Representing Finite Groups

Ambar N. Sengupta

Representing Finite Groups

A Semisimple Introduction

 Springer

Ambar N. Sengupta
Department of Mathematics
Louisiana State University
Baton Rouge Louisiana
USA
sengupta@math.lsu.edu

ISBN 978-1-4899-9808-8   ISBN 978-1-4614-1231-1 (eBook)
DOI 10.1007/978-1-4614-1231-1
Springer New York Dordrecht Heidelberg London

Mathematics Subject Classification (2010): 20C05, 20C30, 20C35, 16D60, 51F25

Printed on acid-free paper

Springer is part of Springer Science+Business Media (www.springer.com)

To my mother

Preface

Geometry is nothing but an expression of a symmetry group. Fortunately, geometry escaped this stifling straitjacket description, an urban legend formulation of Felix Klein's Erlangen program. Nonetheless, there is a valuable ge(r)m of truth in this vision of geometry. Arithmetic and geometry have been intertwined since Euclid's development of arithmetic from geometric constructions. A group, in the abstract, is a set of elements, devoid of concrete form, with just one operation satisfying a minimalist set of axioms. Representation theory is the study of how such an abstract group appears in different avatars as symmetries of geometries over number fields or more general fields of scalars. This book is an initiating journey into this subject.

A large part of the route we take passes through the representation theory of semisimple algebras. We will also make a day-tour from the realm of finite groups to look at the representation theory of unitary groups. These are infinite, continuous groups, but their representation theory is intricately interlinked with the representation theory of permutation groups, and hence this detour from the main route of the book seems worthwhile.

Our navigation system is set to avoid speedways as well as slick shortcuts. Efficiency and speed are not high priorities on this journey. For many of the ideas we view the same set of results from several vantage points. Sometimes we pause to look back at the territory covered or to peer into what lies ahead. We stop to examine glittering objects – specific examples – up close.

The role played by the characteristic of the field underlying a representation is described carefully in each result. We stay almost always within the semisimple territory, etched out by the requirement that the characteristic of the field does not divide the number of elements of the group. By not making any special choice for the field $\mathbb{F}$, we are able to see the role of semisimplicity at every stage and in every result.

Authors generally threaten readers with the admonishment that they *must* do the exercises to appreciate the text. This could give rise to insomnia if one wishes to peruse parts of this text at bedtime. However, for daytime readers, there are several exercises to engage in, some of which may call for breaking intellectual sweat, if the eyes glaze over from simply reading.

The style of presentation I have used is unconventional in some ways. Aside from the very informal tone, I have departed from rigid mathematical custom by repeating definitions instead of sending the reader scurrying back and forth to consult them. I have also included all hypotheses (such as those on the ground field $\mathbb{F}$ of a representation) in the statement of every result, instead of stating them at the beginnings of sections or chapters. This should help the reader who wishes to take just a quick look at some result or sees the statement on a sample page online.

For whom is this book intended? For graduate and undergraduate students, for teachers, for researchers, and also for those who simply want to explore this beautiful subject for itself. This book is an introduction to the subject; at the end, or even part way through, the reader will have enough equipment and experience to read more specialized monographs to pursue roads not traveled here.

A disclaimer on originality needs to be made. To the best of my knowledge, there is no result in this book that is not already "known." Mathematical results evolve in form, from original discovery through mutations and cultural forces, and I have added historical remarks or references only for some of the major results. The reader interested in a more thorough historical analysis should consult works by historians of the subject.

Acknowledgment for much is due to many. To friends, family, strangers, colleagues, students, and a large number of fellow travelers in life and mathematics, I owe thanks for comments, corrections, criticism, encouragement and discouragement. Many discussions with Thierry Lévy have influenced my view of topics in representation theory. I have enjoyed many anecdotes shared with me by Hui-Hsiung Kuo on the frustrations and rewards of writing a book. I am grateful to William Adkins, Daniel Cohen, Subhash Chaturvedi, and Thierry Lévy for specific comments and corrections. It is a pleasure to thank Sergio Albeverio for his kind hospitality at the University of Bonn where this work was completed. Comments by referees, ranging from the insightful to the infuriating, led to innumerable improvements in presentation and content. Vaishali Damle, my editor at Springer, was a calm and steady guide all through the process of turning the original rough notes to the final form of the book. Financial support for my research program from Louisiana State University, a Mercator Guest Professorship at the University of Bonn, and US National Science Foundation Grant DMS-0601141 is gratefully acknowledged. Here I need to add the required disclaimer: Any opinions, findings and conclusions or recommendations expressed in this material are those of the author and do not necessarily reflect the views of the National Science Foundation. Beyond all this, I thank Ingeborg for support that can neither be quantified in numbers nor articulated in words.

Contents

Chapter 1

Concepts and Constructs

A group is an abstract mathematical object, a set with elements and an operation satisfying certain axioms. A representation of a group realizes the elements of the group concretely as geometric symmetries. The same group may have many different such representations. A group that arises naturally as a specific set of symmetries may have representations as geometric symmetries at different levels.

In quantum physics, the group of rotations in three-dimensional space gives rise to symmetries of a complex Hilbert space whose rays represent states of a physical system; the same abstract group appears once, classically, in the avatar of rotations in space and then expresses itself at the level of a more "implicate order" in quantum theory as unitary transformations on Hilbert spaces [5].

In this chapter we acquaint ourselves with the basic concepts, defining group representations, irreducibility, and characters. We will work through certain useful standard constructions with representations and explore a few results that follow very quickly from the basic notions.

All through this chapter G denotes a group and $\mathbb{F}$ a field. We will work with vector spaces, usually denoted V, W, or Z, over the field $\mathbb{F}$. *There are no standing hypotheses on G or $\mathbb{F}$*, and any conditions needed will be stated where necessary.

A.N. Sengupta, *Representing Finite Groups: A Semisimple Introduction*,
DOI 10.1007/978-1-4614-1231-1_1, © Springer Science+Business Media, LLC 2012

1.1   Representations of Groups

A representation ρ of a group G on a vector space V associates with each element $g \in G$ a linear map

$$\rho(g) : V \to V : v \mapsto \rho(g)v$$

such that

$$\rho(gh) = \rho(g)\rho(h) \qquad \text{for all } g, h \in G, \text{ and}$$
$$\rho(e) = I, \tag{1.1}$$

where $I : V \to V$ is the identity map and e is the identity element in G. Here our vector space V is over a field $\mathbb{F}$, and we denote by

$$\mathrm{End}_{\mathbb{F}}(V)$$

the set of all endomorphisms of V. A representation ρ of G on V is thus a map

$$\rho : G \to \mathrm{End}_{\mathbb{F}}(V)$$

satisfying (1.1). The homomorphism condition (1.1), applied with $h = g^{-1}$, implies that each $\rho(g)$ is invertible and

$$\rho(g^{-1}) = \rho(g)^{-1} \quad \text{for all } g \in G.$$

A representation ρ of G on V is said to be *faithful* if $\rho(g) \neq I$ when g is not the identity element in G. Thus, a faithful representation ρ provides an isomorphic copy $\rho(G)$ of G sitting inside $\mathrm{End}_{\mathbb{F}}(V)$.

A *complex representation* is a representation on a vector space over the field $\mathbb{C}$ of complex numbers.

The vector space V on which the linear maps $\rho(g)$ operate is the *representation space* of ρ. We will often say "the representation V" instead of "the representation ρ on the vector space V." Sometimes we write V_ρ for the representation space of ρ.

If V is finite-dimensional, then, on choosing a basis $b_1, ..., b_n$, the endomorphism $\rho(g)$ is encoded in the matrix

$$\begin{bmatrix} \rho(g)_{11} & \rho(g)_{12} & \cdots & \rho(g)_{1n} \\ \rho(g)_{21} & \rho(g)_{22} & \cdots & \rho(g)_{2n} \\ \vdots & \vdots & \vdots & \vdots \\ \rho(g)_{n1} & \rho(g)_{n2} & \cdots & \rho(g)_{nn} \end{bmatrix}. \tag{1.2}$$

Indeed, when a fixed basis has been chosen in a context, we will often not make a distinction between $\rho(g)$ and its matrix form.

As an example, consider the group S_n of permutations of $[n] = \{1, ..., n\}$. This group has a natural action on the vector space $\mathbb{F}^n$ by permutation of coordinates:

$$S_n \times \mathbb{F}^n \to \mathbb{F}^n$$
$$\big(\sigma, (v_1, \ldots, v_n)\big) \mapsto R(\sigma)(v_1, \ldots, v_n) \stackrel{\text{def}}{=} (v_{\sigma^{-1}(1)}, \ldots, v_{\sigma^{-1}(n)}). \tag{1.3}$$

Another way to understand this is by specifying

$$R(\sigma)e_j = e_{\sigma(j)} \qquad \text{for all } j \in [n].$$

Here e_j is the jth vector in the standard basis of $\mathbb{F}^n$; it has 1 in the jth entry and 0 in all other entries. Thus, for example, for the representation of S_4 acting on $\mathbb{F}^4$, the matrix for $R((134))$ relative to the standard basis of $\mathbb{F}^4$ is

$$R((134)) = \begin{bmatrix} 0 & 0 & 0 & 1 \\ 0 & 1 & 0 & 0 \\ 1 & 0 & 0 & 0 \\ 0 & 0 & 1 & 0 \end{bmatrix}.$$

For a transposition $(j\,k)$, we have

$$R((j\,k))e_j = e_k, \qquad R((j\,k))e_k = e_j,$$
$$R((j\,k))e_i = e_i \text{if } i \notin \{j, k\}.$$

We can think of $R((j\,k))$ geometrically as reflection across the hyperplane $\{v \in \mathbb{F}^n : v_j = v_k\}$. Writing a general permutation $\sigma \in S_n$ as a product of transpositions, $R(\sigma)$ is a product of such reflections. The determinant

$$\epsilon(\sigma) = \det R(\sigma) \tag{1.4}$$

is -1 on transpositions, and hence is just the *signature* or *sign* of σ, being $+1$ if σ is a product of an even number of transpositions and -1 otherwise. The signature map ϵ is itself a representation of S_n, a one-dimensional representation, when each $\epsilon(\sigma)$ is viewed as the linear map $\mathbb{F} \to \mathbb{F} : c \mapsto \epsilon(\sigma)c$.

Exercise 1.3 develops the idea contained in the representation R a step further to explore a way to construct more representations of S_n.

The term "representation" will, for us, always mean representation on a vector space. However, we will occasionally notice that a particular complex representation ρ on a vector space V has a striking additional feature: there is a basis in V relative to which all the matrices $\rho(g)$ have integer entries, or all entries lie inside some other subring of $\mathbb{C}$. This is a glimpse at another territory: representations on modules over rings. We will not explore this theory, but will cast an occasional glance at it.

1.2   Representations and Their Morphisms

If ρ_1 and ρ_2 are representations of G on vector spaces V_1 and V_2 over a field $\mathbb{F}$, and

$$T : V_1 \to V_2$$

is a linear map such that

$$\rho_2(g) \circ T = T \circ \rho_1(g) \qquad \text{for all } g \in G, \tag{1.5}$$

then we consider T to be a *morphism* from representation ρ_1 to representation ρ_2. For instance, the identity map $I : V_1 \to V_1$ is a morphism from ρ_1 to itself. Condition (1.5) is also described by saying that T is an *intertwining operator* between representations ρ_1 and ρ_2 or from ρ_1 to ρ_2.

The composition of two morphisms is clearly also a morphism, and the inverse of an invertible morphism is again a morphism. An invertible morphism of representations is called an *isomorphism* or *equivalence* of representations. Thus, representations ρ_1 and ρ_2 are equivalent if there is an invertible intertwining operator from one to the other.

1.3   Direct Sums and Tensor Products

If ρ_1 and ρ_2 are representations of G on vector spaces V_1 and V_2, respectively, over a field $\mathbb{F}$, then we have the direct sum

$$\rho_1 \oplus \rho_2$$

representation on $V_1 \oplus V_2$:

$$(\rho_1 \oplus \rho_2)(g) = (\rho_1(g), \rho_2(g)) \in \mathrm{End}_{\mathbb{F}}(V_1 \oplus V_2). \tag{1.6}$$

If bases are chosen in V_1 and V_2, then the matrix for $(\rho_1 \oplus \rho_2)(g)$ is block-diagonal, with the blocks $\rho_1(g)$ and $\rho_2(g)$ on the diagonal:

$$g \mapsto \begin{bmatrix} \rho_1(g) & 0 \\ 0 & \rho_2(g) \end{bmatrix}.$$

This notion clearly generalizes to a direct sum (or product) of any family of representations.

We also have the *tensor product* $\rho_1 \otimes \rho_2$ of the representations, acting on $V_1 \otimes V_2$, specified through

$$(\rho_1 \otimes \rho_2)(g) = \rho_1(g) \otimes \rho_2(g). \tag{1.7}$$

1.4 Change of Field

There is a more subtle operation on vector spaces, involving change of the ground field over which the vector spaces are defined. Let V be a vector space over a field $\mathbb{F}$ and let $\mathbb{F}_1 \supset \mathbb{F}$ be a field that contains $\mathbb{F}$ as a subfield. Then V specifies a vector space over $\mathbb{F}_1$:

$$V_{\mathbb{F}_1} = \mathbb{F}_1 \otimes_{\mathbb{F}} V. \tag{1.8}$$

Here we have, on the surface, a tensor product of two vector spaces over $\mathbb{F}$: the field $\mathbb{F}_1$, treated as a vector space over the subfield. But $V_{\mathbb{F}_1}$ acquires the structure of a vector space over $\mathbb{F}_1$ by the multiplication rule

$$c(a \otimes v) = (ca) \otimes v$$

for all $c, a \in \mathbb{F}_1$ and $v \in V$. More concretely, if $V \neq \{0\}$ has a basis B, then $V_{\mathbb{F}_1}$ can be taken to be the vector space over $\mathbb{F}_1$ with the same set B as a basis but now using coefficients from the field $\mathbb{F}_1$.

Now suppose ρ is a representation of a group G on a vector space V over $\mathbb{F}$. Then a representation $\rho_{\mathbb{F}_1}$ on $V_{\mathbb{F}_1}$ arises as follows:

$$\rho_{\mathbb{F}_1}(g)(a \otimes v) = a \otimes \rho(g)v \tag{1.9}$$

for all $a \in \mathbb{F}_1$, $v \in V$, and $g \in G$.

To get a concrete feel for $\rho_{\mathbb{F}_1}$, let us look at the matrix form. Choose a basis $b_1, ..., b_n$ for V, assumed to be finite-dimensional and nonzero. Then, almost by definition, this is also a basis for $V_{\mathbb{F}_1}$, only now with scalars to be drawn from $\mathbb{F}_1$. Thus, *the matrix for $\rho_{\mathbb{F}_1}(g)$ is exactly the same as the matrix for $\rho(g)$* for every $g \in G$. The difference is only that we should think of this matrix now as a matrix over $\mathbb{F}_1$ whose entries happen to lie in the subfield $\mathbb{F}$.

This raises a fundamental question: given a representation ρ, is it possible to find a basis of the vector space such that all entries of all the matrices $\rho(g)$ lie in some proper subfield of the field we started with? A deep result of Brauer [7] shows that all irreducible complex representations of a finite group can be realized over a field obtained by adjoining suitable roots of unity to the field $\mathbb{Q}$ of rationals.

1.5   Invariant Subspaces and Quotients

Let ρ be a representation of a group G on a vector space V over a field $\mathbb{F}$.

A subspace $W \subset V$ is said to be *invariant* under ρ if

$$\rho(g)W \subset W \text{ for all } g \in G.$$

In this case, restricting the action of ρ to W,

$$\rho|W : g \mapsto \rho(g)|W \in \text{End}_{\mathbb{F}}(W)$$

is a representation of G on W. It is a *subrepresentation* of ρ. Put another way, the inclusion map

$$W \to V : w \mapsto w$$

is a morphism from $\rho|W$ to ρ.

If W is invariant, then a representation on the quotient space

$$V/W$$

is obtained by setting

$$\rho_{V/W}(g) : v + W \mapsto \rho(g)v + W \qquad \text{for all } v \in V, \tag{1.10}$$

for all $g \in G$.

1.6   Dual Representations

For a vector space V over a field $\mathbb{F}$, let V' be the dual space of all linear mappings of V into $\mathbb{F}$:

$$V' = \operatorname{Hom}_{\mathbb{F}}(V, \mathbb{F}). \tag{1.11}$$

If ρ is a representation of a group G on V, the *dual representation* ρ' on V' is defined as follows:

$$\rho'(g)f = f \circ \rho(g)^{-1} \qquad \text{for all } g \in G, \text{ and } f \in V'. \tag{1.12}$$

It is readily checked that this does indeed specify a representation ρ' of G on V'.

The *adjoint* of $A \in \operatorname{End}_{\mathbb{F}}(V)$ is the element $A' \in \operatorname{End}_{\mathbb{F}}(V')$ given by

$$A'f = f \circ A. \tag{1.13}$$

Thus,

$$\rho'(g) = \rho(g^{-1})' \tag{1.14}$$

for all $g \in G$.

Suppose now that V is finite-dimensional. For a basis $b_1, \ldots, b_n$ of V, the corresponding *dual basis* in V' consists of the sequence of elements $b'_1, \ldots, b'_n \in V'$ specified by the requirement

$$b'_j(b_k) = \delta_{jk} \overset{\text{def}}{=} \begin{cases} 1 & \text{if } j = k, \\ 0 & \text{if } j \neq k. \end{cases} \tag{1.15}$$

It is a pleasant little exercise to check that $b'_1, \ldots, b'_n$ do indeed form a basis of V'; a consequence is that V' is also finite dimensional and

$$\dim_{\mathbb{F}} V' = \dim_{\mathbb{F}} V,$$

under the assumption that this is finite.

Proceeding further with the finite basis $b_1, \ldots, b_n$ of V, for any $A \in \operatorname{End}_{\mathbb{F}}(V)$, the matrix of A' relative to the dual basis $\{b'_i\}$ is related to the matrix of A relative to $\{b_i\}$ as follows:

$$\begin{aligned} A'_{jk} &\overset{\text{def}}{=} (A'b'_k)(b_j) \\ &= b'_k(Ab_j) \\ &= A_{kj}. \end{aligned} \tag{1.16}$$

Thus, the matrix for A' is the *transpose* of the matrix for A. For this reason, the adjoint A' is also denoted as A^{t} or A^{tr}:

$$A^{\mathrm{t}} = A^{\mathrm{tr}} = A'.$$

From all this we see that the matrix for $\rho'(g)$ is the transpose of the matrix for $\rho(g^{-1})$:

$$\rho'(g) = \rho(g^{-1})^{\mathrm{tr}} \qquad \text{for all } g \in G, \text{ as matrices,} \qquad (1.17)$$

relative to dual bases.

Here we have an illustration of the interplay between a vector space V and its dual V'. The *annihilator* W^0 in V' of a subspace W of V is

$$W^0 = \{f \in V' : f(u) = 0 \quad \text{for all } u \in W\}. \qquad (1.18)$$

This is clearly a subspace in V'. Running in the opposite direction, for any subspace N of V' we have its annihilator in V:

$$N_0 = \{u \in V : f(u) = 0 \quad \text{for all } f \in N\}. \qquad (1.19)$$

The association $W \mapsto W^0$, from subspaces of V to subspaces of V', reverses inclusion and has some other nice features that we package into the following lemma:

Lemma 1.1 *Let V be a vector space over a field $\mathbb{F}$, W and Z be subspaces of V, and N be a subspace of V'. Then*

$$(W^0)_0 = W, \qquad (1.20)$$

and $W^0 \subset Z^0$ if and only if $Z \subset W$. If $A \in \mathrm{End}_{\mathbb{F}}(V)$ maps W into itself, then A' maps W^0 into itself. If A' maps N into itself, then $A(N_0) \subset N_0$. If $\iota : W \to V$ is the inclusion map, and

$$r : V' \to W' : f \mapsto f \circ \iota$$

is the restriction map, then r induces an isomorphism of vector spaces

$$r_* : V'/W^0 \to W' : f + W^0 \mapsto r(f). \qquad (1.21)$$

When V is finite-dimensional,

$$\begin{aligned} \dim W^0 &= \dim V - \dim W \\ \dim N_0 &= \dim V - \dim N. \end{aligned} \qquad (1.22)$$

Proof. Clearly $W \subset (W^0)_0$. Now consider a vector $v \in V$ outside the subspace W. Choose a basis B of V with $v \in B$ and such that B contains a basis of W. Let f be the linear functional on V, for which $f(y)$ is equal to 0 on all vectors $y \in B$ except for $y = v$, on which $f(v) = 1$; then $f \in W^0$ is not 0 on v, and so v is not in $(W^0)_0$. Hence, $(W^0)_0 \subset W$. This proves (1.20).

The mappings $M \to M^0$ and $L \mapsto L_0$ are clearly inclusion-reversing. If $W^0 \subset Z^0$, then $(W^0)_0 \supset (Z^0)_0$, and so $Z \subset W$.

If $A(W) \subset W$ and $f \in W^0$, then $A'f = f \circ A$ is 0 on W, and so $A'(W^0) \subset W^0$. Similarly, if $A'(N) \subset N$ and $v \in N_0$, then for any $f \in N$ we have

$$f(Av) = (A'f)(v) = 0,$$

which means $Av \in N_0$.

Now, turning to the restriction map r, first observe that $\ker r = W^0$. Next, if $f \in W'$, then choose a basis of W and extend it to a basis of V, and define $f_1 \in V'$ by requiring it to agree with f on the basis vectors in W and setting it to 0 on all basis vectors outside W; then $r(f_1) = f$. Thus, r is a surjection onto W', and so induces the isomorphism (1.21).

We will prove the dimension result (1.22) using bases, just to illustrate working with dual bases. Choose a basis $b_1, \ldots, b_m$ of W and extend this to a basis $b_1, \ldots, b_n$ of the full space V (so $0 \le m \le n$). Let $\{b'_j\}$ be the dual basis in V'. Then $f \in V'$ lies in W^0 if and only if $f(b_i) = 0$ for $i \in \{1, \ldots, m\}$, and this, in turn, is equivalent to f lying in the span of b'_i for $i \in \{m+1, \ldots, n\}$. Thus, a basis of W^0 is formed by $b'_{m+1}, \ldots, b'_n$, and this proves the first equality in (1.22). The second equality in (1.22) now follows by viewing the finite-dimensional vector space V as the dual of V' (see the discussion below). $\boxed{\text{QED}}$

The mapping

$$V' \times V \to \mathbb{F} : (f, v) \mapsto f(v)$$

specifies the linear functional f on V when f is held fixed, and specifies a linear functional v_* on V' when v is held fixed:

$$v_* : V' \to \mathbb{F} : f \mapsto f(v).$$

The map

$$V \to (V')' : v \mapsto v_* \tag{1.23}$$

is clearly linear as well as injective. If V is finite-dimensional, then V' and hence $(V')'$ both have the same dimension as V, and this forces the injective linear map $v \mapsto v_*$ to be an isomorphism. Thus, *a finite-dimensional vector space V is isomorphic to its double dual $(V')'$* via the natural isomorphism (1.23).

When working with a vector space and its dual, there is a visually appealing notation due to Dirac often used in quantum physics. (If you find this notation irritating, you will be relieved to hear that we will use this notation very rarely, mainly in a couple of sections in Chap. 7.) A vector in V is denoted

$$|v\rangle$$

and is called a "ket," whereas an element of the dual V' is denoted

$$\langle f|$$

and is called a "bra." The evaluation of the bra on the ket is then, conveniently, the "bra-ket"

$$\langle f|v\rangle \in \mathbb{F}.$$

If $|b_1\rangle, \ldots, |b_n\rangle$ is a basis of V, then the dual basis is denoted $\langle b_1|, \ldots, \langle b_n| \in V'$; hence,

$$\langle b_j|b_k\rangle = \delta_{jk} \stackrel{\text{def}}{=} \begin{cases} 1 & \text{if } j = k; \\ 0 & \text{if } j \neq k. \end{cases} \tag{1.24}$$

There is one small spoiler: the notation $\langle b_j|$ wrongly suggests that it is determined solely by the vector $|b_j\rangle$, when in fact one needs the full basis $|b_1\rangle, \ldots, |b_n\rangle$ to give meaning to it.

1.7 Irreducible Representations

A representation ρ on a vector space V is *irreducible* if $V \neq 0$ and the only invariant subspaces of V are 0 and V. The representation ρ is *reducible* if V is 0 or has a proper nonzero invariant subspace.

A starter example of an irreducible representation of the symmetric group S_n can be extracted from the representation R of S_n as a reflection group in the n-dimensional space we looked at in (1.3). Let $\mathbb{F}$ be a field. For any $\sigma \in S_n$, the linear map $R(\sigma) : \mathbb{F}^n \to \mathbb{F}^n$ is specified by

$$R(\sigma)e_j = e_{\sigma(j)} \qquad \text{for all } j \in \{1, \ldots, n\},$$

where $e_1, ..., e_n$ is the standard basis of $\mathbb{F}^n$. In terms of coordinates, R is specified by

$$S_n \times \mathbb{F}^n \to \mathbb{F}^n : (\sigma, v) \mapsto R(\sigma)v = v \circ \sigma^{-1}, \tag{1.25}$$

where $v \in \mathbb{F}^n$ is to be thought of as a map $v : \{1, ..., n\} \to \mathbb{F} : j \mapsto v_j$. The subspaces

$$E_0 = \{(v_1, ..., v_n) \in \mathbb{F}^n : v_1 + \cdots + v_n = 0\} \tag{1.26}$$

and

$$D = \{(v, v, ..., v) : v \in \mathbb{F}\} \tag{1.27}$$

are clearly invariant subspaces. Thus, the representation R is reducible if $n \geq 2$. If $n1_\mathbb{F} \neq 0$ in $\mathbb{F}$, then the subspaces D and E_0 have in common only the zero vector, and provide a decomposition of $\mathbb{F}^n$ into a direct sum of proper nonzero invariant subspaces. In fact, in this case R is restricted to irreducible representations on the subspaces D and E_0 (work this out in Exercise 1.2.)

As we will see later in Theorem 3.4 (Maschke's theorem), for a finite group G, for which $|G|1_\mathbb{F} \neq 0$ in the field $\mathbb{F}$, every representation is a direct sum of irreducible representations.

A one-dimensional representation is automatically irreducible. Our definitions allow the zero space $V = \{0\}$ to be a reducible representation space as well, and we have to try to be careful everywhere to exclude, or include, this case as necessary.

Even with the little technology at hand, we can prove something interesting:

Theorem 1.1 *Let V be a finite-dimensional representation of a group G, and let us equip V' with the dual representation. Then V is irreducible if and only if V' is irreducible.*

Proof. By Lemma 1.1, if W is an invariant subspace of V, then the annihilator W^0 is an invariant subspace of V'. If W is a proper nonzero, invariant subspace of V, then, by Lemma 1.1 and the first dimensional identity in (1.22), W^0 is also a proper nonzero invariant subspace of V'. In the other direction, for any subspace $N \subset V'$, the annihilator N_0 is invariant as a subspace of V if N is invariant in V'. Comparing dimensions by using the second dimensional identity in (1.22), we find N_0 is a proper nonzero invariant subspace of V if N is a proper nonzero invariant subspace of V'. $\boxed{\text{QED}}$

Here is another little useful observation:

Proposition 1.1 *Any irreducible representation of a finite group is finite dimensional.*

Proof. Let ρ be an irreducible representation of the finite group G on a vector space V. Pick any nonzero $v \in V$ and observe that the linear span of the finite set $\{\rho(g)v : g \in G\}$ is a nonzero invariant subspace of V and so, by irreducibility, must be all of V. Thus, V is finite-dimensional. $\boxed{\text{QED}}$

1.8   Schur's Lemma

The following fundamental result of Schur [69, Sect. 2.I] is called *Schur's lemma.* We will revisit and reformulate it several times.

Theorem 1.2 *A morphism between irreducible representations is either an isomorphism or 0. In more detail, if ρ_1 and ρ_2 are irreducible representations of a group G on vector spaces V_1 and V_2, over an arbitrary field $\mathbb{F}$, and if $T : V_1 \to V_2$ is a linear map for which*

$$T\rho_1(g) = \rho_2(g)T \qquad \text{for all } g \in G, \tag{1.28}$$

then T is either invertible or 0.

If ρ is an irreducible representation of a group G on a finite-dimensional vector space V over an algebraically closed field $\mathbb{F}$ and $S : V \to V$ is a linear map for which

$$S\rho(g) = \rho(g)S \qquad \text{for all } g \in G, \tag{1.29}$$

then $S = cI$ for some scalar $c \in \mathbb{F}$.

Proof. Let ρ_1, ρ_2, and T be as stated. From the intertwining property (1.28), it follows readily that $\ker T$ is invariant under the action of the group. Then, by irreducibility of ρ_1, it follows that $\ker T$ is either $\{0\}$ or V_1. So, if $T \neq 0$, then T is injective. Next, applying the same reasoning to $\operatorname{Im} T \subset V_2$, we see that if $T \neq 0$, then T is surjective. Thus, either $T = 0$ or T is an isomorphism.

Now suppose $\mathbb{F}$ is algebraically closed, V is finite-dimensional, and $S : V \to V$ is an intertwining operator from the irreducible representation ρ on V to itself. The polynomial equation in λ given by

$$\det(S - \lambda I) = 0$$

has a solution $\lambda = c \in \mathbb{F}$. Then $S - cI \in \text{End}_{\mathbb{F}}(E)$ is not invertible. Note that $S - cI$ intertwines ρ with itself (i.e., (1.29) holds with $S - cI$ in place of S). So, by the first half of the result, $S - cI$ is 0. Hence, $S = cI$. $\boxed{\text{QED}}$

We will repeat the argument used above in proving that $S = cI$ a couple of times again later.

Since the conclusion of Schur's lemma for the algebraically closed case is so powerful, it is meaningful to isolate it as a hypothesis, or concept, in itself. A field $\mathbb{F}$ is called a *splitting field* for a finite group G if for every irreducible representation ρ of G the only intertwining operators between ρ and itself are the scalar multiples cI of the identity map $I : V_\rho \to V_\rho$.

Schur's lemma is the Incredible Hulk of representation theory. Despite its innocent face-in-the-crowd appearance, it rises up with enormous power to overcome countless challenges. We will see many examples of this, but for now Theorem 1.3 illustrates a somewhat off-label use of Schur's lemma to prove a simple but significant result first established by Wedderburn. A *division algebra* $\mathbb{D}$ over a field $\mathbb{F}$ is an $\mathbb{F}$-algebra with a multiplicative identity $1_{\mathbb{D}} \neq 0$ in which every nonzero element has a multiplicative inverse. As with any $\mathbb{F}$-algebra, the field $\mathbb{F}$ can be viewed as a subset of $\mathbb{D}$ by identifying $c \in \mathbb{F}$ with $c1_{\mathbb{D}} \in \mathbb{D}$; then $\mathbb{F}$, so viewed, lies in the center of $\mathbb{D}$ because

$$(c1_{\mathbb{D}})d = c(1_{\mathbb{D}}d) = cd = c(d1_{\mathbb{D}}) = d(c1_{\mathbb{D}})$$

for all $c \in \mathbb{F}$ and $d \in \mathbb{D}$.

Theorem 1.3 *If $\mathbb{D}$ is a finite-dimensional division algebra over an algebraically closed field $\mathbb{F}$, then $\mathbb{D} = \mathbb{F}1_{\mathbb{D}}$.*

Proof. View $\mathbb{D}$ as a vector space over the field $\mathbb{F}$, and consider the representation l of the multiplicative group $\mathbb{D}^{\times} = \{d \in \mathbb{D} : d \neq 0\}$ on $\mathbb{D}$ given by

$$l(u) \stackrel{\text{def}}{=} l_u : \mathbb{D} \to \mathbb{D} : v \mapsto uv$$

for all $u \in \mathbb{D}^{\times}$. This is an irreducible representation since for any nonzero $u_1, u \in \mathbb{D}$ we have $l(u_1 u^{-1})u = u_1$, which implies that any nonzero invariant subspace of $\mathbb{D}$ contains every nonzero $u_1 \in \mathbb{D}$ and hence is all of $\mathbb{D}$. Next, for any $c \in \mathbb{D}$, the map

$$r_c : \mathbb{D} \to \mathbb{D} : v \mapsto vc$$

is $\mathbb{F}$-linear and commutes with the action of each l_u:

$$l_u r_c v = uvc = r_c l_u v \qquad \text{for all } v \in \mathbb{D} \text{ and } u \in \mathbb{D}^{\times}.$$

Then by Schur's lemma (second part of Theorem 1.2), there is a $c_0 \in \mathbb{F}$ such that $r_c v = c_0 v$ for all $v \in \mathbb{D}$; taking $v = 1_{\mathbb{D}}$ shows that $c = c_0 1_{\mathbb{D}}$. $\boxed{\text{QED}}$

The preceding proof can, of course, be stripped of its use of Schur's lemma: for any $c \in \mathbb{D}$, the $\mathbb{F}$-linear map $r_c : \mathbb{D} \to \mathbb{D} : v \mapsto cv$, with $\mathbb{D}$ viewed as a finite-dimensional vector space over the algebraically closed field $\mathbb{F}$, has an eigenvalue $c_0 \in \mathbb{F}$, which means that there is a nonzero $y \in \mathbb{D}$ for which

$$y(c - c_0 1_{\mathbb{D}}) = r_c y - y c_0 = r_c y - c_0 y = 0,$$

which implies that $c = c_0 1_{\mathbb{D}} \in \mathbb{F}$. This beautiful argument (with left multiplication instead of r_c) was shared with me anonymously.

1.9   The Frobenius–Schur Indicator

A bilinear mapping

$$S : V \times W \to \mathbb{F},$$

where V and W are vector spaces over the field $\mathbb{F}$, is said to be *nondegenerate* if

$$
\begin{aligned}
S(v, w) = 0 \quad &\text{for all } w \text{ implies that } v = 0; \\
S(v, w) = 0 \quad &\text{for all } v \text{ implies that } w = 0.
\end{aligned}
\tag{1.30}
$$

The following result of Frobenius and Schur [36, Sect. 3] is an illustration of the power of Schur's lemma.

Theorem 1.4 *Let ρ be an irreducible representation of a group G on a finite-dimensional vector space V over an algebraically closed field $\mathbb{F}$. Then there exists an element c_ρ in $\mathbb{F}$ whose value is 0 or ± 1,*

$$c_\rho \in \{0, 1, -1\},$$

such that the following holds: if

$$S : V \times V \to \mathbb{F}$$

is bilinear and satisfies

$$S(\rho(g)v, \rho(g)w) = S(v, w) \qquad \text{for all } v, w \in V, \text{ and } g \in G, \tag{1.31}$$

then

$$S(v, w) = c_\rho S(w, v) \qquad \text{for all } v, w \in V. \tag{1.32}$$

If ρ is not equivalent to the dual representation ρ', then $c_\rho = 0$, and thus, in this case, the only G-invariant bilinear form on the representation space of ρ is 0.

If ρ is equivalent to ρ', then $c_\rho \neq 0$ and there is a nondegenerate bilinear S, invariant under the action of G as in (1.31), and all nonzero bilinear S satisfying (1.31) are nondegenerate and multiples of each other. Thus, if there is a nonzero bilinear form on V that is invariant under the action of G, then that form is nondegenerate and either symmetric or skew-symmetric.

When the group G is finite, every irreducible representation is finite-dimensional and so the finite-dimensionality hypothesis is automatically satisfied. The assumption that the field $\mathbb{F}$ is algebraically closed may be replaced by the requirement that it be a splitting field for G. The scalar c_ρ is called the *Frobenius–Schur indicator* of ρ. We will eventually obtain a simple formula expressing c_ρ in terms of the character of ρ; fast-forward to (7.113) for this.

Proof. Define $S_l, S_r : V \to V'$, where V' is the dual vector space to V, by

$$\begin{aligned} S_l(v) &: w \mapsto S(v, w) \\ S_r(v) &: w \mapsto S(w, v), \end{aligned} \tag{1.33}$$

for all $v, w \in V$. The invariance condition (1.31) translates to

$$\begin{aligned} S_l \rho(g) &= \rho'(g) S_l \\ S_r \rho(g) &= \rho'(g) S_r, \end{aligned} \tag{1.34}$$

for all $g \in G$, where ρ' is the dual representation on V' given by $\rho'(g)\phi = \phi \circ \rho(g)^{-1}$. Now recall from Theorem 1.1 that ρ' is irreducible, since ρ is irreducible. Then by Schur's lemma, the intertwining condition (1.34) implies that S_l is either 0 or an isomorphism.

If $S_l = 0$, then $S = 0$, and so the claim (1.32) holds on taking $c_\rho = 0$ for the case where ρ is not equivalent to its dual.

Next, suppose ρ is equivalent to ρ'. Schur's lemma and the intertwining conditions (1.34) imply that S_l is either 0 or an isomorphism. The same holds for S_r. Thus, if $S \neq 0$, then S_l and S_r are both isomorphisms, and a look at (1.30) shows that S is nondegenerate. Moreover, Schur's lemma also implies that S_l is a scalar multiple of S_r; thus, there exists $k_S \in \mathbb{F}$ such that

$$S_l = k_S S_r. \qquad (1.35)$$

Note that since S is not 0, the scalar k_S is uniquely determined by S, but, at least at this stage, could potentially depend on S. The equality (1.35) spells out that

$$S(v, w) = k_S S(w, v) \qquad \text{for all } v, w \in V,$$

and so, applying this twice, we have

$$S(v, w) = k_S S(w, v) = k_S^2 S(v, w)$$

for all $v, w \in V$. Since S is not 0, it follows that $k_S^2 = 1$ and so $k_S \in \{1, -1\}$. It just remains to be shown that k_S is independent of the choice of S. Suppose $T : V \times V \to \mathbb{F}$ is also a nonzero G-invariant bilinear map. Then the argument used above for S_l and S_r when applied to S_l and T_l implies that there is a scalar $k_{ST} \in \mathbb{F}$ such that

$$T = k_{ST} S.$$

Then

$$\begin{aligned}
T(v, w) &= k_{ST} S(v, w) \\
&= k_{ST} k_S S(w, v) = k_S k_{ST} S(w, v) \qquad (1.36) \\
&= k_S T(w, v)
\end{aligned}$$

for all $v, w \in V$, which shows that $k_T = k_S$. Thus, we can set c_ρ to be k_S for any choice of nonzero G-invariant bilinear $S : V \times V \to \mathbb{F}$.

To finish, observe that $\rho \simeq \rho'$ means that there is a linear isomorphism $T : V \to V'$, which intertwines ρ and ρ'. Take $S(v, w)$ to be $T(v)(w)$ for all $v, w \in V$. Clearly, S is bilinear, G-invariant, and, since T is a bijection, nondegenerate. $\boxed{\text{QED}}$

Exercise 1.18 explores the consequences of the behavior of the bilinear map S in the preceding result.

1.10   Character of a Representation

The *trace* of a square matrix is the sum of the diagonal entries. The *trace* of an endomorphism $A \in \mathrm{End}_{\mathbb{F}}(V)$, where V is a finite-dimensional vector space, is the trace of the matrix of A relative to any basis of V:

$$\operatorname{Tr} A = \sum_{j=1}^{n} A_{jj}, \tag{1.37}$$

where $[A_{jk}]$ is the matrix of A relative to a basis $b_1, \ldots, b_n$. It is a basic but remarkable fact that the trace is independent of the choice of basis used in (1.37). Closely related to this is the fact that the trace is invariant under conjugation:

$$\operatorname{Tr}\left(CAC^{-1}\right) = \operatorname{Tr} A \tag{1.38}$$

for all $A, C \in \operatorname{End}_{\mathbb{F}}(V)$, with C invertible. More generally,

$$\operatorname{Tr}\left(AB\right) = \operatorname{Tr}\left(BA\right) \tag{1.39}$$

for all $A, B \in \operatorname{End}_{\mathbb{F}}(V)$. As we have seen in (1.16), the matrix of the adjoint A' is just the transpose of the matrix of A, relative to dual bases, and so the trace, being the sum of the common diagonal terms, is the same for both A and A':

$$\operatorname{Tr} A = \operatorname{Tr} A'. \tag{1.40}$$

Proofs of these results and more information on the trace are given in Sects. 12.11 and 12.12.

The *character* χ_ρ of a representation of a group G on a finite-dimensional vector space V is the function on G given by

$$\chi_\rho(g) \stackrel{\text{def}}{=} \operatorname{Tr}\rho(g) \qquad \text{for all } g \in G. \tag{1.41}$$

For the simplest representation, where $\rho(g)$ is the identity I on V for all $g \in G$, the character is the constant function with value $\dim_{\mathbb{F}} V$. (For the case of $V = \{0\}$, we can set the character to be 0.)

It may seem odd to single out the trace, and not, say, the determinant or some other such natural function of $\rho(g)$. But observe that if we know the trace of $\rho(g)$, with g running over *all* the elements of G, then we know the traces of $\rho(g^2)$, $\rho(g^3)$, etc., which means that we know the traces of all powers of $\rho(g)$ for every $g \in G$. This is clearly a lot of information about a matrix. As we shall see later in Proposition 1.6, if the group G is finite and $|G|1_{\mathbb{F}} \neq 0$, then, by extending the field if necessary, we can write $\rho(g)$ as a diagonal matrix, say, with diagonal entries $\lambda_1, \ldots, \lambda_d$, with respect to some basis (generally dependent on g). The traces

$$\operatorname{Tr} \rho(g)^k = \sum_{i=1}^{d} \lambda_i^k, \tag{1.42}$$

for $k \in [d]$, determine the elementary symmetric sums

$$e_1 = \sum_{i=1}^{d} \lambda_i, \; e_2 = \sum_{1 \leq i < j \leq d} \lambda_i \lambda_j, \ldots, e_d = \prod_{i=1}^{d} \lambda_i,$$

through the Newton–Girard formulas (Exercise 1.22), provided $d! 1_{\mathbb{F}} \neq 0$ in the field $\mathbb{F}$. Now $\lambda_1, \ldots, \lambda_d$ are the roots of the equation

$$X^d - e_1 X^{d-1} + \cdots + (-1)^d e_d = \prod_{i=1}^{d} (X - \lambda_i) = 0.$$

Thus, if G is finite and $\mathbb{F}$ is algebraically closed and of characteristic 0, the traces (1.42) determine the diagonal form of $\rho(g)$, up to basis change. For a good *computational* procedure for determining the diagonal form of $\rho(g)$ work out Exercise 1.21 after reading Sect. 1.11.

Thus, with suitable conditions on G and on $\mathbb{F}$, knowledge of the character of ρ specifies each $\rho(g)$ up to basis change. In other words, under some simple assumptions, if ρ_1 and ρ_2 are finite-dimensional nonzero representations with the same character, then for each g there are bases relative to which the matrix of $\rho_1(g)$ is the same as the matrix of $\rho_2(g)$. This leaves open the possibility, however, that the special choice of bases might depend on g. Remarkably, this is not so! As we will see much later, in Theorem 7.2, the character determines the representation up to equivalence. For now we will be satisfied with a simple observation:

Proposition 1.2 *If ρ_1 and ρ_2 are equivalent representations of a group G on finite-dimensional vector spaces, then*

$$\chi_{\rho_1}(g) = \chi_{\rho_2}(g) \qquad \text{for all } g \in G. \tag{1.43}$$

Proof. Let $v_1, \ldots, v_d$ be a basis for the representation space V for ρ_1 (if this space is $\{0\}$, then the result is obviously and trivially true, and so we discard this case). Then in the representation space W for ρ_2, the vectors $w_i = Tv_i$ form a basis, where T is any isomorphism $V \to W$. We take for T an isomorphism that intertwines ρ_1 and ρ_2:

$$\rho_2(g) = T\rho_1(g)T^{-1} \qquad \text{for all } g \in G.$$

Then for any $g \in G$, the matrix for $\rho_2(g)$ relative to the basis $w_1, \ldots, w_d$ is the same as the matrix for $\rho_1(g)$ relative to the basis $v_1, \ldots, v_d$. Hence, the trace of $\rho_2(g)$ equals the trace of $\rho_1(g)$. $\boxed{\text{QED}}$

The following observations are readily checked by using bases:

Proposition 1.3 *If ρ_1 and ρ_2 are representations of a group on finite-dimensional vector spaces, then*

$$\begin{aligned}
\chi_{\rho_1 \oplus \rho_2} &= \chi_{\rho_1} + \chi_{\rho_2}, \\
\chi_{\rho_1 \otimes \rho_2} &= \chi_{\rho_1} \chi_{\rho_2}.
\end{aligned} \tag{1.44}$$

Let us work out the character of the representation R of the permutation group S_n on $\mathbb{F}^n$, and on the subspaces D and E_0 given in (1.26) and (1.27), discussed in Sect. 1.7. Here $\mathbb{F}$ is a field and $n \geq 2$ is an integer for which $n1_\mathbb{F} \neq 0$. Recall that for $\sigma \in S_n$, and any standard-basis vector e_j of $\mathbb{F}^n$,

$$R(\sigma)e_j \overset{\text{def}}{=} e_{\sigma(j)}.$$

Hence,

$$\chi_R(\sigma) = |\{j \in [n] : \sigma(j) = j\}| = \text{number of fixed points of } \sigma. \tag{1.45}$$

Now consider the restriction R_D of this action to the "diagonal" subspace $D = \mathbb{F}(e_1 + \cdots + e_n)$. Clearly, $R_D(\sigma)$ is the identity map for every $\sigma \in S_n$, and so the character of R_D is given by

$$\chi_D(\sigma) = 1 \qquad \text{for all } \sigma \in S_n.$$

Then the character χ_0 of the representation $R_0 = R(\cdot)|E_0$ is given by

$$\chi_0(\sigma) = \chi_R(\sigma) - \chi_D(\sigma) = |\{j : \sigma(j) = j\}| - 1. \tag{1.46}$$

Characters can become confusing when working with representations over different fields at the same time. Fortunately, there is no confusion in the simplest natural situation:

Proposition 1.4 *If ρ is a representation of a group G on a finite-dimensional vector space V over a field $\mathbb{F}$, and $\rho_{\mathbb{F}_1}$ is the corresponding representation on $V_{\mathbb{F}_1}$, where $\mathbb{F}_1$ is a field containing $\mathbb{F}$ as a subfield, then*

$$\chi_{\rho_{\mathbb{F}_1}} = \chi_\rho. \tag{1.47}$$

Proof. As seen in Sect. 1.4, $\rho_{\mathbb{F}_1}$ has exactly the same matrix as ρ, relative to suitable bases; hence, the characters are the same as well. $\boxed{\text{QED}}$

If ρ_1 is a one-dimensional representation of a group G, then, for each $g \in G$, the operator $\rho_1(g)$ is simply multiplication by a scalar, denoted again by $\rho_1(g)$. Then the character of ρ_1 is ρ_1 itself! In the converse direction, if χ is a homomorphism of G into the multiplicative group of invertible elements in the field, then χ provides a one-dimensional representation.

1.11 Diagonalizability

Let G be a finite group and ρ be a representation of G on a finite-dimensional vector space V over a field $\mathbb{F}$. Remarkably, under some mild conditions on the field $\mathbb{F}$ as described below in Proposition 1.5, every element $\rho(g)$ can be expressed as a diagonal matrix relative to some basis (depending on g) in V, with the diagonal entries being roots of unity in $\mathbb{F}$:

$$\rho(g) = \begin{bmatrix} \zeta_1(g) & 0 & 0 & \ldots & 0 \\ 0 & \zeta_2(g) & 0 & \ldots & 0 \\ \vdots & \vdots & \vdots & \cdots & \vdots \\ 0 & 0 & 0 & \cdots & \zeta_d(g) \end{bmatrix},$$

where each $\zeta_j(g)$, when raised to the $|G|$th power, gives 1, and d is the dimension of V.

An mth root of unity in a field $\mathbb{F}$ is an element $\zeta \in \mathbb{F}$ for which $\zeta^m = 1$. There are m distinct mth roots of unity in an extension of $\mathbb{F}$ if and only if m is not divisible by the characteristic of $\mathbb{F}$ (Theorem 12.6).

Proposition 1.5 *Suppose $\mathbb{F}$ is a field that contains m distinct mth roots of unity for some $m \in \{1, 2, 3, ..\}$. If $V \neq 0$ is a vector space over $\mathbb{F}$ and $S : V \to V$ is a linear map for which $S^m = I$, then there is a basis of V relative to which the matrix for S is diagonal and each diagonal entry is an mth root of unity.*

Proof. Let $\eta_1, ..., \eta_m$ be the distinct elements of $\mathbb{F}$ for which the polynomial $X^m - 1$ factors as

$$X^m - 1 = (X - \eta_1)...(X - \eta_m).$$

Then

$$(S - \eta_1 I)...(S - \eta_m I) = S^m - I = 0.$$

A result from linear algebra (Theorem 12.9) assures us that V has a basis with respect to which the matrix for S is diagonal, with entries drawn from the η_i. $\boxed{\text{QED}}$

As a consequence, we have

Proposition 1.6 *Suppose G is a group in which $g^m = e$ for all $g \in G$ for some positive integer m; for instance, G is finite of order m. Let $\mathbb{F}$ be a field containing m distinct mth roots of unity. Then, for any representation ρ of G on a vector space $V_\rho \neq 0$ over $\mathbb{F}$, for each $g \in G$ there is a basis of V_ρ with respect to which the matrix of $\rho(g)$ is diagonal and the diagonal entries are each mth roots of unity in $\mathbb{F}$.*

When the representation space is finite-dimensional, this gives us an unexpected and intriguing piece of information about characters:

Theorem 1.5 *Suppose G is a group in which $g^m = e$ for all $g \in G$ for some positive integer m; for instance, G may be finite of order m. Let $\mathbb{F}$ be a field containing m distinct mth roots of unity. Then the character χ of any representation of G on a finite-dimensional vector space over $\mathbb{F}$ is a sum of mth roots of unity.*

A form of this result was proved by Maschke [58], and raised the question as to when there is a basis of the vector space relative to which all $\rho(g)$ have entries in some number field generated by a root of unity.

There is a way to bootstrap our way to a stronger form of the preceding result. Suppose that it is not the field $\mathbb{F}$, but rather an extension, a larger field $\mathbb{F}_1 \supset \mathbb{F}$, which contains m distinct mth roots of unity; for instance, $\mathbb{F}$ might be the reals $\mathbb{R}$ and $\mathbb{F}_1$ is the field $\mathbb{C}$. The representation space V can be dressed up to $V_1 = \mathbb{F}_1 \otimes_\mathbb{F} V$, which is a vector space over $\mathbb{F}_1$, and then a linear map $T : V \to V$ produces an $\mathbb{F}_1$-linear map

$$T_1 : V_1 \to V_1 : 1 \otimes v \mapsto 1 \otimes Tv. \tag{1.48}$$

If B is a basis of V, then $\{1 \otimes w : w \in B\}$ is a basis of V_1, and the matrix of T_1 relative to this basis is the same as the matrix of T relative to B, and so

$$\operatorname{Tr} T_1 = \operatorname{Tr} T. \tag{1.49}$$

(We have seen this before in (1.47).) Consequently, if in Theorem 1.5 we require simply that there be an extension field of $\mathbb{F}$ in which there are m distinct mth roots of unity and ρ is a finite-dimensional representation over $\mathbb{F}$, then the values of the character χ_ρ are again sums of mth roots of unity in $\mathbb{F}_1$ (which, themselves, need not lie in $\mathbb{F}$).

Suppose the field $\mathbb{F}$ has an automorphism, call it *conjugation*,

$$\mathbb{F} \to \mathbb{F} : z \mapsto \overline{z}$$

that takes each root of unity to its inverse; let us call self-conjugate elements *real*. For instance, if $\mathbb{F}$ is a subfield of $\mathbb{C}$, then the usual complex conjugation provides such an automorphism. Then, under the hypotheses of Proposition 1.6, for each $g \in G$ and representation ρ of G on a finite-dimensional vector space $V_\rho \neq 0$, there is a basis of V_ρ relative to which the matrix of $\rho(g)$ is diagonal with entries along the diagonal being roots of unity; hence, $\rho(g^{-1})$, relative to the same basis, has a diagonal matrix, with the diagonal entries being the conjugates of those for $\rho(g)$. Hence,

$$\chi_\rho(g^{-1}) = \overline{\chi_\rho(g)}. \tag{1.50}$$

In particular, if an element of G is conjugate to its inverse, then the value of any character on such an element is real. In the symmetric group S_n, every element is conjugate to its own inverse, and so the characters of all complex representations of S_n are *real-valued*. This is an amazing, specific result about a familiar concrete group that falls out immediately from some of the simplest general observations. Later, with greater effort, it will become clear that, in fact, the characters of S_n have integer values!

1.12   Unitarity

Suppose now that our field $\mathbb{F}$ is a subfield of $\mathbb{C}$, the field of complex numbers, and G is a finite group.

Consider any Hermitian inner product $\langle \cdot, \cdot \rangle$ on V, a vector space over $\mathbb{F}$. This is a map

$$V \times V \to \mathbb{F} : (v, w) \mapsto \langle v, w \rangle$$

such that

$$\begin{aligned}
\langle av_1 + v_2, w \rangle &= \overline{a}\langle v_1, w \rangle + \langle v_2, w \rangle, \\
\langle v, aw_1 + w_2 \rangle &= a\langle v, w_1 \rangle + \langle v, w_2 \rangle, \\
\langle v, v \rangle &\geq 0 \quad \text{(in particular, } \langle v, v \rangle \text{ is real)}, \\
\langle v, v \rangle &= 0 \quad \text{if and only if } v = 0
\end{aligned} \tag{1.51}$$

for all $v, w, v_1, v_2, w_1, w_2 \in V$ and $a \in \mathbb{F}$. The norm $\|v\|$ of any $v \in V$ is defined by

$$\|v\| = \sqrt{\langle v, v \rangle}. \tag{1.52}$$

Note that in (1.51) we used the complex conjugation $z \mapsto \overline{z}$. If $\mathbb{F}$ is the field $\mathbb{R}$ of real numbers, then the conjugation operation is just the identity map. If B is a basis of V, then

$$\left\langle \sum_{u \in B} a_u u, \sum_{u \in B} b_u u \right\rangle \overset{\text{def}}{=} \sum_{u \in B} \overline{a}_u b_u, \tag{1.53}$$

for $a_u, b_u \in \mathbb{F}$ with all except finitely many of these being 0, specifies a Hermitian inner product on V.

It is a bit sad that, contrary to what we are using, mathematical convention has chosen $\langle v, w \rangle$ to be linear in v and conjugate-linear in w. Our convention, followed more in the physics literature, appears especially sensible in the bra-ket notation, making

$$\langle v| : |w\rangle \mapsto \langle v|w\rangle$$

linear. Moreover, when V is finite-dimensional, any linear mapping $L : V \to \mathbb{C}$ can be expressed as

$$L(w) = \langle v, w \rangle \qquad \text{for all } w \in V, \tag{1.54}$$

for a choice of $v \in V$ uniquely determined by L, our convention for inner products making it possible to write w to the right of the operation in both sides of the equality in (1.54).

Let us modify the inner product so that it sees all $\rho(g)$ equally; this is done by averaging:

$$\langle v, w \rangle_0 = \frac{1}{|G|} \sum_{g \in G} \langle \rho(g)v, \rho(g)w \rangle \tag{1.55}$$

for all $v, w \in V$. Then it is clear that

$$\langle \rho(h)v, \rho(h)w \rangle_0 = \langle v, w \rangle_0$$

for all $h \in G$ and all $v, w \in V$. You can quickly check through all the properties needed to certify $\langle \cdot, \cdot \rangle_0$ as an inner product on V.

Thus we have proved

Proposition 1.7 *Let G be a finite group and ρ be a representation of G on a vector space V over a subfield $\mathbb{F}$ of $\mathbb{C}$. Then there is a Hermitian inner product $\langle \cdot, \cdot \rangle_0$ on V such that for every $g \in G$ the operator $\rho(g)$ is unitary in the sense that*

$$\langle \rho(g)v, \rho(g)w \rangle_0 = \langle v, w \rangle_0 \qquad \text{for all } v, w \in V \text{ and } g \in G.$$

In matrix algebra one knows that a unitary matrix can be diagonalized by choosing a suitable orthonormal basis in the space. Then our result here provides an alternative way to understand Proposition 1.6.

1.13   Rival Reads

There are many books on representation theory, even for finite groups, ranging from elementary introductions to extensive expositions. An encyclopedic, yet readable, volume is the work of Curtis and Reiner [15]. The book by Burnside [9], from the early years of the theory, is still worth exploring, as is the book by Littlewood [57]. Among modern books, the book by Weintraub [78] provides an efficient and extensive development of the theory, especially the arithmetic aspects of the theory and the behavior of representations under change of the ground field. The book by Serre [73] is a classic. With a very different flavor, the book by Simon [74] is a fast-paced exposition and crosses the bridge from finite to compact groups. For the representation theory of compact groups, for which there is a much larger library of literature, we recommend the book by Hall [41]. Another introduction which bridges finite and compact groups, and explores some of the noncompact group $SL_2(\mathbb{R})$ as well, is the slim volume by Thomas [75]. Returning to finite groups, the books by Alperin and Bell [1] and James and Liebeck [49] offer introductions with a view to understanding the structure of finite groups. The book by Hill [44] is an elegant and readable introduction which pauses to examine many enlightening examples. Fulton and Harris [38] present general group representation theory with a strong algebraic flavor. Lang's *Algebra* [55] includes a rapid but readable account of finite group representation theory, covering the basics and some deeper results.

1.14 Afterthoughts: Lattices

Logic and geometry interweave in an elegant, and abstract, lattice framework developed by Birkhoff and von Neumann [4] for classical and quantum physics. There is an extensive exposition of this theory, and much more, in Varadarajan [76].

A symmetry transforms one entity to another, preserving certain features of interest. The minimal setting for such a transformation is simply as a mapping of a set into itself. An *action* of a group G on a set S is a mapping

$$G \times S \to S : (g, p) \mapsto L_g(p) = g \cdot p,$$

for which $e \cdot p = p$ for all $p \in S$, where e is the identity in G, and $g \cdot (h \cdot p) = (gh) \cdot p$ for all $g, h \in G$, and $p \in S$. Taking h to be g^{-1} shows that each mapping $L_g : p \mapsto g \cdot p$ is a bijection of S into itself. As a physical example, think of S as the set of states of some physical system; for instance, S could be the phase space of a classical dynamical system. If instead of a single point p of S we consider a subset $A \subset S$, the action of $g \in G$ carries A into the subset $L_g(A)$. Thus,

$$A \mapsto L_g(A)$$

specifies an action of G on the set $\mathcal{P}(S)$ of all subsets of S. Unlike S, the set $\mathcal{P}(S)$ does have some structure: it has a partial ordering given by inclusion $A \subset B$, and there is a minimum element $0 = \emptyset$ and a maximum element $1 = S$. This partial order relation makes $\mathcal{P}(S)$ a *lattice* in the sense that any $A, B \in \mathcal{P}(S)$ have both an infimum $A \wedge B = A \cap B$ and a supremum $A \vee B = A \cup B$. This lattice structure has several additional nice features;

for example, it is *distributive*:

$$(P \cup M) \cap B = (P \cap B) \cup (M \cap B)$$
$$P \cup (M \cap B) = (P \cup M) \cap (P \cup B),$$

(1.56)

for all $P, M, B \in \mathcal{P}(S)$. Moreover, the complementation $A \mapsto A^c$ specified by

$$A \cap A^c = \emptyset \quad \text{and} \quad A \cup A^c = S \tag{1.57}$$

is an order-reversing bijection of $\mathcal{P}(S)$ into itself, and is an *involution*, in the sense that $(A^c)^c = A$ for all $A \in \mathcal{P}(S)$. The action of G on $\mathcal{P}(S)$ clearly preserves the partial order relation and hence the lattice structure, given by the infimum and supremum, as well as complements. Conversely, at least for a finite set S, if a group G acts on $\mathcal{P}(S)$, preserving its partial ordering, then this action arises from an action of G on the underlying set S.

Birkhoff and von Neumann [4] proposed that in quantum theory classical Boolean logic, an example of which is the lattice structure of $\mathcal{P}(S)$, is replaced by a different lattice, encoding the "logic of quantum mechanics." (Deducing nature from 'pure reason' is, of course, not the way we truly arrive at an understanding of nature.) This is a lattice $\mathbb{L}(\mathbb{H})$ of subspaces of a vector space $\mathbb{H}$ over a field $\mathbb{F}$, with additional properties of the lattice being reflected in the nature of $\mathbb{F}$ and an inner product on $\mathbb{H}$. The set of subspaces is ordered by inclusion, the infimum is again the intersection, but the supremum of subspaces $A, B \in \mathbb{L}(\mathbb{H})$ is the minimal subspace in $\mathbb{L}(\mathbb{H})$ containing the sum $A + B$. Unlike the Boolean lattice $\mathcal{P}(S)$, the distributive laws do not hold; a weaker form, the *modular law*, holds:

$$(P + M) \cap B = (P \cap B) + M \qquad \text{if } M \subset B. \tag{1.58}$$

(We will meet this again later in (5.40)).

The construction of the field $\mathbb{F}$ and the vector space $\mathbb{H}$ is part of classical projective geometry. The inner product arises from logical negation, which is expressed as a complementation in $\mathbb{L}(\mathbb{H})$: Birkhoff and von Neumann [4, Appendix] show how a complementation $A \mapsto A^\perp$ in the lattice $\mathbb{L}(\mathbb{H})$ induces, when $\dim \mathbb{H} > 3$, an inner product on $\mathbb{H}$ for which $A^\perp$ is the orthogonal complement of A. In the standard form of quantum theory, $\mathbb{F}$ is the field $\mathbb{C}$ of complex numbers, and $\mathbb{L}(\mathbb{H})$ is the lattice of *closed* subspaces of a Hilbert space $\mathbb{H}$. More broadly, one could consider the scalars to be drawn from a division ring, such as the quaternions.

Subspaces M and N of a Hilbert space $\mathbb{H}$ are *orthogonal* if $\langle x, y \rangle = 0$ for all $x \in M$ and $y \in N$. Let us say that closed subspaces M and N in a Hilbert space $\mathbb{H}$ are *perpendicular* to each other if $M = M_1 + K$ and $N = N_1 + K$, where M_1, N_1, and K_1 are closed subspaces of $\mathbb{H}$ that are orthogonal to each other. Consider now a set $\mathcal{A}$ of closed subspaces of $\mathbb{H}$ such that any two elements of $\mathcal{A}$ are perpendicular to each other or one element is contained in the other, and the closed sum of the subspaces in $\mathcal{A}$ is all of $\mathbb{H}$. Then the set $\mathbb{L}(\mathcal{A})$ of all subspaces that are closures of sums of elements of $\mathcal{A}$ satisfies the *distributive laws*:

$$
\begin{aligned}
a + (b \cap c) &= (a + b) \cap (a + c) \\
a \cap (b + c) &= (a \cap b) + (a \cap c)
\end{aligned}
\tag{1.59}
$$

for $a, b, c \in \mathbb{L}(\mathcal{A})$. Thus, the lattice $\mathbb{L}(\mathcal{A})$ is a *Boolean algebra*, as is the case for a classical physical system, unlike the full lattice $\mathbb{L}(\mathbb{H})$ that describes a quantum system. The simplest instance of these notions is seen for $\mathbb{H} = \mathbb{C}^2$, with two complementary atoms, which are orthogonal one-dimensional subspaces. This is the model Hilbert space of a "single qubit" quantum system.

Aside from the lattice framework, an analytically more useful structure is the algebra of operators obtained as suitable (strong) limits of complex linear combinations of projection operators onto the closed subspaces of $\mathbb{H}$. This is a quantum form of the commutative algebra formed by using only the subspaces in the Boolean algebra $\mathbb{L}(\mathcal{A})$. Specifically, a physical observable is modeled mathematically by a self-adjoint operator A on a Hilbert space H, and the spectral theorem expresses A as an integral:

$$
A = \int_{\mathbb{R}} \lambda \, dP_A(\lambda),
\tag{1.60}
$$

where P_A is the *spectral measure* of A, and the values of P^A form a family of commuting orthogonal projections onto closed subspaces. (For spectral theory and Hilbert spaces, see, e.g., Rudin [67].) The most familiar example of this is when $\mathbb{H}$ is finite-dimensional, in which case the integral in (1.60) is a sum, a linear combination of orthogonal projection operators onto the eigenspaces of A, and (1.60) is the usual diagonalization of the self-adjoint operator A.

A symmetry of the physical system in this framework is an automorphism of the complemented lattice $\mathbb{L}(\mathbb{H})$ and, combining fundamental theorems from projective geometry and a result of Wigner, one realizes such a symmetry by a linear or conjugate-linear unitary mapping $\mathbb{H} \to \mathbb{H}$ (see Varadarajan [76] for details and more on this). If ρ is a unitary representation of a finite group G on a finite-dimensional inner product space $\mathbb{H}$, then $\rho_g : A \mapsto \rho(g)A$, for $A \in \mathbb{L}(\mathbb{H})$, is an automorphism of the complemented lattice $\mathbb{L}(\mathbb{H})$, and thus such a representation ρ of G provides a group of symmetries of a quantum system. The requirement that ρ be a representation may be weakened, requiring only that it be a *projective representation,* where $\rho(g)\rho(h)$ must only be a multiple of $\rho(gh)$, for it to produce a group of symmetries of $\mathbb{L}(\mathbb{H})$.

Exercises

1.1. Let G be a finite group and P be a nonempty set on which G acts; this means that there is a map

$$G \times P \to P : (g, p) \mapsto g \cdot p,$$

for which $e \cdot p = p$ for all $p \in P$, where e is the identity in G, and $g \cdot (h \cdot p) = (gh) \cdot p$ for all $g, h \in G$, and $p \in P$. The set P, along with the action of G, is called a G-set. Now suppose V is a vector space over a field $\mathbb{F}$, with P as a basis. Define, for each $g \in G$, the map $\rho(g) : V \to V$ to be the linear map induced by permutation of the basis elements by the action of g:

$$\rho(g) : V \to V : \sum_{p \in P} a_p p \mapsto \sum_{p \in P} a_p g \cdot p.$$

Show that ρ is a representation of G. Interpret the character value $\chi_\rho(g)$ in terms of the action of g on P if P is finite. Next, if P_1 and P_2 are G-sets with corresponding representations ρ_1 and ρ_2, interpret the representation ρ_{12} corresponding to the natural action of G on the product $P_1 \times P_2$ in terms of the tensor product $\rho_1 \otimes \rho_2$.

1.2. Let $n \geq 2$ be a positive integer, $\mathbb{F}$ be a field in which $n1_\mathbb{F} \neq 0$, and consider the representation R of S_n on $\mathbb{F}^n$ given by

$$R(\sigma)(v_1, ..., v_n) = (v_{\sigma^{-1}(1)}, \ldots, v_{\sigma^{-1}(n)})$$
$$\text{for all } (v_1, \ldots, v_n) \in \mathbb{F}^n \text{ and } \sigma \in S_n.$$

Let
$$D = \{(v, ..., v) : v \in \mathbb{F}\} \subset \mathbb{F}^n$$
and
$$E_0 = \{(v_1, \ldots, v_n) \in \mathbb{F}^n : v_1 + \cdots + v_n = 0\}.$$
Show that

(a) No nonzero vector in E_0 is in D and, in fact, $\mathbb{F}^n$ is the direct sum of E_0 and D (since $n \geq 2$, E_0 does contain a nonzero vector!).

(b) Each vector $e_1 - e_j$ lies in the span of $\{R(\sigma)w : \sigma \in S_n\}$ for any $w \in E_0$.

(c) The restriction R_0 of R to the subspace E_0 is an irreducible representation of S_n.

1.3. Let P_n be the set of all partitions of $[n] = \{1, \ldots, n\}$ into k disjoint nonempty subsets, where $k \in [n]$. For $\sigma \in S_n$ and $p \in P_k$, let $\sigma \cdot p = \{\sigma(B) : B \in p\}$. In this way S_n acts on P_k. Now let V_k be the vector space, over a field $\mathbb{F}$, with basis P_k, and let $R_k : S_n \to \text{End}_{\mathbb{F}}(V_k)$ be the corresponding representation given by the method in Exercise 1.1. What is the relationship of this to the representation R in Exercise 1.2?

1.4. Determine all one-dimensional representations of S_n over any field.

1.5. Prove Proposition 1.3.

1.6. Let $n \in \{3, 4, ...\}$ and $\mathbb{F}$ be a field of characteristic 0. Denote by R_0 the restriction of the representation of S_n on $\mathbb{F}^n$ to the subspace $E_0 = \{x \in \mathbb{F}^n : x_1 + \cdots + x_n = 0\}$. Let ϵ be the one-dimensional representation of S_n on $\mathbb{F}$ given by the signature, where $\sigma \in S_n$ acts by multiplication by the signature $\epsilon(\sigma) \in \{+1, -1\}$. Show that $R_1 = R_0 \otimes \epsilon$ is an irreducible representation of S_n. Show that R_1 is not equivalent to R_0.

1.7. Consider S_3, which is generated by the cyclic permutation $c = (123)$ and the transposition $r = (12)$, subject to the relations

$$c^3 = \iota, \qquad r^2 = \iota, \qquad rcr^{-1} = c^2.$$

Let $\mathbb{F}$ be a field. The group S_3 acts on $\mathbb{F}^3$ by permutation of coordinates, and preserves the subspace $E_0 = \{(x_1, x_2, x_3) : x_1 + x_2 + x_3 = 0\}$; the restriction of the action to E_0 is a two-dimensional representation R_0 of S_3. Work out the matrices for $R_0(\cdot)$ relative to the basis $u_1 = (1, 0, -1)$ and $u_2 = (0, 1, -1)$ of E_0. Then work out the values of the character χ_0 on all the six elements of S_3. Compute the sum

$$\sum_{\sigma \in S_3} \chi_0(\sigma)\chi_0(\sigma^{-1}).$$

1.8. Consider A_4, the group of even permutations on $\{1, 2, 3, 4\}$, acting through permutation of coordinates of $\mathbb{F}^4$, where $\mathbb{F}$ is a field. Let R_0 be the restriction of this action to the subspace $E_0 = \{(x_1, x_2, x_3, x_4) \in \mathbb{F}^4 : x_1 + x_2 + x_3 + x_4 = 0\}$. Work out the values of the character of R_0 on all elements of A_4.

1.9. Give an example of a representation ρ of a finite group G on a finite-dimensional vector space V over a field of characteristic 0, such that there is an element $g \in G$ for which $\rho(g)$ is not diagonal in any basis of V.

1.10. Explore the validity of the statement of Theorem 1.1 when V is infinite-dimensional.

1.11. Let V and W be finite-dimensional representations of a group G over the same field. Show that (a) $V'' \simeq V$ and (b) $V \simeq W$ if and only if $V' \simeq W'$, where $\simeq$ denotes equivalence of representations.

1.12. Suppose ρ is an irreducible representation of a finite group G on a vector space V over a field $\mathbb{F}$. If $\mathbb{F}_1 \supset \mathbb{F}$ is an extension field of $\mathbb{F}$, is the representation $\rho_{\mathbb{F}_1}$ on $V_{\mathbb{F}_1}$ necessarily irreducible?

1.13. If H is a normal subgroup of a finite group G and ρ is a representation of the group G/H, let ρ_G be the representation of G specified by

$$\rho_G(g) = \rho(gH) \qquad \text{for all } g \in G.$$

Show that ρ_G is irreducible if and only if ρ is irreducible. Work out the character of ρ_G in terms of the character of ρ.

1.14. Let ρ be a representation of a group G on a finite-dimensional vector space $V \neq 0$.

(a) Show that there is a subspace of V on which ρ is restricted to an irreducible representation.

(b) Show that there is a chain of subspaces $V_1 \subset V_2 \subset \cdots \subset V_m = V$ such that (i) each V_j is invariant under the action of $\rho(G)$, (ii) the representation $\rho|V_1$ is irreducible, and (iii) the representation obtained from ρ on the quotient V_j/V_{j-1} is irreducible for each $j \in \{2, ..., m\}$.

1.15. Let ρ be a representation of a group G on a vector space V over a field $\mathbb{F}$ and suppose $b_1, ..., b_n$ is a basis of V. There is then a representation τ of G on $\mathrm{End}_{\mathbb{F}}(V)$ given by

$$\tau(g)A = \rho(g) \circ A \qquad \text{for all } g \in G \text{ and } A \in \mathrm{End}_{\mathbb{F}}(V).$$

Let

$$S : \mathrm{End}_{\mathbb{F}}(V) \to V \oplus \cdots \oplus V : A \mapsto (Ab_1, \ldots, Ab_n).$$

Show that S is an equivalence from τ to $\rho \oplus \cdots \oplus \rho$ (n-fold direct sum of ρ with itself).

1.16. Let ρ_1 and ρ_2 be representations of a group G on vector spaces V_1 and V_2, respectively, over a common field $\mathbb{F}$. For $g \in G$, let $\rho_{12}(g) :$ $\mathrm{Hom}(V_1, V_2) \to \mathrm{Hom}(V_1, V_2)$ be given by

$$\rho_{12}(g)T = \rho_2(g)T\rho_1(g)^{-1}.$$

Show that ρ_{12} is a representation of G. Taking V_1 and V_2 to be finite-dimensional, show that this representation is equivalent to the tensor product representation $\rho_1' \otimes \rho_2$ on $V_1' \otimes V_2$.

1.17. Let ρ be a representation of a group G on a finite-dimensional vector space V over a field $\mathbb{F}$. There is then a representation σ of $G \times G$ on $\mathrm{End}_{\mathbb{F}}(V)$ given by

$$\sigma(g,h)A = \rho(g) \circ A \circ \rho(h)^{-1} \qquad \text{for all } g \in G \text{ and } A \in \mathrm{End}_{\mathbb{F}}(V).$$

Let

$$B : V' \otimes V \to \mathrm{End}_{\mathbb{F}}(V) \to \langle f| \otimes |v\rangle \mapsto |v\rangle\langle f|,$$

where $|v\rangle\langle f|$ is the map $V \to V$ carrying any vector $|w\rangle \in V$ to $\langle f|w\rangle|v\rangle$. Show that B is an equivalence from σ to the representation θ of $G \times G$ on $V' \otimes V$ specified by

$$\theta(g, h)\langle f| \otimes |v\rangle = \rho'(h)\langle f| \otimes \rho(g)|v\rangle,$$

where ρ' is the dual representation on V'.

1.18. Let G be a group and ρ be an irreducible representation of G on a finite-dimensional complex vector space V. Assume that there is a Hermitian inner product $\langle \cdot, \cdot \rangle$ on V that is invariant under G, thus making ρ a unitary representation. Assume, moreover, that there is a nonzero symmetric bilinear mapping

$$S : V \times V \to \mathbb{C}$$

that is G-invariant:

$$S\big(\rho(g)v, \rho(g)w\big) = S(v, w) \quad \text{for all } v, w \in V \text{ and } g \in G.$$

For $v \in V$, let $S_*(v)$ be the unique element of V for which

$$\langle S_*(v), w \rangle = S(v, w) \quad \text{for all } w \in V. \tag{1.61}$$

(a) Check that $S_* : V \to V$ is *conjugate*-linear, in the sense that

$$S_*(av + w) = \overline{a}S_*(v) + S_*(w)$$

for all $v, w \in V$ and $a \in \mathbb{C}$. Consequently, S_*^2 is linear. Check that

$$S_*(\rho(g)v) = \rho(g)(S_*v)$$

and

$$S_*^2\rho(g) = \rho(g)S_*^2$$

for all $g \in G$ and $v \in V$.

(b) Show from the symmetry of S that S_*^2 is a Hermitian operator:

$$\langle S_*^2 w, v \rangle = \langle S_*v, S_*w \rangle = \langle w, S_*^2 v \rangle$$

for all $v, w \in V$.

(c) Since S_*^2 is Hermitian, there is an orthonormal basis B of V relative to which S_*^2 has all off-diagonal entries as 0. Show that all the diagonal entries are positive.

(d) Let S_0 be the unique linear operator $V \rightarrow V$ that, relative to the basis B in (c), has a matrix for which all of-diagonal entries are 0 and for which the diagonal entries are the positive square roots of the corresponding entries for the matrix of S_*^2. Thus, $S_0 = (S_*^2)^{1/2}$ in the sense that $S_0^2 = S_*^2$ and S_0 is Hermitian and positive: $\langle S_0 v, v \rangle \geq 0$ with equality if and only if $v = 0$. Show that

$$S_0 \rho(g) = \rho(g) S_0 \text{ for all } g \in G,$$

and also that S_0 commutes with S_*.

(e) Let

$$C = S_* S_0^{-1}. \tag{1.62}$$

Check that $C : V \rightarrow V$ is conjugate-linear, $C^2 = I$, the identity map on V, and $C\rho(g) = \rho(g)C$ for all $g \in V$.

(f) By writing any $v \in V$ as

$$v = \frac{1}{2}(v + Cv) + i\frac{1}{2i}(v - Cv),$$

show that

$$V = V_{\mathbb{R}} \oplus iV_{\mathbb{R}},$$

where $V_{\mathbb{R}}$ is the *real vector space* consisting of all $v \in V$ for which $Cv = v$.

(g) Show that $\rho(g)V_{\mathbb{R}} \subset V_{\mathbb{R}}$ for all $g \in G$. Let $\rho_{\mathbb{R}}$ be the representation of G on the real vector space $V_{\mathbb{R}}$ given by the restriction of ρ. Show that ρ is the complexification of $\rho_{\mathbb{R}}$. In particular, there is a basis of V relative to which all matrices $\rho(g)$ have all real entries.

(h) Conversely, show that if there is a basis of V for which all entries of all the matrices $\rho(g)$ are real, then there is a nonzero symmetric G-invariant bilinear form on V.

(i) Prove that for an irreducible complex character χ of a finite group, the Frobenius–Schur indicator has value 0 if the character is not real-valued, has value 1 if the character arises from the complexification of a real representation, and has value -1 if the character is real-valued but does not arise from the complexification of a real representation.

1.19. Let ρ be a representation of a group G on a vector space V over a field $\mathbb{F}$. Show that the subspace $V^{\hat{\otimes}2}$ consisting of symmetric tensors in $V \otimes V$ is invariant under the tensor product representation $\rho \otimes \rho$. Assume that G is finite, containing m elements, and the field $\mathbb{F}$ has characteristic $\neq 2$ and contains m distinct mth roots of unity. Work out the character of the representation ρ_s that is given by the restriction of $\rho \otimes \rho$ to $V^{\hat{\otimes}2}$. (Hint: Diagonalize.)

1.20. Let ρ be an irreducible complex representation of a finite group G on a space of dimension d_ρ and let χ_ρ be its character. If g is an element of G for which $|\chi_\rho(g)| = d_\rho$, show that $\rho(g)$ is of the form cI for some root of unity c.

1.21. Let χ be the character of a representation ρ of a finite group G on a finite-dimensional complex vector space $V \neq 0$. Dixon [24] describes a computationally convenient way to recover the diagonalized form of $\rho(g)$ from the values of χ on the powers of g; in fact, he explains how to recover the diagonalized form of $\rho(g)$, and hence also the value of $\chi(g)$, given only approximate values of the character. Here is a pathway through these ideas:

(a) Suppose U is an $n \times n$ complex diagonal matrix such that $U^d = I$, where d is a positive integer. Let ζ be any dth root of unity. Show that

$$\frac{1}{d} \sum_{k=0}^{d-1} \mathrm{Tr}(U^k)\zeta^{-k}$$

   = the number of times ζ appears on the diagonal of U.

(1.63)

(Hint: If $w^d = 1$, where d is a positive integer, then $1 + w + w^2 + \cdots + w^{d-1}$ is 0 if $w \neq 1$, and is d if $w = 1$.)

(b) If all the values of the character χ are known, use (a) to explain how the diagonalized form of $\rho(g)$ can be computed for every $g \in G$.

(c) Now consider $g \in G$, and let d be a positive integer for which $g^d = e$. Suppose we know the values of χ on the powers of g within an error margin $< 1/2$. In other words, suppose we have complex numbers $z_1, ..., z_d$ with $|z_j - \chi(g^j)| < 1/2$ for all $j \in \{1, ..., d\}$. Show that for any dth root of unity ζ the integer closest to $d^{-1} \sum_{k=1}^{d} z_k \zeta^{-k}$ is the multiplicity of ζ in the diagonalized form of $\rho(g)$. Thus, the values $z_1, ..., z_k$ can be used to compute the diagonalized form of $\rho(g)$ and hence also the exact value of χ on the powers of g. Modify this to allow for approximate values of the powers of ζ as well.

1.22. Let $X, X_1, ..., X_n$ be indeterminates and

$$f(X) = \prod_{j=1}^{n} (X - X_j) = \sum_{k=0}^{n} (-1)^{n-k} E_{n-k} X^k, \qquad (1.64)$$

where E_k is the kth *elementary symmetric polynomial*

$$E_k = \sum_{B \in P_k} \prod_{i \in B} X_i, \qquad (1.65)$$

with P_k being the set of all k-element subsets of $[n] = \{1, ..., n\}$ for $k \in [n]$ and $E_0 = 1$. Using the algebraic identities

$$f'(X) = \sum_{i=1}^{n} \frac{f(X) - f(X_i)}{X - X_i} = \sum_{i=1}^{n} \sum_{j=1}^{n} (-1)^{n-j} E_{n-j} \frac{X^j - X_i^j}{X - X_i}$$

$$= \sum_{k=1}^{n} \left(\sum_{j=k}^{n} (-1)^{n-j} E_{n-j} N_{j-k} \right) X^{k-1}, \qquad (1.66)$$

where, for all $k \in \{0, 1, ..., n\}$, N_k is the kth *power sum*

$$N_k = \sum_{i=1}^{n} X_i^k, \qquad (1.67)$$

and $f'(X) = \sum_{k=1}^{n} (-1)^{n-k} E_{n-k} k X^{k-1}$, show that

$$k E_k + \sum_{j=0}^{k-1} (-1)^{k-j} E_j N_{k-j} = 0. \qquad (1.68)$$

From this show that

$$\det \begin{bmatrix} N_1 & 1 & 0 & 0 & \cdots & 0 \\ N_2 & N_1 & 2 & 0 & \cdots & 0 \\ N_3 & N_2 & N_1 & 3 & \cdots & 0 \\ \vdots & \vdots & \vdots & \vdots & \ddots & \vdots \\ N_{k-1} & N_{k-2} & N_{k-3} & N_{k-4} & \cdots & k-1 \\ N_k & N_{k-1} & N_{k-2} & N_{k-3} & \cdots & N_1 \end{bmatrix} = k! E_k, \qquad (1.69)$$

by expressing the first column as a linear combination of the other columns and a column vector whose entries are all 0 except for the last entry, which is $(-1)^{k-1} k E_k$. This expresses the kth elementary symmetric sum E_k in terms of the power sums as

$$E_k = \frac{1}{k!} \det \begin{bmatrix} N_1 & 1 & 0 & 0 & \cdots & 0 \\ N_2 & N_1 & 2 & 0 & \cdots & 0 \\ N_3 & N_2 & N_1 & 3 & \cdots & 0 \\ \vdots & \vdots & \vdots & \vdots & \ddots & \vdots \\ N_{k-1} & N_{k-2} & N_{k-3} & N_{k-4} & \cdots & k-1 \\ N_k & N_{k-1} & N_{k-2} & N_{k-3} & \cdots & N_1 \end{bmatrix}, \qquad (1.70)$$

which is a polynomial in $N_1, \ldots, N_n$ with coefficients in any field in which $k! \neq 0$. Relations between the elementary symmetric polynomials and the power sums are called *Newton–Girard formulas*.

A Reckoning

He sits in a seamless room
staring
into the depths
of a wall that is not a wall,
opaque,
unfathomable.

Though deep understanding
lies
just beyond that wall,
the vision he desires
can be seen
only from within the room.

Sometimes a sorrow transports
    him
  through the door that is not
  a door,
  down stairs that are not stairs
  to the world beyond the place
  of seeking:

    down fifty steps
hand carved into the mountains
    stony side
    to a goat path that leads to
    switchbacks,
  becoming a trail that becomes
  a road;
    and thus he wanders to the
  town beyond.

Though barely dusk,
  the night lights brighten
  guiding him
    to the well known place of
  respite.

They were boisterous within,
but they respect him as the one
      who seeks,
and so they sit subdued,
waiting,
hoping for the revelation that
    never comes.

Amidst the quiet clinking of glasses
and the softly whispered reverence,
  a woman approaches,
    escorts him to their accustomed
    place.
They speak with words that are
      not words
about ideas that are not ideas
enshrouded by a silence that is
    not silence.

His presence stifles their gaiety,
her gaiety,
and so he soon grows restless
and desires to return to his hope-
      less toil.

The hand upon his cheek,
the tear glistening in her eye,
the whispered words husband mine,
will linger with him
until he once again attains
his room that is not a room.

  As he leaves,
   before the door can slam be-
    hind him,
   he hears their voices
   rise
   once again

in blessed celebration,
   hers distinctly above the others.

But he follows his trail
 and his switchbacks
and his goat path
and the fifty steps
to his seamless world

prepared once again
to let his god
who is not a god
take potshots at his soul.

            Charlie Egedy

Chapter 2

Basic Examples

We will work our way through examples in this chapter, looking at representations and characters of some familiar finite groups. We focus on complex representations, but any algebraically closed field of characteristic zero (e.g., the algebraic closure $\overline{\mathbb{Q}}$ of the rationals) could be substituted for $\mathbb{C}$.

Recall that the character χ_ρ of a finite-dimensional representation ρ of a group G is the function on the group specified by

$$\chi_\rho(g) = \operatorname{Tr} \rho(g). \tag{2.1}$$

Characters are invariant under conjugation, and so χ_ρ takes a constant value $\chi_\rho(C)$ on any conjugacy class C. As we have seen before in (1.50),

$$\chi_\rho(g^{-1}) = \overline{\chi_\rho(g)} \qquad \text{for all } g \in G, \tag{2.2}$$

for any complex representation ρ. We say that a character is *irreducible* if it is the character of an irreducible representation. A *complex character* is the character of a complex representation.

We denote by $\mathcal{R}_G$ a maximal set of inequivalent irreducible complex representations of G. Let $\mathcal{C}_G$ be the set of all conjugacy classes in G. If C is a conjugacy class, then we denote by C^{-1} the conjugacy class consisting of the inverses of the elements in C.

It will be useful to keep at hand some facts (proofs are given in Chap. 7) about complex representations of any finite group G: (a) there are only finitely many inequivalent irreducible complex representations of G and these are all finite-dimensional; (b) two finite-dimensional complex representations

A.N. Sengupta, *Representing Finite Groups: A Semisimple Introduction*,
DOI 10.1007/978-1-4614-1231-1_2, © Springer Science+Business Media, LLC 2012

of G are equivalent if and only if they have the same character; (c) a complex representation of G is irreducible if and only if its character χ_ρ satisfies

$$\sum_{g \in G} |\chi_\rho(g)|^2 = \sum_{C \in \mathcal{C}_G} |C||\chi_\rho(C)|^2 = |G|; \qquad (2.3)$$

and (d) the number of inequivalent irreducible complex representations of G is equal to the number of conjugacy classes in G.

In going through the examples in this chapter, we will sometimes pause to use or verify some standard properties of complex characters of a finite group G (again, proofs are given in Chap. 7). These properties are summarized in the orthogonality relations among complex characters:

$$\sum_{h \in G} \chi_\rho(gh)\chi_{\rho_1}(h^{-1}) = |G|\chi_\rho(g)\delta_{\rho\rho_1},$$

$$\sum_{\rho \in \mathcal{R}_G} \chi_\rho(C')\chi_\rho(C^{-1}) = \frac{|G|}{|C|}\delta_{C'C}, \qquad (2.4)$$

where δ_{ab} is 1 if $a = b$ and is 0 otherwise, the relations above being valid for all $\rho, \rho_1 \in \mathcal{R}_G$, all conjugacy classes $C, C' \in \mathcal{C}_G$, and all elements $g \in G$. Specializing this to specific cases (such as $\rho = \rho_1$ or $g = e$), we have

$$\sum_{\rho \in \mathcal{R}_G} (\dim \rho)^2 = |G|,$$

$$\sum_{\rho \in \mathcal{R}_G} \dim \rho \, \chi_\rho(g) = 0 \qquad \text{if } g \neq e, \qquad (2.5)$$

$$\sum_{g \in G} \chi_{\rho_1}(g)\chi_{\rho_2}(g^{-1}) = |G|\delta_{\rho_1 \rho_2} \dim \rho \qquad \text{for } \rho_1, \rho_2 \in \mathcal{R}_G.$$

2.1 Cyclic Groups

Let us work out all irreducible representations of a cyclic group C_n containing n elements. Being cyclic, C_n contains a *generator* c, which is an element such that C_n consists exactly of the powers $c, c^2, ..., c^n$, where c^n is the identity e in the group. Figure **2.1** displays C_8 as eight equally spaced points around the unit circle in the complex plane.

Let ρ be a representation of C_n on a complex vector space $V \neq 0$. By Proposition 1.6, there is a basis of V relative to which the matrix of $\rho(c)$ is

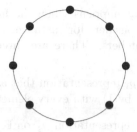

Fig. 2.1 The cyclic group C_8

diagonal, with each diagonal entry being an nth root of unity. If V is of finite dimension d, then

$$\text{matrix of } \rho(c) = \begin{bmatrix} \eta_1 & 0 & 0 & \dots & 0 \\ 0 & \eta_2 & 0 & \dots & 0 \\ \vdots & \vdots & \vdots & \dots & \vdots \\ 0 & 0 & 0 & \dots & \eta_d \end{bmatrix}.$$

Since c generates the full group C_n, the matrix for ρ is diagonal on all the elements c^j in C_n. Thus, V is a direct sum of one-dimensional subspaces, each of which provides a representation of C_n. Of course, any one-dimensional representation is automatically irreducible.

Let us summarize our observations:

Theorem 2.1 *Let C_n be a cyclic group of order $n \in \{1, 2, ...\}$. Every complex representation of C_n is a direct sum of irreducible representations. Each irreducible complex representation of C_n is one-dimensional, specified by the requirement that a generator element $c \in G$ act through multiplication by an nth root of unity. Each nth root of unity provides, in this way, an irreducible complex representation of C_n, and these representations are mutually inequivalent.*

Thus, there are exactly n inequivalent irreducible complex representations of C_n.

Everything we have done here applies for representations of C_n over a field containing n distinct roots of unity.

Let us now look at what happens when the field does not contain the requisite roots of unity. Consider, for instance, the representations of C_3 over the field $\mathbb{R}$ of real numbers. There are three geometrically apparent representations:

1. The one-dimensional ρ_1 representation that associates the identity operator (multiplication by 1) with every element of C_3;

2. The two-dimensional representation ρ_2^+ on $\mathbb{R}^2$ in which c is associated with rotation by 120°;

3. The two-dimensional representation ρ_2^- on $\mathbb{R}^2$ in which c is associated with rotation by −120°.

These are clearly all irreducible. Moreover, any irreducible representation of C_3 on $\mathbb{R}^2$ is clearly either (2) or (3).

Now consider a general real vector space V on which C_3 has a representation ρ. Choose a basis B in V, and let $V_{\mathbb{C}}$ be the complex vector space with B as a basis (put another way, $V_{\mathbb{C}}$ is $\mathbb{C} \otimes_{\mathbb{R}} V$, viewed as a complex vector space). Then ρ gives, naturally, a representation of C_3 on $V_{\mathbb{C}}$. Then $V_{\mathbb{C}}$ is a direct sum of complex one-dimensional subspaces, each invariant under the action of C_3. Since a complex one-dimensional vector space is a real two-dimensional space, and we have already determined all two-dimensional real representations of C_3, we have finished classifying all real representations of C_3. Too fast, you say? Then proceed to Exercise 2.6.

Finite Abelian groups are products of cyclic groups. This could give the impression that there is nothing very interesting in the representations of such groups. But even a very simple representation can be of great use. For any prime p, the nonzero elements in $\mathbb{Z}_p = \mathbb{Z}/p\mathbb{Z}$ form a group $\mathbb{Z}_p^*$ under multiplication. For any $a \in \mathbb{Z}_p^*$, define

$$\lambda_p(a) = a^{(p-1)/2},$$

this being 1 in the case $p = 2$. Since its square is $a^{p-1} = 1$, $\lambda_p(a)$ is necessarily ± 1. Clearly,

$$\lambda_p : \mathbb{Z}_p^* \to \{1, -1\}$$

is a group homomorphism, and hence gives a one-dimensional representation, which is the same as a one-dimensional character of $\mathbb{Z}_p^*$. The *Legendre symbol* $\left(\frac{a}{p}\right)$ is defined for any integer a by

$$\left(\frac{a}{p}\right) = \begin{cases} \lambda_p(a \bmod p) & \text{if } a \text{ is coprime to } p \\ 0 & \text{if } a \text{ is divisible by } p. \end{cases}$$

The celebrated law of quadratic reciprocity, conjectured by Euler and Legendre and proved first, and many times over, by Gauss, states that

$$\left(\frac{p}{q}\right)\left(\frac{q}{p}\right) = (-1)^{(p-1)/2}(-1)^{(q-1)/2},$$

if p and q are odd primes. For an extension of these ideas using the character theory of general finite groups, see the article by Duke and Hopkins [25].

2.2   Dihedral Groups

The dihedral group D_n, for n any positive integer, is a group of $2n$ elements generated by two elements c and r, where c has order n, r has order 2, and conjugation by r turns c into c^{-1}:

$$c^n = e, \qquad r^2 = e, \qquad rcr^{-1} = c^{-1}. \tag{2.6}$$

Geometrically, think of c as counterclockwise rotation in the plane by the angle $2\pi/n$ and r as reflection across a fixed line through the origin. The distinct elements of D_n are

$$e, c, c^2, \ldots, c^{n-1}, r, cr, c^2r, \ldots, c^{n-1}r.$$

This geometric view of D_n, illustrated in Fig. **2.2**, immediately yields a real two-dimensional representation: let c act on $\mathbb{R}^2$ through counterclockwise rotation by angle $2\pi/n$ and let r act through reflection across the x-axis. Relative to the standard basis of $\mathbb{R}^2$ these two linear maps have the following matrix forms:

$$\rho(c) = \begin{bmatrix} \cos(2\pi/n) & -\sin(2\pi/n) \\ \sin(2\pi/n) & \cos(2\pi/n) \end{bmatrix}, \qquad \rho(r) = \begin{bmatrix} 1 & 0 \\ 0 & -1 \end{bmatrix}.$$

It instructive to see what happens when we complexify and take this representation over to $\mathbb{C}^2$. Choose in $\mathbb{C}^2$ the basis given by eigenvectors of $\rho(c)$:

$$b_1 = \begin{pmatrix} 1 \\ -i \end{pmatrix} \quad \text{and} \quad b_2 = \begin{pmatrix} 1 \\ i \end{pmatrix}.$$

Fig. 2.2 The dihedral group D_4

Then

$$\rho_{\mathbb{C}}(c)b_1 = \eta b_1 \quad \text{and} \quad \rho_{\mathbb{C}}(c)b_2 = \eta^{-1}b_2,$$

where $\eta = e^{2\pi i/n}$, and

$$\rho_{\mathbb{C}}(r)b_1 = b_2 \quad \text{and} \quad \rho_{\mathbb{C}}(r)b_2 = b_1.$$

Thus, relative to the basis given by b_1 and b_2, the matrices of $\rho_{\mathbb{C}}(c)$ and $\rho_{\mathbb{C}}(r)$ are

$$\begin{bmatrix} \eta & 0 \\ 0 & \eta^{-1} \end{bmatrix} \quad \text{and} \quad \begin{bmatrix} 0 & 1 \\ 1 & 0 \end{bmatrix}.$$

Switching our perspective from the standard basis to that given by b_1 and b_2 produces a two-dimensional complex representation ρ_1 on $\mathbb{C}^2$ given by

$$\rho_1(c) = \begin{bmatrix} \eta & 0 \\ 0 & \eta^{-1} \end{bmatrix}, \qquad \rho_1(r) = \begin{bmatrix} 0 & 1 \\ 1 & 0 \end{bmatrix}. \tag{2.7}$$

Having been obtained by a change of basis, this representation is equivalent to the representation $\rho_{\mathbb{C}}$, which in turn is the complexification of the rotation–reflection real representation ρ of the dihedral group on $\mathbb{R}^2$.

More generally, we have the representation ρ_m specified by requiring

$$\rho_m(c) = \begin{bmatrix} \eta^m & 0 \\ 0 & \eta^{-m} \end{bmatrix}, \qquad \rho_m(r) = \begin{bmatrix} 0 & 1 \\ 1 & 0 \end{bmatrix}$$

for any $m \in \mathbb{Z}$; of course, to avoid repetition, we may focus on $m \in \{1, 2, ..., n-1\}$. The values of ρ_m on all elements of D_n are given by

$$\rho_m(c^j) = \begin{bmatrix} \eta^{mj} & 0 \\ 0 & \eta^{-mj} \end{bmatrix}, \qquad \rho_m(c^j r) = \begin{bmatrix} 0 & \eta^{mj} \\ \eta^{-mj} & 0 \end{bmatrix}.$$

(Having written this, we notice that this representation makes sense over any field $\mathbb{F}$ containing nth roots of unity. However, we stick to the ground field $\mathbb{C}$, or at least $\mathbb{Q}$ with any primitive nth root of unity adjoined.)

Clearly, ρ_m repeats itself when m changes by multiples of n. Thus, we need only focus on $\rho_1, ..., \rho_{n-1}$.

Is ρ_m reducible? Yes if, and only if, there is a nonzero vector $v \in \mathbb{C}^2$ fixed by $\rho_m(r)$ and $\rho_m(c)$. Being fixed by $\rho_m(r)$ means that such a vector must be a multiple of $(1, 1)$ in $\mathbb{C}^2$. But $\mathbb{C}(1, 1)$ is also invariant under $\rho_m(c)$ if and only if η^m is equal to η^{-m}.

Thus, ρ_m for $m \in \{1, ..., n-1\}$ is irreducible if $n \neq 2m$ and is reducible if $n = 2m$.

Are we counting things too many times? Indeed, the representations ρ_m are not all inequivalent. Interchanging the two axes converts ρ_m into $\rho_{-m} = \rho_{n-m}$. Thus, we can narrow our focus to ρ_m for $1 \leq m < n/2$.

We have now identified $n/2 - 1$ irreducible two-dimensional complex representations if n is even, and $(n-1)/2$ irreducible two-dimensional complex representations if n is odd.

The character χ_m of ρ_m is obtained by taking the trace of ρ_m on the elements of the group D_n:

$$\chi_m(c^j) = \eta^{mj} + \eta^{-mj}, \qquad \chi_m(c^j r) = 0.$$

Now consider a one-dimensional complex representation θ of D_n. First, from $\theta(r)^2 = 1$, we see that $\theta(r) = \pm 1$. If we apply θ to the relation that rcr^{-1} equals c^{-1}, it follows that $\theta(c)$ must also be ± 1. But then, from $c^n = e$, it follows that $\theta(c)$ can be -1 only if n is even. Thus, we have the one-dimensional representations specified by

$$\begin{aligned} \theta_{+,\pm}(c) &= 1, \quad \theta_{+,\pm}(r) = \pm 1 \qquad \text{if } n \text{ is even or odd,} \\ \theta_{-,\pm}(c) &= -1, \quad \theta_{-,\pm}(r) = \pm 1 \qquad \text{if } n \text{ is even.} \end{aligned} \tag{2.8}$$

This gives us four one-dimensional complex representations if n is even, and two if n is odd. (Indeed, the reasoning here works for any ground field.)

Thus, for n is even we have identified a total of $3 + n/2$ irreducible representations, and for n is odd we have identified $(n+3)/2$ irreducible representations.

As noted in the first equation in (2.5), the sum $\sum_{\chi \in \mathcal{R}_G} d_\chi^2$ over all distinct complex irreducible characters of a finite group G is the total number of elements in G. In this case the sum should be $2n$. Working out the sum over all the irreducible characters χ we have determined, we obtain

$$\left(\frac{n}{2} - 1\right) 2^2 + 4 = 2n \qquad \text{for even } n;$$
$$\left(\frac{n-1}{2}\right) 2^2 + 2 = 2n \qquad \text{for odd } n. \tag{2.9}$$

Thus, our list of irreducible complex representations contains all irreducible complex representations, up to equivalence.

Our next objective is to work out all complex characters of D_n. Since characters are constant on conjugacy classes, let us first determine the conjugacy classes in D_n.

Since rcr^{-1} is c^{-1}, it follows that

$$r(c^j r)r^{-1} = c^{-j}r = c^{n-j}r.$$

This already indicates that the conjugacy class structure is different for n is even and n is odd. In fact, notice that conjugating $c^j r$ by c results in increasing j by 2:

$$c(c^j r)c^{-1} = c^{j+1}cr = c^{j+2}r.$$

If n is even, the conjugacy classes are:

$$\{e\}, \{c, c^{n-1}\}, \{c^2, c^{n-2}\}, ..., \{c^{n/2-1}, c^{n/2+1}\}, \{c^{n/2}\},$$
$$\{r, c^2 r, ..., c^{n-2}r\}, \{cr, c^3 r, ..., c^{n-1}r\}. \tag{2.10}$$

Note that there are $3 + n/2$ conjugacy classes, and this exactly matches the number of inequivalent irreducible complex representations obtained earlier.

To see how this plays out in practice, let us look at D_4. Our analysis shows that there are five conjugacy classes:

$$\{e\}, \{c, c^3\}, \{c^2\}, \{r, c^2 r\}, \{cr, c^3 r\}.$$

There are four one-dimensional complex representations $\theta_{\pm,\pm}$, and one irreducible two-dimensional complex representation ρ_1 specified through

$$\rho_1(c) = \begin{bmatrix} i & 0 \\ 0 & -i \end{bmatrix}, \qquad \rho_1(r) = \begin{bmatrix} 0 & 1 \\ -1 & 0 \end{bmatrix}.$$

Table **2.1** contains the *character table* of D_4, listing the values of the irreducible complex characters of D_4 on the various conjugacy classes. The latter

Table 2.1 Complex irreducible characters of D_4

	1	2	1	2	2
	e	c	c^2	r	cr
$\theta_{+,+}$	1	1	1	1	1
$\theta_{+,-}$	1	1	1	-1	-1
$\theta_{-,+}$	1	-1	1	1	-1
$\theta_{-,-}$	1	-1	1	-1	1
χ_1	2	0	-2	0	0

Table 2.2 Complex irreducible characters of $D_3 = S_3$

	1	2	3
	e	c	r
$\theta_{+,+}$	1	1	1
$\theta_{+,-}$	1	1	-1
χ_1	2	-1	0

are displayed in a row (second from top), each conjugacy class identified by an element it contains; above each conjugacy class we have listed the number of elements it contains. Each row in the main body of the table displays the values of a character on the conjugacy classes.

The case for odd n proceeds similarly. Take, for instance, $n = 3$. The group D_3 is generated by elements c and r subject to the relations

$$c^3 = e, \quad r^2 = e, \quad rcr^{-1} = c^{-1}.$$

The conjugacy classes are

$$\{e\}, \{c, c^2\}, \{r, cr, c^2r\}$$

The irreducible complex representations are $\theta_{+,+}$, $\theta_{+,-}$, ρ_1. Their values are displayed in Table **2.2**, where the first row displays the number of elements in the conjugacy classes listed (by choice of an element) in the second row. The dimensions of the representations can be read off from the first column in the main body of the table. Observe that the sum of the squares of the dimensions of the representations of S_3 listed in the table is

$$1^2 + 1^2 + 2^2 = 6,$$

which is exactly the number of elements in D_3. This verifies the first property listed earlier in (2.5).

Table **2.3** Conjugacy classes in S_4

Number of elements	1	6	8	6	3
Conjugacy class	ι	(12)	(123)	(1234)	(12)(34)

2.3   The Symmetric Group S_4

The symmetric group S_3 is isomorphic to the dihedral group D_3, and we have already determined the irreducible representations of D_3 over the complex numbers. Let us turn now to the symmetric group S_4, which is the group of permutations of $\{1, 2, 3, 4\}$. Geometrically, this is the group of rotational symmetries of a cube.

Two elements of S_4 are conjugate if and only if they have the same cycle structure; thus, for instance, (134) and (213) are conjugate, and these are not conjugate to (12)(34). The following elements belong to all the distinct conjugacy classes:

$$\iota, \quad (12), \quad (123), \quad (1234), \quad (12)(34),$$

where ι is the identity permutation. The conjugacy classes, each identified by one element they contain, are listed with the number of elements in each conjugacy class in Table **2.3**.

There are two one-dimensional complex representations of S_4 we are familiar with: the trivial one, associating 1 with every element of S_4, and the signature representation ϵ whose value is $+1$ on even permutations and -1 on odd ones.

We also have seen a three-dimensional irreducible complex representation of S_4; recall the representation R of S_4 on $\mathbb{C}^4$ given by permutation of coordinates:

$$(x_1, x_2, x_3, x_4) \mapsto (x_{\sigma^{-1}(1)}, \dots, x_{\sigma^{-1}(4)})$$

Equivalently,

$$R(\sigma)e_j = e_{\sigma(j)} \qquad \text{for } j \in \{1, 2, 3, 4\},$$

where $e_1, \dots, e_4$ are the standard basis vectors of $\mathbb{C}^4$. The three-dimensional subspace

$$E_0 = \{(x_1, x_2, x_3, x_4) \in \mathbb{C}^4 : x_1 + x_2 + x_3 + x_4 = 0\}$$

Table 2.4 The characters χ_R and χ_0 on conjugacy classes

Conjugacy class	ι	(12)	(123)	(1234)	(12)(34)
χ_R	4	2	1	0	0
χ_0	3	1	0	-1	-1
χ_1	3	-1	0	1	-1

is mapped into itself by the action of R, and the restriction to E_0 gives an irreducible representation R_0 of S_4. In fact,

$$\mathbb{C}^4 = E_0 \oplus \mathbb{C}(1,1,1,1)$$

decomposes the space $\mathbb{C}^4$ into complementary invariant, irreducible subspaces. The subspace $\mathbb{C}(1,1,1,1)$ carries the trivial representation (all elements act through the identity map). Examining the effect of the group elements on the standard basis vectors, we can work out the character of R. For instance, $R((12))$ interchanges e_1 and e_2, and leaves e_3 and e_4 fixed, and so its matrix is

$$\begin{bmatrix} 0 & 1 & 0 & 0 \\ 1 & 0 & 0 & 0 \\ 0 & 0 & 1 & 0 \\ 0 & 0 & 0 & 1 \end{bmatrix}$$

and the trace is

$$\chi_R((12)) = 2.$$

Subtracting the trivial character, which is 1 on all elements of S_4, we obtain the character χ_0 of the representation R_0. All this is displayed in the first three rows in Table **2.4**.

We can create another three-dimensional complex representation R_1 by tensoring R_0 with the signature ϵ:

$$R_1 = R_0 \otimes \epsilon.$$

The character χ_1 of R_1 is then written down by taking products, and is displayed in the fourth row in Table **2.4**.

Since R_0 is irreducible and R_1 acts by a simple ± 1 scaling of R_0, it is clear that R_1 is also irreducible. Thus, we now have two one-dimensional complex representations and two three-dimensional complex irreducible representations. The sum of the squares of the dimensions is

$$1^2 + 1^2 + 3^2 + 3^2 = 20.$$

From the first relation in (2.5) we know that the sum of the squares of the dimensions of all the inequivalent irreducible complex representations is $|S_4| = 24$. Thus, looking at the equation

$$24 = 1^2 + 1^2 + 3^2 + 3^2 + ?^2,$$

we see that we are missing a two-dimensional irreducible complex representation R_2. Leaving the entries for this blank, we have Table **2.5**.

Table 2.5 Character table for S_4 with a missing row

	1	6	8	6	3
	ι	(12)	(123)	(1234)	(12)(34)
Trivial	1	1	1	1	1
ϵ	1	-1	1	-1	1
χ_0	3	1	0	-1	-1
χ_1	3	-1	0	1	-1
χ_2	2	?	?	?	?

As an illustration of the power of character theory, let us work out the character χ_2 of this "missing" representation R_2, without even bothering to search for the representation itself. Recall from (2.5) the relation

$$\sum_{\rho} (\dim \rho)\, \chi_\rho(\sigma) = 0, \qquad \text{if } \sigma \neq \iota,$$

Table 2.6 Character table for S_4

	1	6	8	6	3
	ι	(12)	(123)	(1234)	(12)(34)
Trivial	1	1	1	1	1
ϵ	1	−1	1	−1	1
χ_0	3	1	0	−1	−1
χ_1	3	−1	0	1	−1
χ_2	2	0	−1	0	2

where the sum runs over a maximal set of inequivalent irreducible complex representations of S_4 and σ is any element of S_4. This means that *the vector formed by the first column* in the main body of the table (i.e., the column for the conjugacy class $\{\iota\}$) *is orthogonal to the vectors* formed by the columns *for the other conjugacy classes*. Using this we can work out the entries missing from the character table. For instance, taking $\sigma = (12)$, we have

$$2\chi_2((12)) + 3 * \underbrace{(-1)}_{\chi_1((12))} + 3 * 1 + 1 * (-1) + 1 * 1 = 0,$$

which yields

$$\chi_2((12)) = 0.$$

For $\sigma = (123)$, we have

$$2\chi_2((123)) + 3 * \underbrace{0}_{\chi_1((123))} + 3 * 0 + 1 * 1 + 1 * 1 = 0,$$

which produces

$$\chi_2((123)) = -1.$$

Filling in the entire last row of the character table in this way produces the Table **2.6**.

Just to be sure that the indirectly detected character χ_2 is irreducible, let us run the check given in (2.3) for irreducible complex characters: the sum of the quantities $|C||\chi_2(C)|^2$ over all the conjugacy classes C should be 24. Indeed, we have

$$\sum_C |C||\chi_2(C)|^2 = 1*2^2 + 6*0^2 + 8*(-1)^2 + 6*0^2 + 3*2^2 = 24 = |S_4|,$$

a pleasant proof of the power of the theory and tools promised to be developed in the chapters ahead.

2.4   Quaternionic Units

Before moving on to general theory in the next chapter, let us look at another example which produces a little surprise. The unit quaternions

$$1, -1, i, -i, j, -j, k, -k$$

form a group Q under multiplication. We can take

$$-1, i, j, k$$

as generators, with the relations

$$(-1)^2 = 1, \ i^2 = j^2 = k^2 = -1, \ ij = k.$$

The conjugacy classes are

$$\{1\}, \{-1\}, \{i, -i\}, \{j, -j\}, \{k, -k\}.$$

We can spot the one-dimensional representations as follows. Since

$$ijij = k^2 = -1 = i^2 = j^2,$$

the value of any one-dimensional representation τ on -1 must be 1 because

$$\tau(-1) = \tau(ijij) = \tau(i)\tau(j)\tau(i)\tau(j) = \tau(i^2 j^2) = \tau(1) = 1, \qquad (2.11)$$

and then the values on i and j must each be ± 1. (For another formulation of this argument, see Exercise 4.6.) A little thought shows that

Table 2.7 Character table for Q, missing the last row

	1	2	1	2	2
	1	i	-1	j	k
$\chi_{+1,+1}$	1	1	1	1	1
$\chi_{+1,-1}$	1	1	1	-1	-1
$\chi_{-1,+1}$	1	-1	1	1	-1
$\chi_{-1,-1}$	1	-1	1	-1	1
χ_2	2	?	?	?	?

Table 2.8 Character table for Q

	1	2	1	2	2
	1	i	-1	j	k
$\chi_{+,+}$	1	1	1	1	1
$\chi_{+,-}$	1	1	1	-1	-1
$\chi_{-,+}$	1	-1	1	1	-1
$\chi_{-,-}$	1	-1	1	-1	1
χ_2	2	0	-2	0	0

$(\tau(i), \tau(j))$ could be taken to be any of the four possible values $(\pm 1, \pm 1)$ and this would specify a one-dimensional representation τ. Thus, there are four one-dimensional representations. Given that Q contains eight elements, writing this as a sum of squares of dimensions of irreducible complex representations, we have

$$8 = 1^2 + 1^2 + 1^2 + 1^2 + ?^2$$

Clearly, what we are missing is an irreducible complex representation of dimension 2. The incomplete character table is displayed in Table **2.7**.

Remarkably, everything here, with the potential exception of the missing last row, is identical to the information in Table **2.1** for the dihedral group D_4. Then, since the last row is entirely determined by the information available, the entire character table for Q must be identical to that of D_4. Thus the complete character table for Q is as shown in Table **2.8**.

A guess at this stage would be that Q must be isomorphic to D_4, a guess bolstered by the observation that certainly the conjugacy classes look much the same, in terms of the number of elements at least. But this guess is shown to be invalid upon second thought: the dihedral group D_4 has four elements r, cr, $c^2 r$, and $c^3 r$ each of order 2, whereas the only element of order 2 in Q is -1. So we have an interesting observation here: *two nonisomorphic groups can have identical character tables*!

2.5   Afterthoughts: Geometric Groups

In closing this chapter, let us note some important classes of finite groups, although we will not explore their representations specifically.

The group Q of special quaternions we studied in Sect. 2.4 is a particular case of a more general setting. Let V be a finite-dimensional real vector space equipped with an inner product $\langle \cdot, \cdot \rangle$. There is then the *Clifford algebra* $C_{real,d}$, which is an associative algebra over $\mathbb{R}$, with a unit element 1, whose elements are linear combinations of formal products $v_1...v_m$ (with this being 1 if $m = 0$), linear in each $v_i \in V$, with the requirement that

$$vw + wv = -2\langle v, w \rangle 1 \qquad \text{for all } v, w \in V.$$

If e_1, ..., e_d form an orthonormal basis of V, then the products $\pm e_{i_1}...e_{i_k}$, for $k \in \{0, ..., d\}$, form a group Q_d under the multiplication operation of the algebra $C_{real,d}$. When $d = 2$, we write $i = e_1$, $j = e_2$, and $k = e_1 e_2$, and obtain $Q_2 = \{1, -1, i, -i, j, -j, k, -k\}$, the quaternionic group.

In chemistry one studies *crystallographic groups*, which are finite subgroups of the group of Euclidean motions in $\mathbb{R}^3$. *Reflection groups* are groups generated by reflections in Euclidean spaces. Let V be a finite-dimensional real vector space with an inner product $\langle \cdot, \cdot \rangle$. If w is a unit vector in V, then the reflection r_w across the hyperplane

$$w^{\perp} = \{v \in \mathbb{R}^n : \langle v, w \rangle = 0\}$$

takes w to $-w$ and holds all vectors in the "mirror" $w^{\perp}$ fixed; thus,

$$r_w(v) = v - 2\langle v, w \rangle w \qquad \text{for all } v \in V. \tag{2.12}$$

If r_1 and r_2 are reflections across planes $w_1^{\perp}$ and $w_2^{\perp}$, where w_1 and w_2 are unit vectors in V with angle $\theta = \cos^{-1}\langle w_1, w_2 \rangle \in [0, \pi]$ between them, then, geometrically,

$$r_1^2 = r_2^2 = I,$$
$$r_1 r_2 = r_2 r_1 \quad \text{if } \langle w_1, w_2 \rangle = 0, \tag{2.13}$$
$$r_1 r_2 = \text{rotation by angle } 2\theta \text{ in the } w_1\text{--}w_2 \text{ plane.}$$

An abstract *Coxeter group* is a group generated by a family of elements r_i of order 2, with the restriction that certain pair products $r_i r_j$ also have finite order. Of course, for such a group to be finite, every pair product $r_i r_j$ needs to have finite order. An important class of finite Coxeter groups is formed by the *Weyl groups* that arise in the study of Lie algebras. Consider a very special type of Weyl group: the group generated by reflections across the hyperplanes $(e_j - e_k)^\perp$, where $e_1, \ldots, e_n$ form the standard basis of $\mathbb{R}^n$, and j and k are distinct elements running over $[n]$. We can recognize this as essentially the symmetric group S_n, realized geometrically through the faithful representation R in (1.3). From this point of view, S_n can be viewed as being generated by elements $r_1, \ldots, r_{n-1}$, with r_i standing for the transposition $(i\,i+1)$, satisfying the relations

$$r_j^2 = \iota \qquad \text{for all } j \in [n-1],$$

$$r_j r_{j+1} r_j = r_{j+1} r_j r_{j+1} \qquad \text{for all } j \in [n-2], \qquad (2.14)$$

$$r_j r_k = r_j r_k \qquad \text{for all } j, k \in [n-1] \text{ with } |j - k| \geq 2,$$

where ι is the identity element. It would seem to be more natural to write the second equation as $(r_j r_{j+1})^3 = \iota$, which would be equivalent provided each r_j^2 is ι. However, just holding on to the second and third equations generates another important class of groups, the *braid groups* B_n, where B_n is generated abstractly by elements $r_1, \ldots, r_{n-1}$ subject to just the second and third conditions in (2.14). Thus, there is a natural surjection $B_n \to S_n$ mapping r_i to $(i\,i+1)$ for each $i \in [n-1]$.

If $\mathbb{F}$ is a subfield of a field $\mathbb{F}_1$, such that $\dim_{\mathbb{F}} \mathbb{F}_1 < \infty$, then the set of all automorphisms σ of the field $\mathbb{F}_1$ for which $\sigma(c) = c$ for all $c \in \mathbb{F}$ is a finite group under composition. This is the *Galois group* of $\mathbb{F}_1$ over $\mathbb{F}$; the classical case is where $\mathbb{F}_1$ is defined by adjoining to $\mathbb{F}$ roots of polynomial equations over $\mathbb{F}$. Morally related to these ideas are fundamental groups of surfaces; an instance of this, the fundamental group of a compact oriented surface of genus g, is the group with $2g$ generators $a_1, b_1, \ldots, a_g, b_g$ satisfying the constraint

$$a_1 b_1 a_1^{-1} b_1^{-1} \ldots a_g b_g a_g^{-1} b_g^{-1} = e. \qquad (2.15)$$

Such equations, with a_i and b_j represented in more concrete groups, have come up in two- and three-dimensional gauge theories. Far earlier, in his first major work developing character theory, Frobenius [29] studied the number of solutions of equations of this and related types, with each a_i and b_j represented in some finite group. In Sect. 7.9 we will study the Frobenius formula for counting the number of solutions of the equation

$$s_1 \ldots s_m = e$$

for $s_1, \ldots, s_m$ running over specified conjugacy classes in a finite group G. In the case $G = S_n$, restricting the s_i to run over transpositions, a result of Hurwitz relates this number to counting n-sheeted Riemann surfaces with m branch points (see Curtis [14] for related history).

Exercises

2.1. Work out the character table for D_5.

2.2. Consider the subgroup of S_4 given by

$$V_4 = \{\iota, (12)(34), (13)(24), (14)(23)\}.$$

Being a union of conjugacy classes, V_4 is a normal subgroup of S_4. Now view S_3 as the subgroup of S_4 consisting of the permutations that fix 4. Thus, $V_4 \cap S_3 = \{\iota\}$. Show that the mapping

$$S_3 \to S_4/V_4 : \sigma \mapsto \sigma V_4$$

is an isomorphism. Obtain an explicit form of a two-dimensional irreducible complex representation of S_4 for which the character is χ_2 as given in Table **2.6**.

2.3. In S_3 there is the cyclic group C_3 generated by (123), which is a normal subgroup. The quotient $S_3/C_3 \simeq S_2$ is a two-element group. Work out the one-dimensional representation of S_3 that arises from this by the method in Exercise 2.2.

2.4. Construct a two-dimensional irreducible representation of S_3, over any field $\mathbb{F}$ in which $3 \neq 0$, using matrices that have integer entries.

2.5. The alternating group A_4 consists of all even permutations in S_4. It is generated by the elements

$$c = (123), \quad x = (12)(34), \quad y = (13)(24), \quad z = (14)(23)$$

satisfying the relations

$$cxc^{-1} = z, \quad cyc^{-1} = x, \quad czc^{-1} = y, \quad c^3 = \iota, \quad xy = yx = z.$$

Table 2.9 Character table for A_4

	1	3	4	4
	ι	(12)(34)	(123)	(132)
ψ_0	1	1	1	1
ψ_1	1	1	ω	ω^2
ψ_2	1	1	ω^2	ω
χ_1	?	?	?	?

(a) Show that the conjugacy classes are

$$\{\iota\},\ \{x,y,z\},\ \{c,cx,cy,cz\},\ \{c^2,c^2x,c^2y,c^2z\}.$$

Note that c and c^2 are in different conjugacy classes in A_4, even though in S_4 they are conjugate.

(b) Show that the group A_4 generated by all commutators $aba^{-1}b^{-1}$ is $V_4 = \{\iota, x, y, z\}$, which is just the set of commutators in A_4.

(c) Check that there is an isomorphism given by

$$C_3 \mapsto A_4/V_4 : c \mapsto cV_4.$$

(d) Obtain three one-dimensional representations of A_4.

(e) The group $A_4 \subset S_4$ acts by permutation of coordinates on $\mathbb{C}^4$ and preserves the three-dimensional subspace $E_0 = \{(x_1, ..., x_4) : x_1 + \cdots + x_4 = 0\}$. Work out the character χ_3 of this representation of A_4.

(f) Work out the full character table for A_4, by filling in the last row in Table **2.9**.

2.6. Let V be a real vector space and $T : V \to V$ be a linear mapping with $T^m = I$ for some positive integer m. Choose a basis B of V and let $V_{\mathbb{C}}$

be the complex vector space with basis B. Define the *conjugation* map $C : V_{\mathbb{C}} \to V_{\mathbb{C}} : v \mapsto \bar{v}$ by

$$C\left(\sum_{b \in B} v_b b\right) = \sum_{b \in B} \overline{v_b} b,$$

where each $v_b \in \mathbb{C}$, and on the right we just have the ordinary complex conjugates $\overline{v_b}$. Show that

$$x = \tfrac{1}{2}(v + Cv) \text{ and } y = -\tfrac{i}{2}(v - Cv)$$

are in V for every $v \in V_{\mathbb{C}}$. If $v \in V_{\mathbb{C}}$ is an eigenvector of T, show that T maps the subspace $\mathbb{R}x + \mathbb{R}y$ of V spanned by x and y into itself.

2.7. Work out an irreducible representation of the group

$$Q = \{1, -1, i, -i, j, -j, k, -1\}$$

of unit quaternions on $\mathbb{C}^2$, by associating suitable 2×2 matrices with the elements of Q.

Chapter 3

The Group Algebra

The simplest meaningful object we can construct out of a field $\mathbb{F}$ and a group G is a vector space over $\mathbb{F}$, with the elements of G as basis. A typical element of this vector space is a linear combination

$$a_1 g_1 + \cdots + a_n g_n,$$

where $g_1, ..., g_n$ are the elements of G, and $a_1, ..., a_n$ are drawn from $\mathbb{F}$. This vector space, denoted $\mathbb{F}[G]$, is endowed with a natural representation ρ_{reg} of the group G, specified by

$$\rho_{\text{reg}}(g)(a_1 g_1 + \cdots + a_n g_n) = a_1 g g_1 + \cdots + a_n g g_n.$$

Put another way, the elements of the group G form a basis of $\mathbb{F}[G]$, and the action of G simply permutes this basis by left-multiplication.

The representation ρ_{reg} on $\mathbb{F}[G]$ is the mother of all irreducible representations: if the group G is finite and $|G|1_{\mathbb{F}} \neq 0$, then the representation ρ_{reg} on $\mathbb{F}[G]$ decomposes as a direct sum of irreducible representations of G, and

every irreducible representation of G is equivalent to one of the representations appearing in the decomposition of ρ_{reg}.

This result, and much more, will be proved in Chap. 4, where we will examine the representation ρ_{reg} in detail. For now, in this chapter, we will introduce $\mathbb{F}[G]$ officially, and establish some of its basic features.

Beyond being a vector space, $\mathbb{F}[G]$ is also an *algebra*: there is a natural multiplication operation in $\mathbb{F}[G]$ arising from the multiplication of the elements of the group G. We will explore this algebra structure in a specific

A.N. Sengupta, *Representing Finite Groups: A Semisimple Introduction,*   59
DOI 10.1007/978-1-4614-1231-1_3, © Springer Science+Business Media, LLC 2012

example, with G being the permutation group S_3, and draw some valuable lessons and insights from this example. We will also prove a wonderful structural property of $\mathbb{F}[G]$ called *semisimplicity* that is at the heart of the decomposability of representations of G into irreducible ones.

3.1   Definition of the Group Algebra

It is time to delve into the formal definition of the *group algebra*

$$\mathbb{F}[G],$$

where G is a group and $\mathbb{F}$ a field. As a set, this consists of all formal linear combinations

$$a_1 g_1 + \cdots + a_n g_n,$$

where $g_1, ..., g_n$ are elements of G, and $a_1, ..., a_n \in \mathbb{F}$. We add and multiply these new objects in the only natural way that is sensible. For example,

$$(2g_1 + 3g_2) + (-4g_1 + 5g_3) = (-2)g_1 + 3g_2 + 5g_3$$

and

$$(2g_1 - 4g_2)(g_4 + g_3) = 2g_1 g_4 + 2g_1 g_3 - 4g_2 g_4 - 4g_2 g_3.$$

Officially, $\mathbb{F}[G]$ consists of all maps

$$x : G \mapsto \mathbb{F} : g \mapsto x_g$$

such that x_g is 0 for all except finitely many $g \in G$; thus, $\mathbb{F}[G]$ is the direct sum of copies of the field $\mathbb{F}$, one copy for each element of G. In the case of interest to us, G is finite and $\mathbb{F}[G]$ is simply the set of all $\mathbb{F}$-valued functions on G.

It turns out to be very convenient, indeed intuitively crucial, to write $x \in \mathbb{F}[G]$ in the form

$$x = \sum_{g \in G} x_g g.$$

To avoid clutter we usually write $\sum_g$ when we mean $\sum_{g \in G}$.

Addition and multiplication, as well as multiplication by elements $c \in \mathbb{F}$, are defined in the obvious way:

$$\sum_g x_g g + \sum_g y_g g = \sum_g (x_g + y_g)g,$$

$$\sum_g x_g g \sum_h y_h h = \sum_g \left(\sum_h x_h y_{h^{-1}g}\right) g, \qquad (3.1)$$

$$c\sum_g x_g g = \sum_g cx_g g.$$

It is readily checked that $\mathbb{F}[G]$ is an *algebra* over $\mathbb{F}$: it is a ring as well as an $\mathbb{F}$-module, and the multiplication

$$\mathbb{F}[G] \times \mathbb{F}[G] \to \mathbb{F}[G] : (x, y) \mapsto xy$$

is $\mathbb{F}$-bilinear, associative, and has a nonzero multiplicative identity element $1e$, where e is the identity in G.

Sometimes it is useful to think of G as a subset of $\mathbb{F}[G]$, by identifying $g \in G$ with the element $1g \in \mathbb{F}[G]$. But the multiplicative unit $1e$ in $\mathbb{F}[G]$ will also be denoted 1, and in this way $\mathbb{F}$ may be viewed as a subset of $\mathbb{F}[G]$:

$$\mathbb{F} \to \mathbb{F}[G] : c \mapsto ce.$$

Occasionally we will also work with $R[G]$, where R is a commutative ring such as $\mathbb{Z}$. This is defined just as $\mathbb{F}[G]$ is, except that the field $\mathbb{F}$ is replaced by the ring R, and $R[G]$ is an algebra over the ring R.

3.2   Representations of G and $\mathbb{F}[G]$

The algebra $\mathbb{F}[G]$ has a very useful feature: any representation

$$\rho : G \to \mathrm{End}_{\mathbb{F}}(E)$$

defines, in a unique way, a representation of the algebra $\mathbb{F}[G]$ in terms of operators on E. More specifically, for each element

$$x = \sum_g x_g g \in \mathbb{F}[G]$$

we have an endomorphism

$$\rho(x) \stackrel{\text{def}}{=} \sum_g x_g \rho(g) \in \text{End}_{\mathbb{F}}(E). \tag{3.2}$$

This induces an $\mathbb{F}[G]$-module structure on E:

$$\left(\sum_g x_g g \right) v = \sum_g x_g \rho(g) v. \tag{3.3}$$

It is very useful to look at representations in this way.

Put another way, we have an extension of ρ to an algebra homomorphism

$$\rho : \mathbb{F}[G] \rightarrow \text{End}_{\mathbb{F}}(E) : \sum_g a_g g \mapsto \sum_g a_g \rho(g). \tag{3.4}$$

Thus, *a representation of G specifies a module over the ring $\mathbb{F}[G]$.* Conversely, if E is an $\mathbb{F}[G]$-module, then we have a representation of G on E, by restricting multiplication to the elements in $\mathbb{F}[G]$ that are in G.

In summary, representations of G on vector spaces over $\mathbb{F}$ correspond naturally to $\mathbb{F}[G]$-modules. Depending on the context, it is sometimes useful to think in terms of representations of G and sometimes in terms of $\mathbb{F}[G]$-modules.

A subrepresentation or invariant subspace corresponds to a submodule, and direct sums of representations correspond to direct sums of modules. A morphism of representations corresponds to an $\mathbb{F}[G]$-linear map, and an isomorphism, or equivalence, of representations is an isomorphism of $\mathbb{F}[G]$-modules.

An irreducible representation corresponds to a *simple* module, which is a nonzero module with no proper nonzero submodules.

Here is Schur's lemma (Theorem 1.2) in module language:

Theorem 3.1 *Let G be a finite group and $\mathbb{F}$ be a field. Suppose E and F are simple $\mathbb{F}[G]$-modules and $T : E \rightarrow F$ is an $\mathbb{F}[G]$-linear map. Then either T is 0 or T is an isomorphism of $\mathbb{F}[G]$-modules. If, moreover, $\mathbb{F}$ is algebraically closed, then any $\mathbb{F}[G]$-linear map $S : E \rightarrow E$ is of the form $S = \lambda I$ for some scalar $\lambda \in \mathbb{F}$.*

Sometimes the following version is also useful:

Theorem 3.2 *Let G be a finite group and $\mathbb{F}$ be a field. Suppose E is a simple $\mathbb{F}[G]$-module, F is an $\mathbb{F}[G]$-module, and $T : E \to F$ is an $\mathbb{F}[G]$-linear surjective map. Then either $F = 0$ or T is an isomorphism of $\mathbb{F}[G]$-modules.*

Proof. The kernel $\ker T$, as a submodule of the simple module E, is either 0 or E. So if $F \neq 0$, then $T \neq 0$ and so $\ker T = 0$, in which case the surjective map T is an isomorphism. $\boxed{\text{QED}}$

3.3   The Center

A natural first question about an algebra is whether it has an interesting *center*. By *center* of an algebra we mean the set of all elements in the algebra that commute with every element of the algebra.

It is easy to determine the center Z of the group algebra $\mathbb{F}[G]$ of a group G over a field $\mathbb{F}$. An element

$$x = \sum_{h \in G} x_h h$$

belongs to the center if and only if it commutes with every $g \in G$:

$$g x g^{-1} = x,$$

which expands to

$$\sum_{h \in G} x_h g h g^{-1} = \sum_{h \in G} x_h h.$$

Thus, x lies in Z if and only if

$$x_{g^{-1}hg} = x_h \qquad \text{for every } g, h \in G. \tag{3.5}$$

This means that the function $g \mapsto x_g$ is constant on conjugacy classes in G. Thus, x is in the center if and only if it can be expressed as a linear combination of the elements

$$z_C = \sum_{g \in C} g, \qquad \text{where } C \text{ is a finite conjugacy class in } G. \tag{3.6}$$

We are primarily interested in finite groups, and then the added qualifier of finiteness of the conjugacy classes is not needed.

If C and C' are distinct conjugacy classes, then z_C and $z_{C'}$ are sums over disjoint sets of elements of G, and so the collection of all such z_C is linearly independent. This yields a simple but important result:

Theorem 3.3 *Suppose G is a finite group, $\mathbb{F}$ is a field, and let $z_C \in \mathbb{F}[G]$ be the sum of all the elements in a conjugacy class C in G. The center Z of $\mathbb{F}[G]$ is a vector space over $\mathbb{F}$ and the elements z_C, with C running over all conjugacy classes of G, form a basis of Z. In particular, the dimension of the center of $\mathbb{F}[G]$ is equal to the number of conjugacy classes in G.*

The center Z of $\mathbb{F}[G]$ is, of course, also an algebra in its own right. Since we have a handy basis, consisting of the vectors z_C, of Z, we can get a full grip on the algebra structure of Z by working out all the products between the basis elements z_C. There is one simple, yet remarkable fact here:

Proposition 3.1 *Suppose G is a finite group and C_1, ..., C_s are all the distinct conjugacy classes in G. For each $j \in [s]$, let $z_j \in \mathbb{Z}[G]$ be the sum of all the elements of C_j. Then for any $l, n \in [s]$, the product $z_l z_n$ is a linear combination of the vectors z_m with coefficients that are nonnegative integers. Specifically,*

$$z_l z_n = \sum_{C \in \mathcal{C}} \kappa_{l,mn} z_m, \tag{3.7}$$

where $\kappa_{l,mn}$ counts the number of solutions of the equation $c = ab$, for any fixed $c \in C_m$ with a, b running over C_l and C_n, respectively:

$$\kappa_{l,mn} = |\{(a,b) \in C_l \times C_n \,|\, c = ab\}| \tag{3.8}$$

for any fixed $c \in C_m$.

The numbers $\kappa_{l,mn}$ are sometimes called the *structure constants* of the group G. As we shall see in Sect. 7.6, these constants can be used to work out all the irreducible characters of the group.

Proof. Note first that $c = ab$ if and only if $(gag^{-1})(gbg^{-1}) = gcg^{-1}$ for every $g \in G$, and so the number $\kappa_{l,mn}$ is completely specified by the conjugacy class C_m in which c lies in the definition (3.8). In the product $z_l z_n$, the coefficient of $c \in C_m$ is clearly $\kappa_{l,mn}$. $\boxed{\text{QED}}$

If you wish, you can leap ahead to Sect. 3.5 and then proceed to the next chapter.

3.4   Deconstructing $\mathbb{F}[S_3]$

To get a hands-on feel for the group algebra we will work out the structure of the group algebra $\mathbb{F}[S_3]$, where $\mathbb{F}$ is a field in which $6 \neq 0$; thus, the characteristic of the field is not 2 or 3. The reason for imposing this condition will become clear as we proceed. We will work through this example slowly, avoiding fast tricks/tracks, and it will serve us well later. The method we use will introduce and highlight many key ideas and techniques that we will use later to analyze the structure of $\mathbb{F}[G]$ for general finite groups, and also for general algebras.

From what we have learned in the preceding section, the center Z of $\mathbb{F}[S_3]$ is a vector space with a basis constructed from the conjugacy classes of S_3. These classes are

$$\{\iota\}, \{c, c^2\}, \{r, cr, c^2r\},$$

where $r = (12)$ and $c = (123)$. The center Z has the basis

$$\iota, \quad C = c + c^2, \quad R = r + cr + c^2r.$$

Table **3.1** shows the multiplicative structure of Z. Notice that the structure constants of S_3 can be read from this table.

Table 3.1 Multiplication in the center of $\mathbb{F}[S_3]$

	1	C	R
1	1	C	R
C	C	$2 + C$	$2R$
R	R	$2R$	$3 + 3C$

The structure of the algebra $\mathbb{F}[G]$, for any finite group G, can be probed by means of *idempotent* elements. An element $u \in \mathbb{F}[G]$ is an *idempotent* if

$$u^2 = u.$$

Idempotents u and v are called *orthogonal* if uv and vu are 0. In this case, $u + v$ is also an idempotent:

$$(u + v)^2 = u^2 + uv + vu + v^2 = u + 0 + 0 + v.$$

Clearly, 0 and 1 are idempotent. But what is really useful is to find a maximal set of orthogonal idempotents $u_1, \ldots, u_m$ in the center Z that are not 0 or 1, and have the spanning property

$$u_1 + \cdots + u_m = 1. \tag{3.9}$$

An idempotent in an algebra which lies in the center of the algebra is called a *central idempotent*.

The spanning condition (3.9) for the central idempotents u_i implies that any element $a \in \mathbb{F}[G]$ can be decomposed as

$$a = a1 = au_1 + \cdots + au_m,$$

and the orthogonality property, along with the centrality of the idempotents u_j, shows that

$$au_j au_k = aau_j u_k = 0 \qquad \text{for } j \neq k.$$

In view of this, the map

$$I : \mathbb{F}[G]u_1 \times \cdots \times \mathbb{F}[G]u_m \to \mathbb{F}[G] : (a_1, \ldots, a_m) \mapsto a_1 + \cdots + a_m$$

is an *isomorphism of algebras*, in the sense that it is a bijection, and preserves multiplication and addition:

$$
\begin{aligned}
I(a_1 + a_1', \ldots, a_m + a_m') &= I(a_1, \ldots, a_m) + I(a_1', \ldots, a_m'), \\
I(a_1 a_1', \ldots, a_m a_m') &= I(a_1, \ldots, a_m) I(a_1', \ldots, a_m').
\end{aligned}
\tag{3.10}
$$

All this is verified easily. The multiplicative property and the injectivity of I follow from the orthogonality and centrality of the idempotents u_1, ..., u_m.

Thus, the isomorphism I decomposes $\mathbb{F}[G]$ into a product of the smaller algebras $\mathbb{F}[G]u_j$. Notice that within the algebra $\mathbb{F}[G]u_j$ the element u_j plays the role of the multiplicative unit.

Now we are motivated to search for central idempotents in $\mathbb{F}[S_3]$. Using the basis of Z given by $1 = \iota$, C, R, we consider

$$u = x1 + yC + zR$$

with $x, y, z \in \mathbb{F}$. We are going to do this by brute force; later, in Theorem 7.11, we will see how the character table of a group can be used systematically to obtain the central idempotents in the group algebra. The condition for idempotence, $u^2 = u$, leads to three (quadratic) equations in the three unknowns x, y, and z. The solutions lead to the following elements:

$$u_1 = \frac{1}{6}(1 + C + R), \quad u_2 = \frac{1}{6}(1 + C - R), \quad u_3 = \frac{1}{3}(2 - C),$$

$$u_1 + u_2 = \frac{1}{3}(1 + C), \quad u_2 + u_3 = \frac{1}{6}(5 - C - R), \quad u_3 + u_1 = \frac{1}{6}(5 - C + R).$$

$$(3.11)$$

The division by 6 is the reason for the condition that $6 \neq 0$ in $\mathbb{F}$. We check readily that u_1, u_2, and u_3 are orthogonal; for instance,

$$(1 + C + R)(1 + C - R) = 1 + 2C + C^2 - R^2 = 1 + 2C + 2 + C - 3 - 3C = 0.$$

For now, as an aside, we can observe that there are idempotents in $\mathbb{F}[S_3]$ that are not central; for instance,

$$\frac{1}{2}(1 + r) \qquad \text{and} \qquad \frac{1}{2}(1 - r)$$

are readily checked to be orthogonal idempotents, adding up to 1, but they are not central.

Thus, we have a decomposition of $\mathbb{F}[S_3]$ into a product of smaller algebras:

$$\mathbb{F}[S_3] \simeq \mathbb{F}[S_3]u_1 \times \mathbb{F}[S_3]u_2 \times \mathbb{F}[S_3]u_3. \qquad (3.12)$$

Simple calculations show that

$$cu_1 = u_1 \qquad \text{and} \qquad ru_1 = u_1,$$

which imply that $\mathbb{F}[S_3]u_1$ is simply the one-dimensional space generated by u_1:

$$\mathbb{F}[S_3]u_1 = \mathbb{F}u_1.$$

In fact, what we see is that left-multiplication by elements of S_3 on $\mathbb{F}[S_3]u_1$ is a one-dimensional representation of S_3, the trivial one.

Next,

$$cu_2 = u_2 \qquad \text{and} \qquad ru_2 = -u_2,$$

which imply that $\mathbb{F}[S_3]u_2$ is also one-dimensional:

$$\mathbb{F}[S_3]u_2 = \mathbb{F}u_2.$$

Moreover, multiplication on the left by elements of S_3 on $\mathbb{F}[S_3]u_2$ gives a one-dimensional representation ϵ of S_3, this time the one given by the parity: on even permutations ϵ is 1, and on odd permutations it is -1.

We know that the full space $\mathbb{F}[S_3]$ has a basis consisting of the six elements of S_3. Thus,

$$\dim \mathbb{F}[S_3]u_3 = 6 - 1 - 1 = 4.$$

We can see this more definitively by working out the elements of $\mathbb{F}[S_3]u_3$. For this we should resist the thought of simply multiplying each element of $\mathbb{F}[S_3]$ by u_3; this might not be a method that would give any general insights which would be meaningful for groups other than S_3. Instead, observe that

$$\text{an element } x \in \mathbb{F}[S_3] \text{ lies in } \mathbb{F}[S_3]u_3 \text{ if and only if } xu_3 = x. \qquad (3.13)$$

This follows readily from the idempotence of u_3. Then, taking an element

$$x = \alpha + \beta c + \gamma c^2 + \theta r + \phi cr + \psi c^2 r \in \mathbb{F}[S_3],$$

we can work out what the condition $xu_3 = x$ says about the coefficients α, β, ..., $\psi \in \mathbb{F}$:

$$\begin{aligned} \alpha + \beta + \gamma &= 0, \\ \theta + \phi + \psi &= 0. \end{aligned} \qquad (3.14)$$

This leaves four (linearly) independent elements among the six coefficients $\alpha, ..., \psi$, verifying again that $\mathbb{F}[S_3]u_3$ is four-dimensional. Dropping α and θ as coordinates, we can write $x \in \mathbb{F}[S_3]u_3$ as

$$x = \beta(c - 1) + \gamma(c^2 - 1) + \phi(c - 1)r + \psi(c^2 - 1)r. \qquad (3.15)$$

With this choice, we see that

$$c - 1, \ (c^2 - 1), \ (c - 1)r, \ (c^2 - 1)r \text{ form a basis of } \mathbb{F}[S_3]u_3. \qquad (3.16)$$

Another choice would be to "split the difference" between the multipliers 1 and r, and bring in the two elements

$$r_+ = \frac{1}{2}(1 + r) \qquad \text{and} \qquad r_- = \frac{1}{2}(1 - r).$$

The nice thing about these elements is that they are idempotents, and we will use them again shortly. So we have another choice of basis for $\mathbb{F}[S_3]u_3$:

$$b_1^+ = (c - 1)r_+, \ b_2^+ = (c^2 - 1)r_+, \ b_1^- = (c - 1)r_-, \ b_2^- = (c^2 - 1)r_-. \qquad (3.17)$$

How does the representation ρ_{reg}, restricted to $\mathbb{F}[S_3]u_3$, look relative to this basis? Simply eyeballing the vectors in the basis, we can see that the first two span a subspace invariant under left-multiplication by all elements of S_3, and the span of the last two vectors is also invariant. For the subspace spanned by the b_j^+, the matrices for left-multiplication by c and r are given by

$$c \mapsto \begin{bmatrix} -1 & -1 \\ 1 & 0 \end{bmatrix}, \qquad r \mapsto \begin{bmatrix} 0 & 1 \\ 1 & 0 \end{bmatrix}. \tag{3.18}$$

This representation is irreducible: clearly, any vector fixed (or taken to its negative) by the action of r would have to be a multiple of $(1, 1)$, and the only such multiple fixed by the action of c is the zero vector. Observe that the character χ_2 of this representation is specified on the conjugacy classes by

$$\chi_2(c) = -1, \qquad \chi_2(r) = 0.$$

For the subspace spanned by the vectors b_j^-, these matrices are given by

$$c \mapsto \begin{bmatrix} -1 & -1 \\ 1 & 0 \end{bmatrix}, \qquad r \mapsto \begin{bmatrix} 0 & -1 \\ -1 & 0 \end{bmatrix} \tag{3.19}$$

At first it is not obvious how this relates to (3.18). However, we can use a new basis given by

$$B_1^- = \frac{1}{2}b_1^- - b_2^-, \qquad B_2^- = b_1^- - \frac{1}{2}b_2^-,$$

and with respect to this basis, the matrices for the left-multiplication action of c and r are given again by exactly the same matrices as in (3.18):

$$cB_1^- = -B_1^- + B_2^-, \qquad cB_2^- = -B_1^-.$$

Thus, we have a decomposition of $\mathbb{F}[S_3]u_3$ into subspaces

$$\mathbb{F}[S_3]u_3 = (\text{span of } b_1^+, b_2^+) \oplus (\text{span of } B_1^-, B_2^-),$$

each of which carries the same representation of S_3, specified as in (3.18). Observe that from the way we constructed the invariant subspaces,

$$\text{span of } b_1^+, b_2^+ = \mathbb{F}[S_3]u_3 r_+ \qquad \text{and} \qquad \text{span of } B_1^-, B_2^- = \mathbb{F}[S_3]u_3 r_-.$$

Thus, we have a clean and complete decomposition of $\mathbb{F}[S_3]$ into subspaces

$$\mathbb{F}[S_3] = \mathbb{F}[S_3]u_1 \oplus \mathbb{F}[S_3]u_2 \oplus (\mathbb{F}[S_3]y_1 \oplus \mathbb{F}[S_3]y_2), \qquad (3.20)$$

where

$$y_1 = \frac{1}{2}(1+r)u_3, \quad y_2 = \frac{1}{2}(1-r)u_3. \qquad (3.21)$$

Each of these subspaces carries a representation of S_3 given by multiplication on the left; moreover, each of these is an irreducible representation.

Having done all this, we still do not have a complete analysis of the structure of $\mathbb{F}[S_3]$ as an *algebra*. What remains is to analyze the structure of the smaller algebra

$$\mathbb{F}[S_3]u_3.$$

Perhaps we should try our idempotent trick again? Clearly

$$v_1 = \frac{1}{2}(1+r)u_3 \quad \text{and} \qquad v_2 = \frac{1}{2}(1-r)u_3 \qquad (3.22)$$

are orthogonal idempotents and add up to u_3.

In the absence of centrality, we cannot use our previous method of identifying the algebra with products of certain subalgebras. However, we can do something similar, using the fact that v_1 and v_2 are orthogonal idempotents in $\mathbb{F}[S_3]u_3$ whose sum is u_3, which is the multiplicative identity in this algebra $\mathbb{F}[S_3]u_3$. We can decompose any $x \in \mathbb{F}[S_3]u_3$ as

$$x = (y_1 + y_2)x(y_1 + y_2) = y_1xy_1 + y_1xy_2 + y_2xy_1 + y_2xy_2. \qquad (3.23)$$

Let us write

$$x_{jk} = y_jxy_k. \qquad (3.24)$$

Observe next that for $x, w \in \mathbb{F}[S_3]u_3$, the product xw decomposes as

$$xw = (x_{11} + x_{12} + x_{21} + x_{22})(w_{11} + w_{12} + w_{21} + w_{22}) = \sum_{j,k=1}^{2} \left(\sum_{m=1}^{2} x_{jm}w_{mk} \right).$$

Using the orthogonality of the idempotents y_1 and y_2, we have

$$(xw)_{jk} = y_j(xw)y_k = \sum_{m=1}^{2} x_{jm}w_{mk}.$$

Does this remind us of something? Yes, it is matrix multiplication! Thus, the association

$$x \mapsto \begin{bmatrix} x_{11} & x_{12} \\ x_{21} & x_{22} \end{bmatrix} \tag{3.25}$$

preserves multiplication. Clearly, it also preserves/respects addition, and multiplication by scalars (elements of $\mathbb{F}$). Thus, we have identified $\mathbb{F}[S_3]u_3$ as an algebra of matrices.

However, something is not clear yet: what kind of objects are the entries of the matrix $[x_{jk}]$? Since we know that $\mathbb{F}[S_3]u_3$ is a four-dimensional vector space over $\mathbb{F}$, it seems that the entries of the matrix ought to be scalars drawn from $\mathbb{F}$. To see if or in what way this is true, we need to explore the nature of the quantities

$$x_{jk} = y_j x y_k, \qquad \text{with } x \in \mathbb{F}[S_3]u_3.$$

We have reached the "shut up and calculate" point; for

$$x = \beta(c-1) + \gamma(c^2-1) + \phi(c-1)r + \psi(c^2-1)r,$$

as in (3.15), the matrix $[x_{jk}]$ works out to be

$$
\begin{bmatrix} x_{11} & x_{12} \\ x_{21} & x_{22} \end{bmatrix}
$$
$$
= \begin{bmatrix} -\frac{3}{2}(\beta+\gamma+\phi+\psi)y_1 & (\beta-\gamma-\phi+\psi)\frac{1}{4}(1+r)(c-c^2) \\ (\beta-\gamma-\phi-\psi)\frac{1}{4}(1-r)(c-c^2) & -\frac{3}{2}(\beta+\gamma-\phi-\psi)y_2 \end{bmatrix}.
$$
$$\tag{3.26}$$

Perhaps then we should associate the matrix

$$
\begin{bmatrix} -\frac{3}{2}(\beta+\gamma+\phi+\psi) & (\beta-\gamma-\phi+\psi) \\ (\beta-\gamma-\phi-\psi) & -\frac{3}{2}(\beta+\gamma-\phi-\psi) \end{bmatrix}
$$

with $x \in \mathbb{F}[S_3]u_3$? This would certainly identify $\mathbb{F}[S_3]u_3$, as a vector space, with the vector space of 2×2 matrices with entries in $\mathbb{F}$. But to also properly encode multiplication in $\mathbb{F}[S_3]u_3$ into matrix multiplication we observe, after calculations, that

$$\frac{1}{4}(1+r)(c-c^2)\frac{1}{4}(1-r)(c-c^2) = -\frac{3}{4}y_1.$$

The factor of $-3/4$ can throw things off balance. So we use the mapping

$$x \mapsto \begin{bmatrix} -\frac{3}{2}(\beta + \gamma + \phi + \psi) & -\frac{3}{4}(\beta - \gamma - \phi + \psi) \\ (\beta - \gamma - \phi - \psi) & -\frac{3}{2}(\beta + \gamma - \phi - \psi) \end{bmatrix}. \qquad (3.27)$$

This identifies the algebra $\mathbb{F}[S_3]u_3$ with the algebra of all 2×2 matrices with entries drawn from the field $\mathbb{F}$:

$$\mathbb{F}[S_3]u_3 \simeq \mathrm{Matr}_{2\times2}(\mathbb{F}). \qquad (3.28)$$

Thus, we have completely worked out the structure of the algebra $\mathbb{F}[S_3]$:

$$\mathbb{F}[S_3] \simeq \mathbb{F} \times \mathbb{F} \times \mathrm{Matr}_{2\times2}(\mathbb{F}), \qquad (3.29)$$

where the first two terms arise from the one-dimensional algebras $\mathbb{F}[S_3]u_1$ and $\mathbb{F}[S_3]u_2$.

What are the lessons of this long exercise? Here is a summary, writing A for the algebra $\mathbb{F}[S_3]$:

- We found a basis of the center Z of A consisting of idempotents u_1, u_2, and u_3. Then A is realized as isomorphic to a *product* of smaller algebras:

$$A \simeq Au_1 \times Au_2 \times Au_3.$$

- Au_1 and Au_2 are one-dimensional, and hence carry one-dimensional irreducible representations of $\mathbb{F}[S_3]$ by left-multiplication.

- The subspace Au_3 was decomposed again by the method of idempotents: we found orthogonal idempotents y_1 and y_2, adding up to u_3, and then

$$Au_3 = Ay_1 \oplus Ay_2,$$

with Ay_1 and Ay_2 being irreducible representations of S_3 under left-multiplication.

- The set

$$\{y_j x y_k \mid x \in Au_3\}$$

is a one-dimensional subspace of Ay_k for each $j, k \in \{1, 2\}$.

- There is then a convenient decomposition of each $x \in Au_3$ as

$$x = y_1 x y_1 + y_1 x y_2 + y_2 x y_1 + y_2 x y_2,$$

which suggests the association of a matrix with x:

$$x \mapsto \begin{bmatrix} x_{11} & x_{12} \\ x_{21} & x_{22} \end{bmatrix}.$$

- Au_3, as an algebra, is isomorphic to the algebra $\mathrm{Matr}_{2\times 2}(\mathbb{F})$.

Remarkably, much of this is valid even when we use a general finite group G in place of S_3. Indeed, a lot of it works even for algebras that can be decomposed into a sum of subspaces which are invariant under left-multiplication by elements of the algebra. In Chap. 5 we will traverse this territory.

Let us not forget that all the way through we were dividing by 2 and 3, and indeed even in forming the idempotents, we needed to divide by 6. So for our analysis of the structure of $\mathbb{F}[S_3]$ we needed to assume that 6 is not 0 in the field $\mathbb{F}$. What is special about 6? It is no coincidence that 6 is just the number of elements of S_3. In the more general setting of $\mathbb{F}[G]$, we will need to assume that $|G|1_{\mathbb{F}} \neq 0$ to make progress in understanding the structure of $\mathbb{F}[G]$.

We can also make some other observations that are more specific to S_3. For instance, the representation on each irreducible subspace is given by matrices with *integer* entries! This is not something we can expect to hold for a general finite group. But it does raise a question: perhaps some groups have a kind of "rigidity" that forces irreducible representations to be realizable in suitable integer rings? (Leap ahead to Exercise 6.3 to dip your foot in these waters.)

3.5 When $\mathbb{F}[G]$ Is Semisimple

Closing this chapter, we will prove a fundamental structural property of the group algebra $\mathbb{F}[G]$ that will yield a large trove of results about representations of G. This property is semisimplicity.

A module E over a ring is *semisimple* if for any submodule F in E there is a submodule F_c in E such that E is the direct sum of F and F_c. A ring is *semisimple* if it is semisimple as a left module over itself.

If E is the direct sum of submodules F and F_c, then these submodules are said to be *complements* of each other.

Our immediate objective here is to prove Maschke's theorem:

Theorem 3.4 *Suppose G is a finite group and $\mathbb{F}$ is a field whose characteristic is not a divisor of $|G|$. Then every module over the ring $\mathbb{F}[G]$ is semisimple. In particular, $\mathbb{F}[G]$ is semisimple.*

Note the condition that $|G|$ is not divisible by the characteristic of $\mathbb{F}$. We have seen this condition arise in the study of the structure of $\mathbb{F}[S_3]$. In fact, the converse of the Theorem 3.4 also holds: if $\mathbb{F}[G]$ is semisimple, then the characteristic of $\mathbb{F}$ is not a divisor of $|G|$; this is Exercise 3.3.

Proof. Let E be an $\mathbb{F}[G]$-module and F be a submodule. We have then the $\mathbb{F}$-linear inclusion

$$j : F \to E.$$

Since E and F are vector spaces over $\mathbb{F}$, there is an $\mathbb{F}$-linear map

$$P : E \to F$$

satisfying

$$Pj = \mathrm{id}_F. \tag{3.30}$$

(Choose a basis of F and extend this to a basis of E. Then let P be the map that keeps each of the basis elements of F fixed, but maps all the other basis elements to 0.)

All we have to do is modify P to make it $\mathbb{F}[G]$-linear. Observe that the inclusion map j is invariant under "conjugation" by any element of G:

$$gjg^{-1} = j \qquad \text{for all } g \in G.$$

Consequently,

$$gPg^{-1}j = gPjg^{-1} = \mathrm{id}_F \qquad \text{for all } g \in G. \tag{3.31}$$

So we have

$$P_0 j = \mathrm{id}_F,$$

where P_0 is the G-averaged version of P:

$$P_0 = \frac{1}{|G|} \sum_{g \in G} gPg^{-1};$$

here the division makes sense because $|G| \neq 0$ in $\mathbb{F}$. Clearly, P_0 is G-invariant and hence $\mathbb{F}[G]$-linear. Moreover, just as P, the G-averaged version P_0 is also a "projection" onto F in the sense that $P_0 v = v$ for all v in F.

We can decompose any $x \in E$ as

$$x = \underbrace{P_0 x}_{\in F} + \underbrace{x - P_0 x}_{\in F_c}.$$

This shows that E splits as a direct sum of $\mathbb{F}[G]$-submodules:

$$E = F \oplus F_c,$$

where

$$F_c = \ker P_0$$

is also an $\mathbb{F}[G]$-submodule of E. Thus, every submodule of an $\mathbb{F}[G]$-module has a complementary submodule. In particular, this applies to $\mathbb{F}[G]$ itself, and so $\mathbb{F}[G]$ is semisimple. $\boxed{\text{QED}}$

The version above is a long way, in the evolution of the formulation, from Maschke's original result [59], which was reformulated and reproved by Frobenius, Burnside, Schur, and Weyl (see [14, Sect. III.4]).

The map

$$\mathbb{F}[G] \to \mathbb{F}[G] : x \mapsto \hat{x} = \sum_{g \in G} x_g g^{-1} \tag{3.32}$$

turns left into right:

$$\widehat{(xy)} = \hat{y}\hat{x}.$$

This makes every right $\mathbb{F}[G]$-module a left $\mathbb{F}[G]$-module by defining the left-module structure through

$$g \cdot v = vg^{-1},$$

and then every sub-right-module is a sub-left-module. Thus, $\mathbb{F}[G]$, *viewed as a right module over itself*, is also semisimple.

Despite the ethereal appearance of the proof of Theorem 3.4, the argument can be exploited to obtain a slow but sure algorithm for decomposing a representation into irreducible components, at least over an algebraically closed field. If a representation ρ on E is not irreducible, and has a proper nonzero invariant subspace $F \subset E$, then starting with an ordinary linear projection map $P : E \to F$, we obtain a G-invariant one by averaging:

$$P_0 = \frac{1}{|G|} \sum_{g \in G} \rho(g)^{-1} P \rho(g).$$

This provides us with a decomposition

$$E = \ker P_0 + \ker(I - P_0)$$

into complementary, invariant subspaces F and $(I - P_0)(E)$ of *lower* dimension than E, and so repeating this procedure breaks down the original space E into irreducible subspaces. But how do we find the starter projection P? Since we have nothing to go on, we can take any linear map $T : E \to E$, and average it to

$$T_0 = \frac{1}{|G|} \sum_{g \in G} \rho(g)^{-1} T \rho(g).$$

Then we can take a suitable polynomial in T_0 that provides a projection map; specifically, if λ is an eigenvalue of T_0 (and that always exists if the field is algebraically closed), then the projection onto the corresponding eigensubspace is a polynomial in T_0 and hence is also G-invariant. This provides us with P_0, without needing to start with a projection P. There is, however, still something that could throw a spanner in the works: what if T_0 turns out to be just a multiple of the identity I? If this were the case for *every* choice of T, then there would in fact be no proper nonzero G-invariant projection map, and ρ would be irreducible and we could halt the program right there. Still, it seems unpleasant to have to search through *all* endomorphisms of E for some T that would yield a T_0 which is not a multiple of I. Fortunately, we can simply try out all the elements in any basis of $\mathrm{End}_{\mathbb{F}}(E)$, because if all such elements lead to multiples of the identity, then of course ρ must be irreducible.

We can now sketch a first draft of an algorithm for breaking down a given representation into subrepresentations. For convenience, let us assume the field of scalars is $\mathbb{C}$. Let us choose an inner product on E that makes each $\rho(g)$ unitary. Instead of endomorphisms of the N-dimensional space E, we work with $N \times N$ matrices. The usual basis of the space of all $N \times N$ matrices consists of the matrices E_{jk}, where E_{jk} has 1 at the (j, k) position and 0 elsewhere for $j, k \in \{1, \ldots, N\}$. It will be more convenient to work with a basis consisting of Hermitian matrices. To this end, replace, for $j \neq k$, the pair of matrices E_{jk} and E_{kj} by the pair of Hermitian matrices

$$E_{jk} + E_{kj} \qquad \text{and} \qquad i(E_{jk} - E_{kj}).$$

This produces a basis $B_1, \ldots, B_{N^2}$ of the space of $N \times N$ matrices, where each B_j is Hermitian. The sketch algorithm is

- For each $1 \leq k \leq N^2$, work out

$$\frac{1}{|G|} \sum_{g \in G} \rho(g) B_k \rho(g)^{-1}$$

  (which, you can check, is Hermitian) and set T_0 equal to the first such matrix which is not a multiple of the identity matrix I.

- Work out, using a suitable matrix-algebra "subroutine," the projection operator P_0 onto an eigensubspace of T_0.

Obviously, this needs more work to turn it into code. For details and more on computational representation theory. see the articles by Blokker and Flodmark [6] and Dixon [23, 24].

3.6 Afterthoughts: Invariants

Although we are focusing almost entirely on finite-dimensional representations of a group, there are infinite-dimensional representations that are of natural and classic interest. Let ρ be a representation of a finite group G on a finite-dimensional vector space V over a field $\mathbb{F}$. Then each tensor power $V^{\otimes n}$ carries the representation $\rho^{\otimes n}$:

$$\rho^{\otimes n}(g)(v_1 \otimes \cdots \otimes v_n) = \rho(g)v_1 \otimes \cdots \otimes \rho(g)v_n. \tag{3.33}$$

Hence, the tensor algebra

$$T(V) = \bigoplus_{n \in \{0,1,2,\ldots\}} V^{\otimes n} \tag{3.34}$$

carries the corresponding direct sum representation of all the tensor powers $\rho^{\otimes n}$, with $\rho^{\otimes 0}$ being the trivial representation (given by the identity map) on $V^{\otimes 0} = \mathbb{F}$. The group S_n of all permutations of $[n]$ acts naturally on $V^{\otimes n}$ by

$$\sigma \cdot (v_1 \otimes \cdots \otimes v_n) = v_{\sigma^{-1}(1)} \otimes \cdots \otimes v_{\sigma^{-1}(n)}.$$

The subspace of all $x \in V^{\otimes n}$ that are fixed, with $\sigma \cdot x = x$ for all $\sigma \in S_n$, is the *symmetric tensor power* $V^{\hat{\otimes} n}$; for $n = 0$ we take this to be $\mathbb{F}$. Clearly, $\rho^{\otimes n}$ leaves $V^{\hat{\otimes} n}$ invariant, and so the tensor algebra representation has a restriction to a representation on the *symmetric tensor algebra*

$$S(V) = \bigoplus_{n \in \{0,1,2,\dots\}} V^{\hat{\otimes} n}. \tag{3.35}$$

There is a more concrete and pleasant way of working with the symmetric tensor algebra representation. For this it is convenient to work with the dual space V' and the dual representation ρ' on V'. Choosing a basis in V, we denote the dual basis in V' by $X_1,\dots, X_n$, which we could also think of as abstract (commuting) indeterminates. An element of the tensor algebra $S(V')$ is then a finite linear combination of monomials $X_1^{w_1}\dots X_n^{w_n}$ with $(w_1,\dots,w_n) \in \mathbb{Z}_{\geq 0}^n$. Thus, $S(V')$ is identifiable with the polynomial algebra $\mathbb{F}[X_1,\dots,X_n]$. The action by ρ' is specified through

$$gX_j \overset{\text{def}}{=} \rho'(g)X_j = X_j \circ \rho(g)^{-1}.$$

A fundamental task, the subject of *invariant theory*, is to determine the set I_ρ of all polynomials $f \in \mathbb{F}[X_1,\dots,X_n]$ that are fixed by the action of G. Clearly, I_ρ is closed under both addition and multiplication, and also contains all scalars in $\mathbb{F}$. Thus, the invariants form a ring, or, more specifically, an algebra over $\mathbb{F}$. A deep and fundamental result obtained by Noether shows that there is a finite set of generators for this ring.

The most familiar example in this context is the symmetric group S_n acting on polynomials in $X_1,\dots, X_n$ in the natural way specified by $\sigma X_j = X_{\sigma^{-1}(j)}$. The ring of invariants is generated by the elementary symmetric polynomials

$$E_k(X_1,\dots,X_n) = \sum_{B \in P_k} \prod_{j \in B} X_j,$$

where P_k is the set of all k-element subsets of $[n]$, and $k \in \{0,1,\dots,n\}$. Another choice of generators is given by the power sums

$$N_k(X_1,\dots,X_n) = \sum_{j=1}^{n} X_j^k.$$

for $k \in \{0, \ldots, n\}$. The Jacobian

$$\det \begin{bmatrix} \frac{\partial N_1}{\partial X_1} & \cdots & \frac{\partial N_1}{\partial X_n} \\ \vdots & \cdots & \vdots \\ \frac{\partial N_n}{\partial X_1} & \cdots & \frac{\partial N_n}{\partial X_n} \end{bmatrix} = n! \det \begin{bmatrix} 1 & 1 & \cdots & 1 \\ X_1 & X_2 & \cdots & X_n \\ \vdots & \vdots & \cdots & \vdots \\ X_1^{n-1} & X_2^{n-1} & \cdots & X_n^{n-1} \end{bmatrix} \quad (3.36)$$

$$= n! \prod_{1 \le j < k \le n} (X_k - X_j),$$

where in the last step we have the formula for the Vandermonde determinant which we will meet again in other contexts (see Exercise 3.10). The simple observation that the determinant is not identically 0 already has a substantial consequence: the polynomials $N_1, \ldots, N_n$ are algebraically independent. Here is a "one sentence" proof: if f is a polynomial in n variables, of least total degree, for which $f(N_1, \ldots, N_n)$, as a polynomial in the X_i, is 0, then the row vector

$$[\partial_1 f(N_1, \ldots, N_n), \ldots, \partial_n f(N_1, \ldots, N_n)]$$

multiplied on the right by the Jacobian matrix in (3.36) is 0, and so, since the determinant of this matrix is not 0, each $\partial_i f(N_1, \ldots, N_n)$ is 0, from which, by minimality of the degree of f, it follows that f is constant and hence 0. The factorization that takes place in the last step in (3.36) is no coincidence; it is an instance of a deeper fact about reflection groups, of which the symmetric group S_n is an example. (For more on the power sums N_j, see Exercise 1.22.)

The slim but carefully detailed volume of Dieudonné and Carrell [22] and the beautiful text of Neusel [62] are excellent introductions to this subject.

Exercises

3.1. Let G be a finite group, $\mathbb{F}$ be a field, and G^* be the set of all nonzero multiplicative homomorphisms $G \to \mathbb{F}$. For $f \in G^*$, let

$$s_f = \sum_{g \in G} f(g^1) g.$$

Show that $\mathbb{F} s_f$ is an invariant subspace of $\mathbb{F}[G]$. The representation of G on $\mathbb{F} s_f$ given by left-multiplication is f, in the sense that $gv = f(g)v$ for all $g \in G$ and $v \in \mathbb{F} s_f$.

3.2. Show that if G is a finite group containing more than one element, and $\mathbb{F}$ is any field, then $\mathbb{F}[G]$ contains nonzero elements a and b whose product ab is 0.

3.3. Suppose $\mathbb{F}$ is a field of characteristic $p > 0$ and G is a finite group with $|G|$ a multiple of p. Let $s = \sum_{g \in G} g \in \mathbb{F}[G]$. Show that the submodule $\mathbb{F}[G]s$ contains no nonzero idempotent and conclude that $\mathbb{F}[G]s$ has no complementary submodule in $\mathbb{F}[G]$. (Exercise 4.14 pushes this much further.) Thus, $\mathbb{F}[G]$ *is not semisimple if the characteristic of $\mathbb{F}$ is a divisor of $|G|$.*

3.4. For any finite group G and commutative ring R, explain why the *augmentation map*

$$\epsilon : R[G] \to R : \sum_g x_g g \mapsto \sum_g x_g \qquad (3.37)$$

is a homomorphism of rings. Show that $\ker \epsilon$, which is an ideal in $R[G]$, is free as an R-module, with basis $\{g - 1 : g \in G, g \neq e\}$.

3.5. Work out the multiplication table specifying the algebra structure of the center $Z(D_5)$ of the dihedral group D_5. Take the generators of the group to be c and r, satisfying $c^5 = r^2 = e$ and $rcr^{-1} = c^{-1}$. Take as a basis for the center the conjugacy sums 1, $C = c + c^4$, $D = c^2 + c^3$, and $R = (1 + c + c^2 + c^3 + c^4)r$.

3.6. Determine all the central idempotents in the algebra $\mathbb{F}[D_5]$, where D_5 is the dihedral group of order 10, and $\mathbb{F}$ is a field of characteristic 0 containing a square root of 5. Show that some of these form a basis of the center Z of $\mathbb{F}[D_5]$. Then determine the structure of the algebra $\mathbb{F}[D_5]$ as a product of two one-dimensional algebras and two four-dimensional matrix algebras.

3.7. Let G be a finite group and $\mathbb{F}$ be an algebraically closed field in which $|G|1_{\mathbb{F}} \neq 0$. Suppose E is a simple $\mathbb{F}[G]$-module. Fix an $\mathbb{F}$-linear map $P : E \to E$ that is a projection onto a one-dimensional subspace V of E, and let $P_0 = \frac{1}{|G|} \sum_{g \in G} gPg^{-1}$. Show by computing the trace of P_0 and then again by using Schur's lemma (specifically, the second part of Theorem 3.1) that $\dim_{\mathbb{F}} E$ is not divisible by the characteristic of $\mathbb{F}$.

3.8. For $g \in G$, let $R_g : \mathbb{F}[G] \to \mathbb{F}[G] : x \mapsto gx$. Show that

$$\operatorname{Tr}(R_g) = \begin{cases} |G| & \text{if } g = e, \\ 0 & \text{if } g \neq e. \end{cases} \qquad (3.38)$$

3.9. For $g, h \in G$, let $T_{(g,h)} : \mathbb{F}[G] \to \mathbb{F}[G] : x \mapsto gxh^{-1}$. Show that

$$\operatorname{Tr}(T_{(g,h)}) = \begin{cases} 0 & \text{if } g \text{ and } h \text{ are not conjugate,} \\ \frac{|G|}{|C|} & \text{if } g \text{ and } h \text{ belong to the same conjugacy class } C. \end{cases} \qquad (3.39)$$

3.10. Prove the Vandermonde determinant formula:

$$\det \begin{bmatrix} 1 & 1 & \cdots & 1 \\ X_1 & X_2 & \cdots & X_n \\ \vdots & \vdots & \cdots & \vdots \\ X_1^{n-1} & X_2^{n-1} & \cdots & X_n^{n-1} \end{bmatrix} = \prod_{1 \leq j < k \leq n} (X_k - X_j). \qquad (3.40)$$

Chapter 4

More Group Algebra

We are now ready to plunge into a fuller exploration of the group algebra $\mathbb{F}[G]$, where G is a group and $\mathbb{F}$ is a field. Our focus is, of course, on the case where G is finite. For some results, conditions on the field are also needed. However, *all conditions on both G and $\mathbb{F}$ will be spelled out in every result*.

Recall that if G is any group and $\mathbb{F}$ is any field, $\mathbb{F}[G]$ is the vector space, over the field $\mathbb{F}$, with the elements of G forming a basis. Thus, its dimension is $|G|$, the number of elements in G. The typical element of $\mathbb{F}[G]$ is of the form

$$x = \sum_{g \in G} x_g g,$$

with each x_g in $\mathbb{F}$, and $x_g = 0$ except for finitely many g. The subscript $g \in G$ on $\sum$ will be dropped when the group G is known from the context. The multiplication map

$$\mathbb{F}[G] \times \mathbb{F}[G] \to \mathbb{F}[G] : (x, y) \mapsto xy = \sum_g \left(\sum_h x_{gh^{-1}} y_h \right) g$$

is bilinear and associative. Thus, $\mathbb{F}[G]$ is an *algebra* over the field $\mathbb{F}$, with multiplicative identity $1 = 1e$, where e is the identity element of G.

The *regular representation* ρ_{reg} of G associates with each $g \in G$ the map

$$\rho_{\text{reg}}(g) : \mathbb{F}[G] \to \mathbb{F}[G] : x \mapsto gx = \sum_h x_h gh \qquad (4.1)$$

for all elements $x = \sum_{h \in G} x_h h$ in $\mathbb{F}[G]$. It is very useful to view a representation ρ of G on a vector space E as specifying, and specified by, an

A.N. Sengupta, *Representing Finite Groups: A Semisimple Introduction*,
DOI 10.1007/978-1-4614-1231-1_4, © Springer Science+Business Media, LLC 2012

$\mathbb{F}[G]$-module structure on E:

$$\left(\sum_g x_g g\right) v = \sum_{g \in G} x_g \rho(g) v$$

for all $v \in E$ and all $x = \sum_g x_g g \in \mathbb{F}[G]$. With this notation, we can stop writing ρ and write gv instead of $\rho(g)v$. The trade-off between notational ambiguity and clarity is worth it. A subrepresentation then is just a submodule. An irreducible representation E corresponds to a *simple* module, in the sense that $E \neq 0$ and E has no submodules other than 0 and E itself. We will use the terms "irreducible" and "simple" interchangeably in the context of modules.

Inside the algebra $\mathbb{F}[G]$, viewed as a left module over itself, a submodule is a *left ideal*, which means a subset closed under addition and also under multiplication on the left by elements of $\mathbb{F}[G]$. A simple submodule L of $\mathbb{F}[G]$ is thus a *simple left ideal*, in the sense that $L \neq \{0\}$ and that L contains no proper nonzero left ideal.

In this chapter we will study the structure of the regular representation ρ_{reg} on $\mathbb{F}[G]$, and the structure of $\mathbb{F}[G]$ as an algebra. A shorter path to the structure of $\mathbb{F}[G]$ may be charted by reading Sect. 4.2 and then Theorem 4.4, and looking back as needed. If you are eager to hike ahead on your own, you can explore along the path laid out in Exercise 4.15, in which, to add to the adventure, you are not allowed to semisimplify!

4.1   Looking Ahead

Let us take a quick look at the terrain ahead. We will work with a finite group G and a field $\mathbb{F}$ in which $|G|1_{\mathbb{F}} \neq 0$. The significance and endlessly useful consequence of this assumption about $|G|$ is that the algebra $\mathbb{F}[G]$ is *semisimple*. This means that any submodule of $\mathbb{F}[G]$ has a complementary submodule, so their direct sum is all of $\mathbb{F}[G]$. Thus, it is no surprise, as we shall prove in Proposition 4.3, that $\mathbb{F}[G]$ splits into a direct sum of simple left ideals M_j:

$$\mathbb{F}[G] = M_1 \oplus \cdots \oplus M_m.$$

By Schur's lemma (Theorem 3.1) it follows that for any $j, k \in [m]$, either M_j and M_k are isomorphic as $\mathbb{F}[G]$-modules or there is no nonzero module morphism $M_j \to M_k$. Clearly it makes sense then to pick out a maximal set of nonisomorphic simple left ideals $L_1, ..., L_s$, and group the M_j's together according to which L_i they are isomorphic to. This produces a decomposition

$$\mathbb{F}[G] = \underbrace{L_{11} + \cdots + L_{1d_1}}_{A_1} + \cdots + \underbrace{L_{s1} + \cdots + L_{sd_s}}_{A_s}, \qquad (4.2)$$

which is a direct sum, with the first d_1 left ideals being isomorphic to L_1, the next d_2 being isomorphic to L_2, and so on, with the last d_s being isomorphic to L_s. Thus,

$$\mathbb{F}[G] \simeq L_1^{d_1} \oplus \cdots \oplus L_s^{d_s}. \qquad (4.3)$$

We will show that each A_i is a *two-sided ideal*, closed under multiplication both on the left and on the right by elements of $\mathbb{F}[G]$, and contains an element u_i that serves as a multiplicative unit inside A_i. Note that the latter implies that $u_i^2 = u_i$; thus, each u_i is an *idempotent*. Each A_i is therefore an algebra in itself, and it is a *minimal* algebra, in the sense that the only two-sided ideals inside it are 0 and A_i. Furthermore, using Schur's lemma again, we will show that

$$A_j A_k = 0 \qquad \text{if } j \neq k. \qquad (4.4)$$

All this provides an identification of $\mathbb{F}[G]$ with the product of the algebras A_i,

$$\prod_{i=1}^{s} A_i \simeq \mathbb{F}[G], \qquad (4.5)$$

by identifying $(a_1, ..., a_s)$ with the sum $a_1 + \cdots + a_s$. Keep in mind that we are working with a finite group G and a field $\mathbb{F}$ in which $|G|1_{\mathbb{F}} \neq 0$ (which means that the characteristic of $\mathbb{F}$ is not a divisor of $|G|$).

An important result is the realization of $\mathbb{F}[G]$ as an algebra of matrices. We will study this carefully in Sect. 4.5, but for now here is a very brief preview. For each $b \in \mathbb{F}[G]$ we have the map

$$r_b : \mathbb{F}[G] \to \mathbb{F}[G] : x \mapsto xb.$$

A key point here is that r_b is $\mathbb{F}[G]$-linear on viewing $\mathbb{F}[G]$ as a left module over itself. The decomposition of $\mathbb{F}[G]$ as a direct sum in (4.2) provides a matrix for r_b whose entries are $\mathbb{F}[G]$-linear maps $L_{ia} \to L_{jb}$; by Schur's lemma, these are all 0 except, potentially, when $i = j$, in which case, of course,

$L_{ia} \simeq L_{jb} \simeq L_i$. As we will prove later, $\text{End}_{\mathbb{F}[G]}(L_i)$ is a division algebra. Thus, r_b can be displayed as a block-diagonal matrix

$$
b \;\leftrightarrow\; r_b =
\begin{bmatrix}
[b_1] & 0 & 0 & \cdots & 0 \\
0 & [b_2] & 0 & \cdots & 0 \\
\vdots & \vdots & \vdots & \vdots & \vdots \\
0 & 0 & 0 & \cdots & [b_s]
\end{bmatrix}
\tag{4.6}
$$

where the block $[b_i]$ has entries in the division ring $\text{End}_{\mathbb{F}[G]}(L_i)$. This realizes $\mathbb{F}[G]$ as an algebra of block-diagonal matrices, with each block being a matrix with entries in a division algebra (these algebras being different in the different blocks). In the special case where $\mathbb{F}$ is algebraically closed, the division algebras collapse to $\mathbb{F}$ itself, and $\mathbb{F}[G]$ is realized as an algebra of block-diagonal matrices with entries in $\mathbb{F}$.

Decomposing $\mathbb{F}[G]$ into simple left ideals provides a decomposition of the regular representation into irreducible components. The interplay between the regular representation, as given by multiplications on the left, and the representation on $\mathbb{F}[G]$, as given by multiplications on the right, is part of a powerful larger story which we will see recurring later in Schur–Weyl duality.

4.2   Submodules and Idempotents

In this section G is a finite group and $\mathbb{F}$ is any field. Let us begin with a closer look at why idempotents arise in constructing submodules of the group algebra $\mathbb{F}[G]$. Idempotents were introduced and used with great effectiveness by Frobenius in unraveling the structure of $\mathbb{F}[G]$.

Recall that an *idempotent* in any ring is an element v whose square is itself:

$$
v^2 = v.
$$

Idempotents u and v are said to be *orthogonal* if

$$
uv = vu = 0.
$$

The sum of two orthogonal idempotents is clearly again an idempotent. An idempotent is said to be *primitive* or *indecomposable* if it is not zero and cannot be expressed as a sum of two nonzero orthogonal idempotents.

An element v in a left ideal L in $\mathbb{F}[G]$ is called a *generator* if $L = \mathbb{F}[G]v$. Here is a very useful fact:

For an idempotent y, an element x lies in $\mathbb{F}[G]y$ if and only if $xy = x$.
$$(4.7)$$

(You can verify this as a moment's-thought exercise.)

Indecomposability of idempotents translates to indecomposability of the generated left ideals:

Proposition 4.1 *Let G be any finite group and $\mathbb{F}$ be any field. An idempotent $y \in \mathbb{F}[G]$ is indecomposable if and only if $\mathbb{F}[G]y$ cannot be decomposed as a direct sum of two distinct nonzero left ideals in $\mathbb{F}[G]$.*

Proof. Suppose y is an indecomposable idempotent and $\mathbb{F}[G]y$ is the direct sum of left ideals L_1 and M_1. Then

$$y = y_1 + v_1 \tag{4.8}$$

for unique $y_1 \in L_1$ and $v_1 \in M_1$. Since $y_1 \in L_1 \subset \mathbb{F}[G]y$, we can write $y_1 = ay$ for some $a \in \mathbb{F}[G]$ and then, since y is an idempotent, we have $y_1 y = y_1$. Left-multiplying (4.8) by y_1 produces

$$\underbrace{y_1 y}_{=y_1} = \underbrace{y_1 y_1}_{\in L_1} + \underbrace{y_1 v_1}_{\in M_1}$$

and so, by unique decomposition,

$$y_1 = y_1^2 \quad \text{and} \quad y_1 v_1 = 0. \tag{4.9}$$

Then

$$y_1 + v_1 = y = y^2 = y_1^2 + y_1 v_1 + v_1 y_1 + v_1^2 = y_1 + v_1 y_1 + v_1^2,$$

which simplifies to

$$\underbrace{v_1}_{\in M_1} = \underbrace{v_1 y_1}_{\in L_1} + \underbrace{v_1^2}_{\in M_1}.$$

The uniqueness of this decomposition then implies

$$v_1 = v_1^2 \quad \text{and} \quad v_1 y_1 = 0. \tag{4.10}$$

Thus, (4.8) decomposes y into the sum of orthogonal idempotents y_1 and v_1 Since y is indecomposable, at least one of y_1 and v_1 is 0. Say, $v_1 = 0$, but then $y = y_1$, and so $\mathbb{F}[G]y \subset L_1$, which implies $M_1 = 0$. Similarly, if $y_1 = 0$, then $y = v_1$, which implies $\mathbb{F}[G]y \subset M_1$ and so $L_1 = 0$ in this case.

For the converse, suppose

$$y = y_1 + v_1, \tag{4.11}$$

where y_1 and v_1 are nonzero orthogonal idempotents. For any $x \in \mathbb{F}[G]y$, we have $x = ay$ for some $a \in \mathbb{F}[G]$, and then

$$x = xy = \underbrace{xy_1}_{\in \mathbb{F}[G]y_1} + \underbrace{xv_1}_{\in \mathbb{F}[G]v_1}, \tag{4.12}$$

a decomposition which is unique because if $x = u + v$, with $u \in \mathbb{F}[G]y_1$ and $v \in \mathbb{F}[G]v_1$, then $u = xy_1$ and $v = xv_1$ on using orthogonality of y_1 and v_1. Note also, by orthogonality of y_1 and v_1 and (4.11), that $y_1 = y_1 y \in \mathbb{F}[G]y$ and $v_1 = v_1 y \in \mathbb{F}[G]y$. So $\mathbb{F}[G]y$ is the direct sum of the left ideals $\mathbb{F}[G]y_1$ and $\mathbb{F}[G]v_1$. Finally, note that $\mathbb{F}[G]y_1$ contains y_1 and so is not $\{0\}$, and similarly also $\mathbb{F}[G]v_1 \neq \{0\}$. Thus, if the idempotent y is decomposable, then so is the left ideal it generates. $\boxed{\text{QED}}$

With semisimplicity, every left ideal has an idempotent generator:

Proposition 4.2 *Let G be any finite group and $\mathbb{F}$ be a field in which $|G|1_{\mathbb{F}} \neq 0$. If L is a left ideal in the algebra $\mathbb{F}[G]$, then there is an idempotent element $y \in \mathbb{F}[G]$ such that*

$$L = \mathbb{F}[G]y.$$

Proof. By semisimplicity, L has a complementary left ideal L_c such that $\mathbb{F}[G]$ is the direct sum of L and L_c. Decompose $1 \in \mathbb{F}[G]$ as

$$1 = y + z,$$

where $y \in L$ and $z \in L_c$. Then for any $x \in \mathbb{F}[G]$,

$$x = \underbrace{xy}_{\in L} + \underbrace{xz}_{\in L_c},$$

and so x lies in L if and only if x is, in fact, equal to xy. Hence, $L = \mathbb{F}[G]y$, and also y equals yy, which means that y is an idempotent. $\boxed{\text{QED}}$

4.3   Deconstructing $\mathbb{F}[G]$, the Module

Semisimplicity decomposes $\mathbb{F}[G]$ into simple left ideals:

Proposition 4.3 *If G is any finite group and $\mathbb{F}$ is a field in which $|G|1_{\mathbb{F}} \neq 0$, the algebra $\mathbb{F}[G]$, viewed as a left module over itself, decomposes as a direct sum of simple submodules. There are indecomposable orthogonal idempotents $e_1, \ldots, e_m$ in $\mathbb{F}[G]$ such that*

$$1 = e_1 + \cdots + e_m,$$

and the simple left ideals $\mathbb{F}[G]e_1, \ldots, \mathbb{F}[G]e_m$ provide a decomposition of $\mathbb{F}[G]$ as a direct sum:

$$\mathbb{F}[G] = \mathbb{F}[G]e_1 \oplus \cdots \oplus \mathbb{F}[G]e_m.$$

In the language of representations, this decomposes the regular representation into a direct sum of irreducible representations.

Proof. Choose a submodule M_1 in $\mathbb{F}[G]$ that has the smallest nonzero dimension as a vector space over $\mathbb{F}$. Then, of course, M_1 has to be a simple submodule.

Let m be the largest integer m such that there exist simple submodules $M_1, \ldots, M_m$, such that the sum $M = M_1 + \cdots + M_m$ is a direct sum; such an m exists because $\mathbb{F}[G]$ is finite-dimensional as a vector space over $\mathbb{F}$. If M is not all of $\mathbb{F}[G]$, then there is, by semisimplicity, a complementary submodule N that is not zero. Inside N choose a submodule M_{m+1} of smallest positive dimension as the vector space over $\mathbb{F}$. But then M_{m+1} is a simple submodule and the sum $M_1 + \cdots + M_{m+1}$ is direct, which contradicts the definition of m. Hence, M is all of $\mathbb{F}[G]$:

$$\mathbb{F}[G] = M_1 \oplus \cdots \oplus M_m.$$

Splitting the element $1 \in \mathbb{F}[G]$ as a sum of components $e_j \in M_j$, we have

$$1 = e_1 + \cdots + e_m.$$

Then for any $x \in \mathbb{F}[G]$,

$$x = \underbrace{xe_1}_{\in M_1} + \cdots + \underbrace{xe_m}_{\in M_m},$$

and so x lies in M_j if and only if $x = xe_j$ and $xe_k = 0$ for all $k \neq j$. This means, in particular, that

$$e_j^2 = e_j, \quad \text{and} \quad e_j e_k = 0 \quad \text{if } j \neq k,$$

and

$$M_j = \mathbb{F}[G]e_j$$

for all $j, k \in [m]$. $\boxed{\text{QED}}$

Here is another observation for which we use the versatile power of Schur's lemma (Theorem 3.1):

Proposition 4.4 *Let G be a finite group and $\mathbb{F}$ be a field in which $|G|1_\mathbb{F} \neq 0$. View $\mathbb{F}[G]$ as a left module over itself, and let $M_1, \ldots, M_m$ be simple submodules whose direct sum is $\mathbb{F}[G]$. If L is any simple submodule in $\mathbb{F}[G]$, then L is isomorphic to some M_j, and is a subset of the sum of those M_j that are isomorphic to L.*

Proof. Since $\mathbb{F}[G]$ is the direct sum of the submodules M_j, every element $x \in \mathbb{F}[G]$ decomposes uniquely as a sum

$$x = \underbrace{x_1}_{\in M_1} + \cdots + \underbrace{x_m}_{\in M_m},$$

with $x_j \in M_j$ for each $j \in [m]$. We then have the projection maps

$$\pi_j : \mathbb{F}[G] \to M_j : x \mapsto x_j.$$

The uniqueness of the decomposition, along with the fact that $ax_j \in M_j$ for every $a \in \mathbb{F}[G]$, implies that π_j is linear as a map between $\mathbb{F}[G]$-modules:

$$\pi_j(ax + y) = a\pi_j(x) + \pi_j(y)$$

for all $a, x, y \in \mathbb{F}[G]$. Consider now a simple submodule $L \subset \mathbb{F}[G]$. The restriction $\pi_j|L$ is an $\mathbb{F}[G]$-linear map $L \to M_j$. Then by Schur's lemma (Theorem 3.1), this must be either 0 or an isomorphism. Looking at any $x \in L$, as a sum of the components $x_j = \pi_j(x)$, the components that lie in the M_k not isomorphic to L are all zero, and so at least one of the other components must be nonzero when $x \neq 0$. This implies that L is isomorphic to some M_j, and lies inside the sum of those M_j to which it is isomorphic. $\boxed{\text{QED}}$

4.4 Deconstructing $\mathbb{F}[G]$, the Algebra

We turn now to the task of decomposing the algebra $\mathbb{F}[G]$ as a product of smaller, simpler algebras. If S and T are subsets of $\mathbb{F}[G]$, then by ST we mean the set of all elements that are finite sums of products st with $s \in S$ and $t \in T$:

$$ST = \{s_1 t_1 + \cdots + s_k t_k : k \in \{1, 2, ...\}, s_1, ..., s_k \in S, t_1, ..., t_k \in T\}.$$

Thus, with this notation, a subset $J \subset A$ for which $J + J \subset J$ is a left ideal if $AJ \subset J$ is a right ideal if $JA \subset J$, and is a two-sided ideal if $AJA \subset J$.

Let us make a few starter observations about left ideals.

Proposition 4.5 *Let G be a finite group, $\mathbb{F}$ be a field, and L be a simple left ideal in the algebra $A = \mathbb{F}[G]$. Then:*

1. $L = \mathbb{F}[G]u$ *for any nonzero* $u \in L$.

2. *If* $v \in \mathbb{F}[G]$, *then either Lv is 0 or Lv is isomorphic to L, as left $\mathbb{F}[G]$-modules.*

3. *If M is a simple left ideal and $LM \neq 0$, then $M = Lv$ for some $v \in \mathbb{F}[G]$.*

4. *LA, which is the sum of all the right translates Lv, is a two-sided ideal in $\mathbb{F}[G]$.*

5. *If L and M are simple left ideals, and M is not isomorphic to L, then*

$$(LA)(MA) = 0.$$

If, moreover, $|G|1_{\mathbb{F}} \neq 0$, then every simple left ideal M isomorphic to L is of the form Lv for some $v \in \mathbb{F}[G]$.

Notice that for statements 1–4 we do not need the semisimplicity condition that $|G|$ not be divisible by the characteristic of $\mathbb{F}$.

Proof. 1. If $u \in L$ is not zero, then $\mathbb{F}[G]u$ is a nonzero left ideal contained inside the simple left ideal L and hence must be equal to L.

2. For any $v \in \mathbb{F}[G]$, Lv is clearly a left ideal in $\mathbb{F}[G]$. The map

$$f : L \to Lv : a \mapsto av$$

is $\mathbb{F}[G]$-linear, and so Schur's lemma (the version in Theorem 3.2) implies that either $Lv = 0$ or f is an isomorphism of L onto Lv. Thus, either Lv is 0 or Lv is isomorphic, as a left $\mathbb{F}[G]$-module, to L.

3. Next suppose M is also a simple left ideal and $LM \neq 0$. Choose $u \in L$ and $v \in M$ with $uv \neq 0$. Then, by statement 1, $M = \mathbb{F}[G]v$ and so $Lv \subset M$. Since M is simple and Lv, which contains uv, is not 0, we have $M = Lv$.

4. It is clear that LA is both a left ideal and a right ideal.

5. Now suppose L and M are both simple left ideals and $(LA)(MA) \neq 0$. Then $(Lx)(My) \neq 0$ for some $x, y \in \mathbb{F}[G]$. Then $Lx \neq 0$ and $My \neq 0$, and so $Lx \simeq L$ and $My \simeq M$ (where $\simeq$ denotes isomorphism of A-modules), by statement 2. In particular, Lx and My are also simple left ideals. Since $LxMy \neq 0$, it follows from statement 3 that My is a right translate of Lx, which then, by statement 2, implies that $Lx \simeq My$. But, as we have already noted, $Lx \simeq L$ and $My \simeq M$. Hence $L \simeq M$.

We turn finally to proving the stated consequence of semisimplicity of $\mathbb{F}[G]$. By Proposition 4.2, semisimplicity implies $L = \mathbb{F}[G]y$ for an *idempotent* y and so if $f : L \to M$ is an isomorphism of modules, then

$$M = f(L) = f(Ly) = Lf(y),$$

showing that M is a right translate of L. $\boxed{\text{QED}}$

If we add up all the simple left ideals that are isomorphic to a given simple left ideal L, we get

$$\sum_{x \in \mathbb{F}[G]} Lx = L\mathbb{F}[G],$$

a two-sided ideal that is clearly the smallest two-sided ideal containing L. Such two-sided ideals form the key structural pieces in the decomposition of the algebra $\mathbb{F}[G]$.

Theorem 4.1 *Let G be a finite group and $\mathbb{F}$ be a field in which $|G|1_{\mathbb{F}} \neq 0$. Then there are subspaces $A_1, \ldots, A_s \subset \mathbb{F}[G]$ such that each A_j is an algebra under the multiplication operation inherited from $\mathbb{F}[G]$, and the map*

$$I : \prod_{j=1}^{s} A_j \to \mathbb{F}[G] : (a_1, \ldots, a_s) \mapsto a_1 + \cdots + a_s \qquad (4.13)$$

is an isomorphism of algebras. Moreover:

1. *Every simple left ideal is contained inside exactly one of $A_1, \ldots, A_s$.*

2. $A_j A_k = 0$ *if* $j \neq k$.

3. *Each* A_j *is a two-sided ideal in* $\mathbb{F}[G]$.

4. *Each* A_j *is of the form* $\mathbb{F}[G]u_j$, *with* $u_1, \ldots, u_s$ *being nonzero orthogonal central idempotents, and with*

$$u_1 + \cdots + u_s = 1.$$

5. *Every two-sided ideal in* $\mathbb{F}[G]$ *is a sum of some of the* $A_1, \ldots, A_s$.

6. *For every* $j \in [s]$, *the only two-sided ideals of* A_j *are* 0 *and* A_j *itself.*

7. *No* u_j *can be decomposed as a sum of two nonzero central idempotents.*

This is a lot and the proof is lengthy, but not hard. Statements 1–4, and also statement 7, hold even when $\mathbb{F}[G]$ is not semisimple; for this, following an alternative route, you can work through Exercise 4.15. Theorem 4.4 will reconstruct the two-sided ideals A_i by producing the central idempotents u_i first.

Proof. First view $\mathbb{F}[G]$ as a left module over itself. We saw in Proposition 4.3 that $\mathbb{F}[G]$ is a direct sum of a finite set of simple submodules $M_1, \ldots, M_m$:

$$\mathbb{F}[G] = M_1 \oplus \cdots \oplus M_m.$$

Moreover, by Proposition 4.4, every simple submodule is isomorphic to one of these submodules and also lies inside the sum of those M_j to which it is isomorphic. Thus, it makes sense to group together all the M_j that are mutually isomorphic and form their sums.

Let $L_1, \ldots, L_s$ be a maximal set of simple submodules among the M_j such that no two are isomorphic with each other. Now, for each $j \in [s]$, set A_j to be the sum of all those M_i that are isomorphic to L_j. Then $\mathbb{F}[G]$ is the direct sum of the submodules A_j:

$$\mathbb{F}[G] = A_1 \oplus \cdots \oplus A_s. \tag{4.14}$$

Let us keep in mind, from Proposition 4.4, that any simple submodule which is isomorphic to L_j actually lies inside A_j. Thus, A_j is the sum of *all* the simple submodules that are isomorphic to L_j. This, along with the fact that the right side in (4.14) is a *direct* sum, proves statement 1.

By the last statement in Proposition 4.5, every submodule of $\mathbb{F}[G]$ that is isomorphic to L_j is a right translate $L_j y$ of L_j. Conversely every right translate $L_j y$ is either 0 or isomorphic to L_j. Hence, the algebra A_j is the sum of all right translates of L_j:

$$A_j = L_j \mathbb{F}[G].$$

From this it is clear that A_j is also a right ideal. This proves statement 3.

By Proposition 4.5 (statement 5) it follows that

$$A_j A_k = 0 \qquad \text{if } j \neq k.$$

This establishes statement 2.

Thus, if $x, y \in \mathbb{F}[G]$ decompose as

$$x = x_1 + \cdots + x_s, \quad y = y_1 + \cdots + y_s,$$

with $x_j, y_j \in A_j$ for each j, then

$$xy = x_1 y_1 + \cdots + x_s y_s.$$

Let us now express 1 as a sum of components $u_j \in A_j$:

$$1 = u_1 + \cdots + u_s.$$

Since $A_j A_k$ is 0 for $j \neq k$, it follows by working out the product $u_j 1$ that

$$u_j = u_j^2 \quad \text{and} \quad u_j u_k = 0 \qquad \text{for all } j, k \in [s] \text{ with } j \neq k.$$

Thus, the u_j are orthogonal idempotents and add up to 1.

For $x \in \mathbb{F}[G]$ we have

$$x = x1 = xu_1 + \cdots + xu_s,$$

which gives the decomposition of x into the component pieces in the A_j, and also shows that x lies inside A_j if and only if xu_j is x itself; hence,

$$A_j = \mathbb{F}[G]u_j \qquad \text{for all } j \in [s].$$

Clearly, u_j is the multiplicative identity element in A_j, which is thus an algebra in itself. Note that if u_j were 0, then A_j would be 0 and this is impossible because A_j is a sum of simple, hence nonzero, submodules. This completes the proof of statement 4, except for centrality of the u_j, which is shown below.

It is now clear that the mapping

$$I : \prod_{j=1}^{s} A_j \to \mathbb{F}[G] : (a_1, ..., a_s) \mapsto a_1 + \cdots + a_s$$

is an isomorphism of algebras, as stated after (4.13).

Let us check that each u_j is in the center of $\mathbb{F}[G]$. For any $x \in \mathbb{F}[G]$ we have

$$x = 1x = \underbrace{u_1 x}_{\in A_1} + \cdots + \underbrace{u_s x}_{\in A_s} .$$

Comparing with the decomposition "on the left"

$$x = x1 = \underbrace{x u_1}_{\in A_1} + \cdots + \underbrace{x u_s}_{\in A_s}$$

and using the uniqueness of decomposition of $\mathbb{F}[G]$ as a *direct* sum of the A_j, we see that x commutes with each u_j. Hence, $u_1, ..., u_s$ are all central idempotents.

Now consider a two-sided ideal $B \neq 0$ in $\mathbb{F}[G]$. Let $j \in [s]$. The set BA_j, consisting of all sums of elements ba_j with b drawn from B and a_j from A_j, is a two-sided ideal and is clearly contained inside $B \cap A_j$. If BA_j contains a nonzero element x, then, working with a simple left ideal L contained in $\mathbb{F}[G]x \subset BA_j$, it follows that BA_j contains all right translates of L; thus, if $BA_j \neq 0$, then $BA_j \supset A_j$, and hence $BA_j = A_j$. Thus, looking at the decomposition

$$B = BA = BA_1 + \cdots + BA_s,$$

we see that B is the sum of those A_j for which $BA_j \neq 0$. This proves statement 5.

Now we prove statement 6, that the algebra A_j is minimal in the sense that any two-sided ideal in it is either 0 or A_j. Suppose J is a two-sided ideal in the algebra A_j. For any $x \in \mathbb{F}[G]$, and $y \in A_j$, we know that xy equals $x_j y$, where x_j is the component of x in A_j in the decomposition of A as the direct sum of $A_1, ..., A_s$. Consequently, any left ideal within A_j is a left ideal in the full algebra $\mathbb{F}[G]$. Similarly, any right ideal in A_j is a right ideal in

$\mathbb{F}[G]$. Hence, a two-sided ideal J inside the algebra A_j is a two-sided ideal in $\mathbb{F}[G]$ and hence is a sum of some of the two-sided ideals A_i. But these two-sided ideals are complementary and J lies inside A_j; hence, J is equal to A_j.

Finally, we prove statement 7, showing that the central idempotent generators u_j are indecomposable within the class of central idempotents. Suppose

$$u_j = u + v,$$

where u and v are orthogonal central idempotents. Then

$$uu_j = uu + uv = u^2 + 0 = u,$$

and so

$$\mathbb{F}[G]u = \mathbb{F}[G]uu_j \subset \mathbb{F}[G]u_j = A_j.$$

Furthermore, since u is central, the left ideal $\mathbb{F}[G]u$ is also a right ideal. Being a two-sided ideal lying inside A_j it must then be either 0 or A_j itself. If $\mathbb{F}[G]u$ is 0, then $u = 1u$ is 0. If $u \neq 0$, then $\mathbb{F}[G]u = A_j$ and so $u_j = xu$ for some $x \in \mathbb{F}[G]$, and then $v = u_j v = xuv$ is 0. Thus, in the decomposition of u_j into a sum of two central orthogonal idempotents, one of them must be 0.
$\boxed{\text{QED}}$

The next task is to determine the structure of an algebra that does not contain any proper nonzero two-sided ideals. But before turning to that, we note the following uniqueness of the decomposition:

Theorem 4.2 *Let G be a finite group and $\mathbb{F}$ be a field in which $|G|1_{\mathbb{F}} \neq 0$. Suppose $B_1, \ldots, B_r \subset \mathbb{F}[G]$, where each B_j is nonzero, closed under addition and multiplication, and contains no proper nonzero two-sided ideals, and such that*

$$I : B_1 \times \cdots \times B_r \to \mathbb{F}[G] : (b_1, \ldots, b_r) \mapsto b_1 + \cdots + b_r$$

is surjective and preserves addition and multiplication, these operations being defined componentwise for $B_1 \times \cdots \times B_r$. Then $r = s$ and

$$\{B_1, \ldots, B_r\} = \{A_1, \ldots, A_s\},$$

where $A_1, \ldots, A_s$ are the two-sided ideals in $\mathbb{F}[G]$ described in Theorem 4.1.

Proof. The fact that I preserves multiplication implies that

$$B_j B_k = 0 \qquad \text{if } j \neq k.$$

Each B_j is a two-sided ideal in $B = \mathbb{F}[G]$ because

$$BB_j \subset B_1 B_j + \cdots + B_r B_j = 0 + B_j B_j + 0 \subset B_j,$$

and, similarly, $B_j B \subset B_j$.

Then, by Theorem 4.1, each B_j is the sum of some of the two-sided ideals A_i. The condition that B_j contains no proper nonzero two-sided ideal then implies that B_j is equal to some A_i. Hence, I maps

$$\{(0, 0, \ldots, \underbrace{b_j}_{j\text{th position}}, 0, \ldots, 0) : b_j \in B_j\}$$

onto A_i. Now the sets $A_1, \ldots, A_s$ are all distinct. Since the map I is a bijection, it follows that $B_1, \ldots, B_r$ are all distinct. Hence, $r = s$ and $\{B_1, \ldots, B_r\}$ is the same as $\{A_1, \ldots, A_s\}$. $\boxed{\text{QED}}$

4.5 As Simple as Matrix Algebras

We turn now to the determination of the structure of finite-dimensional algebras that contain no proper nonzero two-sided ideals. We will revisit this topic in a more general setting in Sect. 5.6.

Suppose B is a finite-dimensional algebra over a field $\mathbb{F}$ and L is a left ideal in B of minimum positive dimension. Then L, being of minimum dimension, is simple. Let

$$\mathbb{D} = \text{End}_B(L),$$

which is the set of all B-linear maps $f : L \to L$. By Schur's lemma, any such f is either 0 or an isomorphism. Thus, $\mathbb{D}$ is a division ring: it is a ring, with multiplicative identity $(\neq 0)$, in which every nonzero element has a multiplicative inverse. Note that here $\mathbb{D}$ contains $\mathbb{F}$ and is also a vector space over $\mathbb{F}$, necessarily finite-dimensional because it is contained inside the finite-dimensional space $\text{End}_\mathbb{F}(L)$.

Theorem 4.3 *Suppose B is a finite-dimensional algebra over a field $\mathbb{F}$ and assume that the only two-sided ideals in B are 0 and B itself Then B is isomorphic to the algebra*

$$\text{Matr}_{n \times n}(\mathbb{D})$$

*of $n \times n$ matrices over a division algebra $\mathbb{D}$ over $\mathbb{F}$ for some positive integer
n. The division algebra $\mathbb{D}$ can be taken to be $\mathbb{D} = \mathrm{End}_B(L)$, where L is any
simple left ideal in B, with multiplication given by composition in the opposite
order: $f \circ_{op} g = g \circ f$ for $f, g \in \mathrm{End}_B(L)$.*

This fundamental result, evolved in the formulation, grew out of the dis-
sertations of Molien [60] and Wedderburn [77].

To indicate that the multiplication is in the opposite order to the standard
multiplication in $\mathrm{End}_B(L)$, we write

$$\mathbb{D} = \mathrm{End}_B(L)^{\mathrm{opp}}.$$

The appearance of a division ring, as opposed to a field, might seem dis-
appointing. But much of the algebra here is a sharper shadow of synthetic
geometry, a subject nearly lost to mathematical history, where, logically if
not historically, division rings appear more naturally (i.e., from fewer geometric axioms) than fields.

Proof. There are two main steps in realizing B as an algebra of matrices.
First, we will show that B is naturally isomorphic to the algebra $\mathrm{End}_B(B)$
of all B-linear maps $B \to B$, with a little twist applied. Next, we will show
by breaking B up into a direct sum of translates of any simple left ideal that
any element of $\mathrm{End}_B(B)$ can be viewed as a matrix with entries in $\mathbb{D}$.

Any element $b \in B$ specifies a B-linear map

$$r_b : B \to B : x \mapsto xb,$$

and b is recovered from r_b by applying r_b to 1:

$$b = r_b(1).$$

Conversely, if $f \in \mathrm{End}_B(B)$, then

$$f(x) = f(x1) = xf(1) = r_{f(1)}(x) \qquad \text{for all } x \in B.$$

Thus, $b \mapsto r_b$ is a bijection $B \to \mathrm{End}_B(B)$, and is clearly linear over the field
$\mathbb{F}$. Let us look now at how r interacts with multiplication:

$$r_a r_b(x) = r_a(xb) = x(ba) = r_{ba}(x).$$

Thus, the map $b \mapsto r_b$ *reverses* multiplication. Then we have an isomorphism
of algebras

$$B \to \mathrm{End}_B(B)^{\mathrm{opp}},$$

where the superscript indicates that multiplication of endomorphisms should be done in the opposite order to usual.

Now let L be a left ideal in B of minimum positive dimension, as a vector space over $\mathbb{F}$. The minimum dimension condition implies that L is a simple left ideal. Since LB is a nonzero two-sided ideal in B, it is equal to B. But LB is the sum of all right translates Lb with b running over B. If $b \in B$ is such that $Lb \neq 0$, then the mapping between the left B-modules L and Lb given by

$$\phi_b : L \to Lb : x \mapsto xb$$

is clearly B-linear and surjective, and ker ϕ_b, which is not L because $Lb \neq 0$, must be $\{0\}$, and this means that ϕ_b is an isomorphism of B-modules. Thus, any nonzero Lb is a simple left ideal isomorphic as a B-module to L.

Let n be the largest integer for which there exist $b_1, ..., b_n \in B$ such that the sum $Lb_1 + \cdots + Lb_n$ is a direct sum and each Lb_j is nonzero. As noted above, the mapping

$$\phi_j : L \to Lb_j : x \mapsto xb_j \tag{4.15}$$

is an isomorphism of B-modules and Lb_j is a simple left ideal in B.

Note that $n \geq 1$ and also that $n \leq \dim_{\mathbb{F}} B$, which is finite by hypothesis. If $Lb_1 + \cdots + Lb_n$ is not all of LB, then there is some Lb not contained in $S = Lb_1 + \cdots + Lb_n$; but then $Lb \cap S = \{0\}$ by simplicity of Lb and this would contradict the definition of n. Thus,

$$B = LB = Lb_1 \oplus \cdots \oplus Lb_n. \tag{4.16}$$

Putting the maps ϕ_j together yields an isomorphism of left B-modules:

$$\Phi : L^n \to B : (a_1, ..., a_n) \mapsto \phi_1(a_1) + \cdots + \phi_n(a_n).$$

Then any $b \in B$ corresponds to a B-linear map

$$r'_b = \Phi^{-1} \circ r_b \circ \Phi : L^n \to L^n$$

that gives rise to a matrix

$$[b_{jk}]_{1 \leq j, k \leq n},$$

where

$$b_{jk} = p_k \circ r'_b \circ i_j : L \to L,$$

with $p_k : L^n \to L$ being the projection onto the kth component and

$$i_j : L \to L^n : x \mapsto (0, ..., \underbrace{x}_{j\text{th term}}, ..., 0).$$

Note that

$$\sum_{j=1}^{n} i_j p_j = \text{id}_{L^n}.$$

Now we have a key observation: *each component b_{jk} is in $\text{End}_B(L)$, and is thus an element of the division ring $\mathbb{D}$.* Thus, we have associated with each $b \in B$ a matrix $[b_{jk}]$ with entries in $\mathbb{D}$.

If $a, b \in B$, then

$$\begin{aligned}
(ab)_{jk} &= p_k \Phi^{-1} r_{ab} \Phi i_j \\
&= p_k \Phi^{-1} r_b r_a \Phi i_j \\
&= \sum_{l=1}^{n} p_k \Phi^{-1} r_b \Phi i_l p_l \Phi^{-1} r_a \Phi i_j \\
&= \sum_{l=1}^{n} b_{lk} a_{jl} \\
&= \sum_{l=1}^{n} a_{jl} \circ_{\text{op}} b_{lk}.
\end{aligned} \tag{4.17}$$

Thus,

$$[(ab)_{jk}] = [a_{jl}][b_{lk}]$$

as a product of matrices with entries in the ring $\mathbb{D} = \text{End}_B(L)^{\text{opp}}$. It is clear that there is no twist in addition:

$$[(a + b)_{jk}] = [a_{jk}] + [b_{jk}].$$

Thus, the mapping

$$a \mapsto [a_{jk}]$$

preserves addition and multiplication. Clearly, it preserves multiplication by scalars from $\mathbb{F}$, and also carries the multiplicative identity 1 in B to the identity matrix.

If $[c_{jk}]$ is any $n \times n$ matrix with entries in $\mathbb{D}$, then it corresponds to the B-linear mapping

$$L^n \to L^n : (x_1, ..., x_n) \mapsto \left(\sum_{j=1}^{n} c_{j1} x_j, ..., \sum_{j=1}^{n} c_{jn} x_j \right),$$

which, by the identification $L^n \simeq B$, corresponds to an element $f \in \mathrm{End}_B(B)$, which in turn corresponds to the element $c = f(1)$ in B. This recovers c from the matrix $[c_{jk}]$. $\boxed{\text{QED}}$

A ring which is the sum of simple left ideals that are all isomorphic to each other is called a *simple ring*. In the preceding proof, specifically in (4.16), we saw that a finite-dimensional algebra B that contains no proper nonzero two-sided ideals is a simple ring. By a *simple algebra* we mean an algebra which is a simple ring. We study simple rings in Sect. 5.6.

In applying Theorem 4.3 to the algebras A_i contained inside $\mathbb{F}[G]$ (assumed semisimple) as two-sided ideals, we note that a simple left ideal L in A_i is also a simple left ideal when viewed as a subset of $\mathbb{F}[G]$, because if $x \in \mathbb{F}[G]$ is decomposed as $x_1 + \cdots + x_s$, with $x_j \in A_j$ for each j, then

$$xL = (x_1 + \cdots + x_s)L = 0 + x_i L + 0 \subset L,$$

with the last inclusion holding because $x_i \in A_i$ and L is a left ideal in A_i. In fact, essentially the same argument shows that if $f : L \to L$ is linear over A_i, then it is linear over the big algebra $\mathbb{F}[G]$. Thus,

$$\mathrm{End}_{A_i}(L) = \mathrm{End}_{\mathbb{F}[G]}(L)$$

for any minimal two-sided ideal A_i in $\mathbb{F}[G]$ and simple left ideal $L \subset A_i$.

Recall in this context Wedderburn's result (Theorem 1.3): any finite-dimensional division algebra $\mathbb{D}$ over any algebraically closed field $\mathbb{F}$ is equal to $\mathbb{F}$, identified naturally as a subset of $\mathbb{D}$. There is another similar result, also discovered by Wedderburn [77]: *if $\mathbb{F}$ is a finite field, then every finite-dimensional division algebra over $\mathbb{F}$ is a field.*

We have introduced the notion of a splitting field for a group algebra. More generally, a field $\mathbb{F}$ is a *splitting field* for a finite-dimensional $\mathbb{F}$-algebra A if for every simple A-module E, the only A-linear mappings $E \to E$ are of the form cI, where I is the identity mapping on E and $c \in \mathbb{F}$; more compactly, the condition is that $\mathrm{End}_A(E) = \mathbb{F}I$.

4.6   Putting $\mathbb{F}[G]$ Back Together

It is time to look back and see how all the pieces fit together to form the group algebra $\mathbb{F}[G]$.

Theorem 4.4 *Let G be a finite group and $\mathbb{F}$ be any field.*

1. *Then there is a largest integer s for which there exist nonzero central idempotents $u_1, \ldots, u_s$ that add up to 1 and are orthogonal to each other: $u_i u_j = 0$ for all $i, j \in [s]$ with $i \neq j$. For each $i \in [s]$, $\mathbb{F}[G]u_i$ is a two-sided ideal, a subalgebra of $\mathbb{F}[G]$ with u_i as a multiplicative identity, and the mapping*

$$I : \prod_{i=1}^{s} \mathbb{F}[G]u_i \to \mathbb{F}[G] : (a_1, \ldots, a_s) \mapsto a_1 + \cdots + a_s, \qquad (4.18)$$

is an isomorphism of $\mathbb{F}$-algebras, where the left side in (4.18) is the product of the algebras $\mathbb{F}[G]u_i$.

2. *Suppose that $|G|1_{\mathbb{F}} \neq 0$. Then, for each $i \in [s]$, $\mathbb{F}[G]u_i$ is a minimal two-sided ideal in $\mathbb{F}[G]$, and is the direct sum of right translates of a simple left ideal L_i,*

$$\mathbb{F}[G]u_i = \bigoplus_{j=1}^{d_i} L_i a_{ij}, \qquad (4.19)$$

for some positive integer d_i and elements $a_{ij} \in \mathbb{F}[G]$.

3. *If L is any simple left ideal in $\mathbb{F}[G]$, then there is exactly one $i \in [s]$ such that L is isomorphic, as a left $\mathbb{F}[G]$-module, to L_i and then L is a right translate $L_i a$ of L_i; conversely, any nonzero right translate of L_i is a simple left ideal of $\mathbb{F}[G]$ and is isomorphic to L_i.*

4. *Suppose again that $|G|1_{\mathbb{F}} \neq 0$. For any $x \in \mathbb{F}[G]u_i$ let $\rho_i(x) : L_i \to L_i$ be given by $y \mapsto xy$. Then for each $i \in [s]$,*

$$\mathbb{D}_i = \mathrm{End}_{\mathbb{F}[G]}(L_i) \qquad (4.20)$$

is a finite-dimensional division algebra over $\mathbb{F}$, and the map

$$\mathbb{F}[G]u_i \to \mathrm{End}_{\mathbb{D}_i}(L_i) : x \mapsto \rho_i(x) \qquad (4.21)$$

is an isomorphism of $\mathbb{F}$-algebras.

5. If $|G|1_{\mathbb{F}} \neq 0$ and $\mathbb{F}$ is algebraically closed, then $\mathbb{D}_i = \mathbb{F}$, with the latter identified with a subset of $\mathbb{D}_i$ in the natural way, and

$$d_i = \dim_{\mathbb{F}} L_i \quad \text{for all } i \in [s]. \tag{4.22}$$

You can piece together this result by combining Theorem 4.1 with other results we have proven. However, we include a detailed, and hence lengthy, proof, incorporating both ideas we have seen before and some (such as double commutants) that we will explore later. The construction of the two-sided ideals $\mathbb{F}[G]u_i$ given here is different from that in the proof of Theorem 4.1, exploiting the finite dimensionality of $\mathbb{F}[G]$ directly instead of the semisimplicity (which we do need for other parts of this result).

Proof. 1. If $y_1, \ldots, y_m$ are nonzero orthogonal idempotents and $\sum_{i=1}^m c_i y_i = 0$ for some $c_1, \ldots, c_m \in \mathbb{F}$, then multiplying by y_k shows that $c_k y_k = 0$ and so $c_k = 0$; thus, $\{y_1, \ldots, y_m\}$ is linearly independent. Then $m \leq \dim_{\mathbb{F}} \mathbb{F}[G] = |G|$. This proves the existence of a largest integer s as claimed (of course, $s \geq 1$ because 1 itself is a central idempotent). The map I clearly preserves addition and it also preserves multiplication because the central idempotents u_i are orthogonal. Orthogonality also implies that I is injective, whereas surjectivity of I follows from the sum of the u_i being 1.

Note that orthogonality of the central idempotents u_i, along with the fact that the two-sided ideals $\mathbb{F}[G]u_i$ add up to $\mathbb{F}[G]$, implies that any $L \subset \mathbb{F}[G]u_i$ is a left ideal in $\mathbb{F}[G]u_i$ if and only if it is a left (or two-sided) ideal in $\mathbb{F}[G]$.

2. Let L_i be a left ideal in $\mathbb{F}[G]u_i$ of minimum positive dimension $\dim_{\mathbb{F}} L_i$; then L_i is a simple left ideal in $\mathbb{F}[G]$.

Now suppose $|G|1_{\mathbb{F}} \neq 0$. Let B be a two-sided ideal in $\mathbb{F}[G]$ contained inside $\mathbb{F}[G]u_i$. By semisimplicity, there is a left ideal B_c for which

$$B + B_c = \mathbb{F}[G]u_i$$

as a direct sum. Decompose u_i as $y + y_c$, where $y \in B$ and $y_c \in B_c$. Then for any $x \in \mathbb{F}[G]$, we have

$$x u_i = \underbrace{xy}_{\in B} + \underbrace{xy_c}_{\in B_c} = u_i x = \underbrace{yx}_{\in B} + \underbrace{y_c x}_{\in B_c}, \tag{4.23}$$

which implies that $xy = yx$ and $xy_c = yx_c$. Thus, y and y_c are central. Taking $y = yu_i$ for x in (4.23) shows that $y^2 = y$, and taking $x = y_c = y_c u_i$ shows that $y_c^2 = y_c$. Lastly, observe that $yy_c \in B \cap B_c = \{0\}$. Thus, y and y_c are orthogonal central idempotents. Replacing u_i by the two central idempotents y and y_c in the collection $\{u_1, \ldots, u_s\}$ would produce a set of $s + 1$ orthogonal central idempotents adding up to 1; so, in view of the definition of s, either y or y_c is 0. Thus, u_i cannot be decomposed into a sum of two nonzero central orthogonal idempotents, and either B or B_c is $\{0\}$. Thus, $\mathbb{F}[G]u_i$ is a minimal two-sided ideal.

We continue with the hypothesis that $|G|1_{\mathbb{F}} \neq 0$. The sum of all right translates $L_i a$, with a running over $\mathbb{F}[G]u_i$, is the two-sided ideal $L_i \mathbb{F}[G]u_i$ sitting inside $\mathbb{F}[G]u_i$ and so is equal to $\mathbb{F}[G]u_i$. The mapping $r_a|L_i : L_i \to L_i a : x \mapsto xa$ is surjective and, since L_i is a simple left ideal, its kernel is 0 or L_i; thus, $r_a|L_i : L_i \to L_i a$ is an isomorphism if $L_i a \neq 0$. Let M be any simple left ideal of $\mathbb{F}[G]$ lying inside $\mathbb{F}[G]u_i$ and let v be any nonzero element of M. Since $u_i \in \mathbb{F}[G]u_i$, the sum of right translates of L_i, we can write

$$u_i = u_{i1}a_1 + \cdots + u_{ip}a_p$$

for some positive integer p, $u_{i_1}, \ldots, u_{ip} \in L_i$ and $a_1, \ldots, a_p \in \mathbb{F}[G]$. From $u_i v = v \neq 0$, it follows that $u_{ir} a_r v \neq 0$ for some $r \in [p]$. Then the map

$$L_i a_r \to M : x \mapsto xv \qquad (4.24)$$

is a nonzero $\mathbb{F}[G]$-linear map between simple $\mathbb{F}[G]$-modules and hence, by Schur's lemma, is an isomorphism. In particular,

$$M = L_i a_r v \qquad (4.25)$$

is a right translate of L_i. (A form of this argument was used earlier to prove statement 3 in Proposition 4.5.)

Now let d_i be the largest positive integer for which there exist $a_{i1}, \ldots,$ $a_{id_i} \in \mathbb{F}[G]$ such that the sum

$$N = L_i a_{i1} + \cdots + L_i a_{id_i}$$

is a *direct* sum and each $L_i a_{ij}$ is nonzero. Choose a complementary left ideal N_c in $\mathbb{F}[G]u_i$. If $N_c \neq 0$, there is a left ideal of $\mathbb{F}[G]$ of minimum positive dimension lying inside N_c; this left ideal is then a simple left ideal J, necessarily a right translate of L_i, such that $N + J$ is a direct sum. To avoid contradiction, $N = \mathbb{F}[G]u_i$. Thus,

$$\mathbb{F}[G]u_i = L_i a_{i1} + \cdots + L_i a_{id_i} \qquad (4.26)$$

is a direct sum, as claimed.

3. Let $f : L_i \to L_j$ be $\mathbb{F}[G]$-linear, where $i, j \in [s]$. Then for any $y \in L_i$, we have

$$f(y) = f(u_i y) = u_i f(y) \in u_i \mathbb{F}[G]u_j = 0 \quad \text{if } j \neq i.$$

Hence, L_i is not isomorphic to L_j if $j \neq i$. For each $i \in [s]$, let the map $\pi_i : \mathbb{F}[G] \to \mathbb{F}[G]u_i$ be specified by requiring that

$$a = \pi_1(a) + \cdots + \pi_s(a)$$

for all $a \in \mathbb{F}[G]$. Now let L be any simple left ideal in $\mathbb{F}[G]$. If $\pi_i(L) \neq 0$, then, by Schur's lemma, $\pi_i|L$ is an isomorphism of L onto a simple left ideal lying inside $\mathbb{F}[G]u_i$. Then, as seen in the context of (4.24), $\pi_i(L)$ is isomorphic to L_i. Thus, L is isomorphic to L_j if and only if $j = i$. This means $\pi_j = 0$ for $j \neq i$. Hence, $L = \pi_i(L)$ is a right translate of L_i.

4. We have already seen that Schur's lemma implies that $\mathbb{D}_i = \text{End}_{\mathbb{F}[G]}(L_i)$ is a division ring. Since $\dim_{\mathbb{F}} L_i \leq \dim_{\mathbb{F}} \mathbb{F}[G] = |G|$ is finite, $\mathbb{D}_i$ is finite-dimensional as a vector space over $\mathbb{F}$.

It is useful to note that by the decomposition of $\mathbb{F}[G]$ into the sum of the "orthogonal" two-sided ideals $\mathbb{F}[G]u_i$ it follows that

$$\mathbb{D}_i = \text{End}_{\mathbb{F}[G]u_i}(L_i).$$

Our next objective is to prove that $\rho_i : \mathbb{F}[G]u_i \to \text{End}_{\mathbb{D}_i}(L_i)$ is an isomorphism of $\mathbb{F}$-algebras, where $\rho_i(x)y = xy$ for all $x \in \mathbb{F}[G]u_i$ and $y \in L_i$. Here we are viewing L_i as a left $\mathbb{D}_i$-module, where $\mathbb{D}_i = \text{End}_{\mathbb{F}[G]}(L_i)$ acts on L_i in the natural way:

$$fa = f(a) \qquad \text{for all } f \in \mathbb{D}_i \text{ and } a \in L_i.$$

We need to check first that $\rho_i(x)$ is indeed $\mathbb{D}_i$-linear as a map $L_i \to L_i$. For any $x \in \mathbb{F}[G]u_i$ and $f \in \mathbb{D}_i$ we have

$$\rho_i(x)(fa) = xf(a) = f(ax) = f(\rho_i(x)a) = f(\rho_i(x)a) \quad \text{for all } a \in L_i,$$

which means that $\rho_i(x)$ is $\mathbb{D}_i$-linear; thus, $\rho_i(x) \in \mathrm{End}_{\mathbb{D}_i}(L_i)$. It is clear that ρ_i preserves addition and multiplication, and maps the multiplicative identity $u_i \in \mathbb{F}[G]u_i$ to the identity map 1 on L_i. Then $\ker \rho_i$ is a proper two-sided ideal in $\mathbb{F}[G]u_i$ and so, by minimality of the two-sided ideal $\mathbb{F}[G]u_i$, we conclude that $\ker \rho_i = \{0\}$. Next we show that ρ_i is surjective. Consider any $\phi \in \mathrm{End}_{\mathbb{D}_i}(L_i)$; we will show that $\phi = \rho_i(\phi_*)$ for some $\phi_* \in \mathbb{F}[G]u_i$. Note that for any $w \in \mathbb{F}[G]$ the right translation map $r_{wy_i} : L_i \to L_i : v \mapsto vwy_i$ is in $\mathbb{D}_i$, and so

$$\phi(vwy_i) = \phi(r_{wy_i}v) = r_{wy_i}\phi(v) = \phi(v)wy_i \quad \text{for all } v \in L_i.$$

Now write u_i as a sum $b_{i1}y_ia_{i1} + \cdots + b_{id_i}y_ia_{id_i}$, by using (4.26). Then for any $y \in L_i$ we have

$$
\begin{aligned}
\phi(y) &= \phi(b_{i1}y_ia_{i1}y + \cdots + b_{id_i}y_ia_{id_i}y) \\
&= \phi(b_{i1}y_ia_{i1}y) + \cdots + \phi(b_{id_i}y_ia_{id_i}y) \\
&= \phi(b_{i1}y_i)a_{i1}y + \cdots + \phi(b_{id_i}y_i)a_{id_i}y \\
&= \rho_i(\phi_*)y,
\end{aligned}
\tag{4.27}
$$

where $\phi_* = \phi(b_{i1}y_i)a_{i1} + \cdots + \phi(b_{id_i}y_i)a_{id_i}$. This proves that ρ_i is surjective onto $\mathrm{End}_{\mathbb{D}_i}(L_i)$.

5. Finally, suppose $|G|1_{\mathbb{F}} \neq 0$ and $\mathbb{F}$ is algebraically closed. By Schur's lemma (Theorem 3.1), we have $\mathbb{D}_i = \mathbb{F}$. Since

$$\rho_i : \mathbb{F}[G]u_i \to \mathrm{End}_{\mathbb{F}}(L_i)$$

is an isomorphism of vector spaces over $\mathbb{F}$, and

$$\dim_{\mathbb{F}} \mathrm{End}_{\mathbb{F}}(L_i) = \left(\dim_{\mathbb{F}} L_i\right)^2$$

(which follows by realizing each $T \in \mathrm{End}_{\mathbb{F}}(L_i)$ as a square matrix relative to any basis of L_i), we have

$$\left(\dim_{\mathbb{F}} L_i\right)^2 = \dim_{\mathbb{F}}(L_ia_{i1} \oplus \cdots \oplus L_ia_{id_i}) = d_i \dim_{\mathbb{F}}(L_i),$$

because the left ideals L_ia_{ij} are $\mathbb{F}[G]$-isomorphic (and hence also $\mathbb{F}$-isomorphic). We conclude then that $d_i = \dim_{\mathbb{F}} L_i$.

$\boxed{\text{QED}}$

An important discovery of Frobenius [30] emerges naturally from this analysis:

Theorem 4.5 *If G is a finite group, and $\mathbb{F}$ is an algebraically closed field in which $|G|1_{\mathbb{F}} \neq 0$ in $\mathbb{F}$, then*

$$|G| = \sum_{i=1}^{s} d_i^2, \tag{4.28}$$

where $d_i = \dim_{\mathbb{F}} L_i$, and $L_1, ..., L_s$ is a maximal collection of simple left ideals in $\mathbb{F}[G]$ such that no two are isomorphic as $\mathbb{F}[G]$-modules.

Proof. Let $u_1, \ldots, u_s$ be a maximal sequence of nonzero orthogonal central idempotents adding up to 1. Then we know that $\mathbb{F}[G]$ is the direct sum of the two-sided ideals $\mathbb{F}[G]u_i$, and $\mathbb{F}[G]u_i$ is isomorphic to $\mathrm{End}_{\mathbb{F}}(L_i)$, where L_i is any simple left ideal lying in $\mathbb{F}[G]u_i$. Hence,

$$|G| = \dim_{\mathbb{F}} \mathbb{F}[G] = \sum_{i=1}^{s} \dim_{\mathbb{F}} \mathrm{End}_{\mathbb{F}}(L_i) = \sum_{i=1}^{s} d_i^2.$$

QED

Chapter 7 contains more results on the properties of the dimensions d_i.

4.7 The Mother of All Representations

Let ρ be an irreducible representation of a finite group G on a vector space V over a field $\mathbb{F}$. Assume that $|G|1_{\mathbb{F}} \neq 0$ in $\mathbb{F}$. Then $\mathbb{F}[G]$ is semisimple, and so $\mathbb{F}[G]$ is a direct sum of subspaces, each of which is irreducible under ρ_{reg}. In particular,

$$1 = y_1 + \cdots + y_N$$

for some $y_1, ..., y_N$ lying in the distinct irreducible subspaces. For any nonzero $v \in V$ we then have

$$v = y_1 v + \cdots + y_N v,$$

and so at least one of the terms on the right, say, $y_j v$, is nonzero, where y_j lies in a simple submodule $L \subset \mathbb{F}[G]$. Then the map

$$L \to V : x \mapsto \rho(x)v$$

is not zero, and is clearly an intertwining operator from $\rho_{\mathrm{reg}}|L$ to ρ. By Schur's lemma (Theorem 1.2), it is an isomorphism. Thus, we have a remarkable conclusion:

Theorem 4.6 *Suppose G is a finite group and $\mathbb{F}$ is a field in which $|G|1_\mathbb{F} \neq 0$. Then every irreducible representation of G is equivalent to a subrepresentation of the regular representation ρ_{reg} of G on the group algebra $\mathbb{F}[G]$. In particular, every irreducible representation of a finite group is finite-dimensional.*

For an alternative proof, see Exercise 4.1. As we saw before in Proposition 1.1, the finite dimensionality of irreducible representations requires only that G be finite and no restrictions on $\mathbb{F}$ are needed.

Thus, the regular representation is no ordinary representation: it contains the pieces that make up all representations. If you think of what $\mathbb{F}[G]$ is, the vector space with the elements of G as a basis and on which G acts by permutations through multiplication on the left, it is not so surprising that it contains just about all there is to know about the representations of G.

When examining the structure of $\mathbb{F}[G]$, we observed in Proposition 4.4 that if $|G|1_\mathbb{F} \neq 0$, then there are simple left ideals L_1, ..., L_s in $\mathbb{F}[G]$ for some $s \leq \dim_\mathbb{F} \mathbb{F}[G] = |G|$ such that any simple left ideal is isomorphic as an $\mathbb{F}[G]$-module to exactly one of the L_i.

Theorem 4.7 *Suppose G is a finite group and $\mathbb{F}$ is a field in which $|G|1_\mathbb{F} \neq 0$. Then there is a finite number s, and simple left ideals $L_1, \ldots, L_s$ in $\mathbb{F}[G]$ such that every irreducible representation of G is equivalent to the restriction $\rho_{\text{reg}}|L_i$ for exactly one $i \in [s]$. Moreover, if $\mathbb{F}$ is algebraically closed, then*

$$|G| = \sum_{i=1}^{s} d_i^2,$$

where $d_i = \dim_\mathbb{F} L_i$.

A remark about computing representations is in order. Recall the procedure we sketched in Sect. 3.5 for decomposing a representation into irreducible components. If that procedure is applied to the regular representation, where each element of G is represented by a nice permutation matrix, then the algorithm leads to a determination of *all* irreducible complex representations of G.

Theorem 4.8 *Suppose G is a finite group and $\mathbb{F}$ is a field in which $|G|1_\mathbb{F} \neq 0$. Then every $\mathbb{F}[G]$-module E is a direct sum of simple submodules. In other words, every representation of G, on a vector space over the field $\mathbb{F}$, is a direct sum of irreducible representations.*

Proof. We will prove this here under the assumption that E has finite dimension as a vector space over $\mathbb{F}$, which makes it possible to use an inductive argument. (The general case is proved later in Theorem 5.3 using a more sophisticated induction procedure, namely, Zorn's lemma.) If $E = 0$, there is nothing to prove, so suppose $\dim_{\mathbb{F}} E$ is positive but finite. Any submodule of E of minimal positive dimension as a vector space over $\mathbb{F}$ is a simple submodule. So there is a largest positive integer m such that there exist simple submodules $E_1, ..., E_m$ whose sum $F = E_1 + \cdots + E_m$ is a direct sum. If $F \neq E$, then there is a nonzero complementary submodule F_c in E; in other words, a submodule F_c for which E is the direct sum of F and F_c. Inside F_c choose a submodule E_{m+1} of minimal positive dimension (notice that this works because we are working with finite-dimensional vector spaces!). But then the sum $E_1 + \cdots + E_{m+1}$ is a direct sum, contradicting the definition of m. Thus, $F = E$, and so E is a direct sum of irreducible subspaces. $\boxed{\text{QED}}$

4.8 The Center

We will work with a finite group G and any field $\mathbb{F}$.

We know from Proposition 3.3 that the center Z of $\mathbb{F}[G]$ has a basis consisting of the conjugacy class sums

$$z_C = \sum_{g \in C} g,$$

where C runs over all conjugacy classes of G. We will compare this now with what the matrix realization of $\mathbb{F}[G]$ says about Z and draw some interesting conclusions.

Let $A_1, ..., A_s$ be a collection of nonzero two-sided ideals in $\mathbb{F}[G]$ whose direct sum is $\mathbb{F}[G]$ (we will eventually specialize to the case where s is the largest integer for which there is such a finite collection). Then

$$A_j A_k \subset A_k \cap A_k = \{0\} \quad \text{if } j \neq k.$$

Decomposing 1 uniquely as a sum of elements in the A_i, we have

$$1 = u_1 + \cdots + u_s,$$

with $u_i \in A_i$ for each i. Left/right-multiplying by u_i, we have

$$u_i = u_i^2 + 0,$$

which shows that each u_i is an idempotent. Then, multiplying 1 by any $x \in \mathbb{F}[G]$, we have

$$\sum_{i=1}^{s} \underbrace{x u_i}_{\in A_i} = x = \sum_{i=1}^{s} \underbrace{u_i x}_{\in A_i},$$

which shows that (1) each u_i is in the center Z of $\mathbb{F}[G]$, (2) $y u_i = y$ if $y \in A_i$ (and, in particular, $u_i \neq 0$), and (3) $u_i x = 0$ if $x \in A_j$ with $j \neq i$. The idempotents u_i are linearly independent, because if $\sum_{i=1}^{s} c_i u_i = 0$, with coefficients c_i all in $\mathbb{F}$, then multiplying by u_j shows that $c_j u_j = 0$ and hence $c_j = 0$. As seen before,

$$\prod_{i=1}^{s} A_i \to A : (a_1, \ldots, a_s) \mapsto a_1 + \cdots + a_s \qquad (4.29)$$

is an isomorphism of algebras.

Thus, with no assumptions on the field $\mathbb{F}$, we have found a natural set of orthogonal central idempotents $u_1, \ldots, u_s$ that are *linearly independent* over $\mathbb{F}$ and all lie in the center Z. Moreover, from the isomorphism (4.29) it follows that

$$Z = Z(A_1) + \cdots + Z(A_s),$$

where $Z(A_i)$ is the center of A_i.

Now assume that $|G| 1_{\mathbb{F}} \neq 0$ in $\mathbb{F}$. Then we have seen that $A_1, \ldots, A_s$ exist such that A_i is isomorphic to the algebra of $d_i \times d_i$ matrices over a division ring $\mathbb{D}_i$, where d_i is the number of copies of a simple module L_i whose direct sum is isomorphic to A_i. If we now, further, assume that $\mathbb{F}$ is algebraically closed, then the division rings $\mathbb{D}_i$ are all equal to $\mathbb{F}$. Now the center of the algebra of all $d_i \times d_i$ consists just of the scalar matrices (multiples of the identity matrix). From this we see that if $\mathbb{F}$ is algebraically closed and $|G| 1_{\mathbb{F}} \neq 0$ in $\mathbb{F}$, then

$$Z(A_i) = \mathbb{F} u_i.$$

We have thus proved the following proposition:

Proposition 4.6 *Let G be a finite group, $\mathbb{F}$ be any field, and Z be the center of the group algebra $\mathbb{F}[G]$. Let $u_1, \ldots, u_s$ be a maximal string of nonzero central idempotents adding up to 1 in $\mathbb{F}[G]$. Then*

$$s \leq \dim_{\mathbb{F}} Z. \qquad (4.30)$$

If, moreover, $\mathbb{F}$ is algebraically closed and $|G|1_\mathbb{F} \neq 0$, then $u_1, ..., u_s$ form a basis for Z, and so

$$s = \dim_\mathbb{F} Z \quad \text{if } \mathbb{F} \text{ is algebraically closed and } |G|1_\mathbb{F} \neq 0. \tag{4.31}$$

We saw in Theorem 3.3 that the dimension of the center Z, as a vector space over $\mathbb{F}$, is just the number of conjugacy classes in G. Putting this together with the observations we have made in this section, we have a remarkable conclusion:

Theorem 4.9 *Suppose G is a finite group, $\mathbb{F}$ is a field, and Z is the center of the group algebra $\mathbb{F}[G]$. Let s be the number of distinct isomorphism classes of irreducible representations of G, over the field $\mathbb{F}$. Then*

$$s \leq \text{number of conjugacy classes in } G. \tag{4.32}$$

If the field $\mathbb{F}$ is also algebraically closed, and $|G|1_\mathbb{F} \neq 0$, then s equals the number of conjugacy classes in G.

As usual, the condition that $\mathbb{F}$ is algebraically closed can be replaced by the requirement that it be a splitting field for G, since that is what is actually used in the argument. If the characteristic p of the field $\mathbb{F}$ is a divisor of $|G|$ (taking us outside our semisimple comfort zone), then, with $\mathbb{F}$ still being a splitting field for G, the number of distinct isomorphism classes of irreducible representations of G is equal to the number of conjugacy classes of elements whose orders are coprime to p; for a proof, see [66, Theorem 1.5].

4.9 Representing Abelian Groups

Let G be a finite group and $\mathbb{F}$ be an algebraically closed field in which $|G|1_\mathbb{F} \neq 0$. Let $L_1, ..., L_s$ be a maximal set of irreducible, inequivalent representations of G over $\mathbb{F}$. Then the formula

$$|G| = \sum_{i=1}^{s} [\dim_\mathbb{F}(L_i)]^2$$

shows that each L_i is one-dimensional if and only if the number s is equal to $|G|$. Thus, each irreducible representation of G is one-dimensional if and only if the number of conjugacy classes in G equals $|G|$, in other words if each conjugacy class contains just one element. But this means that G is Abelian. We state this formally:

Theorem 4.10 *Assume the ground field* $\mathbb{F}$ *is algebraically closed and* G *is a finite group with* $|G|1_{\mathbb{F}} \neq 0$ *in* $\mathbb{F}$. *All irreducible representations of* G *are one-dimensional if and only if* G *is Abelian.*

If $\mathbb{F}$ is not algebraically closed, then the above result is not true. For example, the representation of the cyclic group $\mathbb{Z}_4$ on $\mathbb{R}^2$ given by rotations, with $1 \in \mathbb{Z}_4$ going to rotation by 90°, is irreducible. If the characteristic of $\mathbb{F}$ is a divisor of $|G|$, so that we are away from our semisimple comfort zone, one can end up with a situation where every irreducible representation of G is one-dimensional even if G is not Abelian; Exercise 4.13 develops an example.

4.10   Indecomposable Idempotents

Before closing our study of $\mathbb{F}[G]$, let us return briefly to one corner that we left unexplored but which will prove useful later. How do we decide if a given idempotent is indecomposable?

In understanding the motivation for some of the results in this section, we will find it useful to keep in mind the case where $\mathbb{F}[G]$ is realized as a matrix algebra. An idempotent is then a projection matrix.

Proposition 4.7 *Let* A *be a finite-dimensional algebra over a field* $\mathbb{F}$; *for instance,* $A = \mathbb{F}[G]$, *where* G *is a finite group. If a nonzero idempotent* $u \in A$ *satisfies the condition*

$$uAu = \mathbb{F}u, \qquad (4.33)$$

then u *is indecomposable.*

Proof. Assume that the idempotent u satisfies (4.33): for every $x \in A$,

$$uxu = \lambda_x u$$

for some $\lambda_x \in \mathbb{F}$. Now suppose u decomposes as

$$u = v + w,$$

where v and w are orthogonal idempotents:

$$v^2 = v, \quad w^2 = w, \quad vw = wv = 0.$$

Now

$$uvu = (v + w)v(v + w) = v + 0 = v,$$

and so, by (4.33), it follows that v is a multiple of u:

$$v = \lambda u \quad \text{for some } \lambda \in \mathbb{F}.$$

Since both u and v are idempotents, it follows that

$$\lambda^2 = \lambda,$$

and so λ is 0 or 1, which means v is 0 or u. Thus, u is indecomposable.
QED

We can take the first step to understanding how inequivalence of simple left ideals reflects on the generators of such ideals:

Theorem 4.11 *Suppose G is a finite group and $\mathbb{F}$ is a field. If y_1 and y_2 are nonzero idempotents in $\mathbb{F}[G]$ for which*

$$y_2 \mathbb{F}[G] y_1 = 0, \tag{4.34}$$

then the left ideals $\mathbb{F}[G]y_1$ and $\mathbb{F}[G]y_2$ are not isomorphic as $\mathbb{F}[G]$-modules.

Proof. Let $f : \mathbb{F}[G]y_1 \to \mathbb{F}[G]y_2$ be $\mathbb{F}[G]$-linear, where y_1 and y_2 are idempotents in $\mathbb{F}[G]$. Then the image $f(y_1)$ is of the form xy_2 for some $x \in \mathbb{F}[G]$, and so

$$f(ay_1) = f(ay_1 y_1) = ay_1 f(y_1) = ay_1 x y_2$$

for all $a \in \mathbb{F}[G]$, and so $f = 0$ if condition (4.34) holds. In particular, $\mathbb{F}[G]y_1$ and $\mathbb{F}[G]y_2$ are not isomorphic as $\mathbb{F}[G]$-modules unless they are both zero.
QED

With semisimplicity thrown in, we have in the converse direction the following theorem:

Theorem 4.12 *Suppose G is a finite group and $\mathbb{F}$ is a field in which $|G|1_{\mathbb{F}} \neq 0$. If y_1 and y_2 are indecomposable idempotents such that the left ideals $\mathbb{F}[G]y_1$ and $\mathbb{F}[G]y_2$ are not isomorphic as $\mathbb{F}[G]$-modules, then*

$$y_2 \mathbb{F}[G] y_1 = 0. \tag{4.35}$$

Proof. By Proposition 4.1, $\mathbb{F}[G]y_2$ and $\mathbb{F}[G]y_1$ are simple modules. Fix any $x \in \mathbb{F}[G]$, and consider the map

$$f : \mathbb{F}[G]y_2 \to \mathbb{F}[G]y_1 : y \mapsto yxy_1,$$

which is clearly $\mathbb{F}[G]$-linear. By Schur's lemma (Theorem 3.1), f is either 0 or an isomorphism. By the hypothesis, f is not an isomorphism, and hence it is 0. In particular, $f(y_2)$ is 0. Thus, y_2xy_1 is 0. $\boxed{\text{QED}}$

In our warm-up exercise (look back at equation (3.24)) decomposing $\mathbb{F}[S_3]$, we found it useful to associate with each $x \in \mathbb{F}[S_3]$ a matrix with entries y_jxy_k, where the y_j are indecomposable idempotents. We also saw there that $\{y_jxy_k : x \in \mathbb{F}[S_3]\}$ is one-dimensional over $\mathbb{F}$. We can now prove this for $\mathbb{F}[G]$, with some assumptions on the field and the group. One way to visualize the following is by thinking of the full algebra $\mathbb{F}[G]$ as a matrix algebra in which the idempotent y_2 is the matrix for a projection operator onto a one-dimensional subspace; then $\{y_2xy_1 : x \in \mathbb{F}[G]\}$ consists of all scalar multiples of y_2.

Theorem 4.13 *Suppose G is a finite group and $\mathbb{F}$ is an algebraically closed field in which $|G|1_\mathbb{F} \neq 0$. If y_1 and y_2 are indecomposable idempotents such that the left ideals $\mathbb{F}[G]y_1$ and $\mathbb{F}[G]y_2$ are isomorphic $\mathbb{F}[G]$-modules, then $\{y_2xy_1 : x \in \mathbb{F}[G]\}$ is a one-dimensional vector space over $\mathbb{F}$.*

Exercise 5.11 provides a more general formulation.

Proof. Let A denote the algebra $\mathbb{F}[G]$. By Schur's lemma (Theorem 3.1), $\mathrm{Hom}_A(Ay_2, Ay_1)$ is a one-dimensional vector space over $\mathbb{F}$. Fix a nonzero $f_0 \in \mathrm{Hom}_A(Ay_2, Ay_1)$. Then $f_0(y_2)$ is of the form x_0y_1 for some $x_0 \in A$, and so

$$f_0(y) = f_0(yy_2y_2) = yy_2f(y_2) = yy_2x_0y_1$$

for all $y \in Ay_2$. Now take any $x \in A$; then the map $Ay_2 \to Ay_1 : y \mapsto yy_2xy_1$ is A-linear, and so is an $\mathbb{F}$-multiple of f_0; in particular, $f(y_2)$ is an $\mathbb{F}$-multiple of $f_0(y_2)$, which just says that y_2xy_1 is an $\mathbb{F}$-multiple of $y_2x_0y_1$. $\boxed{\text{QED}}$

4.11 Beyond Our Borders

Our study of the group algebra $\mathbb{F}[G]$ is entirely focused on the case where the group G is finite. Semisimplicity can play a powerful role even beyond this case, for infinite groups (despite the observation in Exercise 4.12). If our focus does not seem to do full justice to the enduring power of semisimplicity, see Chalabi [11] on group algebras for infinite groups. A comprehensive development of the theory is given in the book by Passman [65].

Our exploration of $\mathbb{F}[G]$ almost always stays within semisimple territory. Modular representation theory, which stays with finite groups but goes deep into fields of finite characteristic, is much harder. To make matters worse for an initiation, books on this subject follow a shock-and-awe style of exposition that leaves the beginner with the wrong impression that this is a subject where "stuff happens," making it hard to discern a coherent structure or philosophy. The works of Puttaswamiah and Dixon [66] and Feit [27] are substantial accounts, but the work of Curtis and Reiner [15], despite its encyclopedic scope, is more readable, as is the concise introduction in the book by Weintraub [78].

There is an entirely different territory to explore when one veers off $\mathbb{F}[G]$ into a "deformation" of its algebraic structure. For instance, consider a finite group W generated by a family of reflections $r_1,..., r_m$ across hyperplanes in some Euclidean space $\mathbb{R}^N$. In the group algebra $\mathbb{F}[W]$, the relations $r_j^2 = 1$ hold. Now consider an algebra $\mathbb{F}[W]_q$, with q being a, possibly formal, parameter, generated by elements $r_1, ..., r_m$ satisfying the relations that the reflections r_j satisfy except that each relation $r_j^2 = 1$ is replaced by a "deformation":

$$r_j^2 = q1 - (1 - q)r_j.$$

When $q = 1$, this reduces to the group algebra $\mathbb{F}[W]$. This leads to the study of Hecke algebras and the general idea of deformation of algebras. This notion of deformation appears in the relationship between certain algebras of functions, or observables, for a classical physical system and algebras for the corresponding observables for the quantum theory of the physical system.

Exercises

4.1. Let G be a group and $\mathbb{F}$ be a field such that the algebra $\mathbb{F}[G]$ is semisimple. Let L be a simple $\mathbb{F}[G]$-module and consider the map $I : \mathbb{F}[G] \to L : x \mapsto xv$ for any fixed nonzero $v \in L$. Using I, and just the fact that every submodule of $\mathbb{F}[G]$ has a complement, produce a submodule of $\mathbb{F}[G]$ that is isomorphic to L.

4.2. Let G be a finite group and $\mathbb{F}$ be a field, and for each $g \in G$ let $R(g) : \mathbb{F}[G] \to \mathbb{F}[G] : x \mapsto gx$ provide the regular representation. Using the elements of G as a basis of $\mathbb{F}[G]$, check that the (a, b)th entry of the matrix for $R(g)$ is

$$R(g)_{ab} \overset{\text{def}}{=} \begin{cases} 1 & \text{if } g = ab^{-1}, \\ 0 & \text{if } g \neq ab^{-1}. \end{cases} \qquad (4.36)$$

Now introduce a variable X_g for each $g \in G$, and verify that the matrix

$$D_G = \sum_{g \in G} R(g) X_g \qquad (4.37)$$

has (a, b)th entry $X_{ab^{-1}}$. The determinant of the matrix D_G was introduced by Dedekind [19] and named the *group determinant*; its factorization, now among the many memes lost to mutations in mathematical evolution, gave rise to the notion of characters of groups. We will return to this in Sect. 7.7. For now show that the group determinant for a cyclic group of order n factors as a product of linear terms:

$$\begin{vmatrix} X_0 & X_{n-1} & X_{n-2} & \cdots & X_1 \\ X_1 & X_0 & X_{n-1} & \cdots & X_2 \\ \vdots & \vdots & \vdots & \cdots & \vdots \\ X_{n-1} & X_{n-2} & X_{n-3} & \cdots & X_0 \end{vmatrix}$$

$$= \prod_{i=1}^{n} \left(X_0 + \eta^j X_1 + \eta^{2j} + \cdots + \eta^{(n-1)j} X_{n-1} \right),$$

$$(4.38)$$

where η is any primitive nth root of unity. The type of determinant on the left in (4.38) is (or, more accurately, was) called a *circulant*.

4.3. Let G be a finite group, and for each $g \in G$ consider indeterminates X_g and Y_g. Explain the the matrix commutation identity

$$[X_{ab^{-1}}]_{a,b \in G} [Y_{b^{-1}a}]_{a,b \in G} = [Y_{b^{-1}a}]_{a,b \in G} [X_{ab^{-1}}]_{a,b \in G}. \qquad (4.39)$$

4.4. Let $C_1, \ldots, C_r$ be the distinct conjugacy classes in G. For each $i \in [r] = \{1, \ldots, r\}$ we have the central element $z_i \in \mathbb{F}[G]$ that is the sum of all the elements of C_i. Recall from (3.7) the structure constants $\kappa_{i,jk}$ of G, specified by requiring that

$$z_i z_k = \sum_{j=1}^{r} \kappa_{i,jk} z_j.$$

Thus, $\kappa_{i,jk}$ is the number of solutions $(a, c) \in C_i \times C_k$, of the equation $a = bc^{-1}$ for any fixed $b \in C_j$. Next let

$$M_i = [\kappa_{i,jk}]_{j,k \in [r]}$$

be the $r \times r$ matrix of the restriction of $R(z_i)$ to the center Z of $\mathbb{F}[G]$, relative to the basis $\{z_j : j \in [r]\}$. Since everything is in the center, the matrices $M_1, \ldots, M_r$ commute with each other. Now attach a variable Y_g to each g but with the condition that $Y_g = Y_h$ if g and h are in the same conjugacy class; also denote this common variable for the conjugacy class C_i as Y_i. Consider the $r \times r$ matrix

$$F_{ZG} = \det \left[\sum_{i=1}^{r} M_i Y_i \right]. \tag{4.40}$$

Explain why F_{ZG} is a product of linear factors of the type $\lambda_1 Y_1 + \cdots + \lambda_m Y_m$.

4.5. Work out all idempotents in the algebra $\mathbb{Z}_2[S_3]$.

4.6. Let τ be a one-dimensional representation of a group G over any field. Show that τ maps all elements of the commutator subgroup (the subgroup generated by $aba^{-1}b^{-1}$ with a, b running over G) to 1. Use this to show that in the case $G = Q = \{\pm 1, \pm i, \pm j, \pm k\}$, the group of unit quaternions, $\tau(-1)$ must be 1 and hence that $\tau(i)$ and $\tau(j)$ must be ± 1. (We saw this earlier in (2.11)).

4.7. Let G be a finite group and $\mathbb{F}$ be an algebraically closed field in which $|G|1_\mathbb{F} \neq 0$. Show that the number of inequivalent one-dimensional representations of G over $\mathbb{F}$ is $|G/G'|$, where G' is the commutator subgroup of G.

4.8. Let G be a cyclic group and $\mathbb{F}$ be algebraically closed, in which $|G|1_\mathbb{F}$ is not 0. Decompose $\mathbb{F}[G]$ as a direct sum of one-dimensional representations of G.

4.9. Let $y = \sum_g y_g g \in \mathbb{Z}[G]$ and suppose that y^2 is a rational multiple of y and $y_e = 1$.

    (a) Show that there is a positive integer γ which is a divisor of $|G|$, and for which $\gamma^{-1}y$ is an idempotent.

(b) Show that the dimension of the representation space for the idempotent $\gamma^{-1}y$ is a divisor of $|G|$.

4.10. Let $\tau : G \to \mathbb{F}^\times$ be a homomorphism of the finite group G into the group of invertible elements of the field $\mathbb{F}$, and assume that the characteristic of $\mathbb{F}$ is not a divisor of $|G|$. Let

$$u_\tau = \frac{1}{|G|} \sum_{g \in G} \tau(g^{-1})g.$$

Show that u_τ is an indecomposable idempotent.

4.11. Let R be a commutative ring, G be a finite group, and y be an element of $R[G]$ for which $gy = y$ for all $g \in G$. Show that $y = y_e s$, where $s = \sum_g g$.

4.12. Show that, for any field $\mathbb{F}$, the ring $\mathbb{F}[G]$ is not semisimple if G is an infinite group.

4.13. Let R be a commutative ring of prime characteristic $p > 0$, G be a group with $|G| = p^n$ for some positive integer n, and E be an $R[G]$-module. Choose a nonzero $v \in E$ and let E_0 be the $\mathbb{Z}$-linear span of $Gv = \{gv : g \in G\}$ in E. Then E_0 is a finite-dimensional vector space over the field $\mathbb{Z}_p$, and so $|E_0| = p^d$, where $d = \dim_{\mathbb{Z}_p} E_0 \geq 1$. By partitioning the set E_0 into the union of disjoint orbits under the action of G, show that there exists a nonzero $w \in E_0$ for which $gw = w$ for all $g \in G$. Now show that if the $R[G]$-module E is simple, then $E = Rw$ and $gv = v$ for all $v \in E$.

4.14. Let $\mathbb{F}$ be a field of characteristic $p > 0$ and G be a group with $|G| = p^n$ for some positive integer n. Prove that $\mathbb{F}[G]$ is indecomposable, and $\mathbb{F}s$, where $s = \sum_g g$, is the unique simple left ideal in $\mathbb{F}[G]$. Show also that $\ker \epsilon$ is the unique maximal ideal in $\mathbb{F}[G]$, where $\epsilon : \mathbb{F}[G] \to \mathbb{F} : \sum_g x_g g \mapsto \sum_g x_g$. In the converse direction, prove that if $\mathbb{F}$ has characteristic $p > 0$ and G is a finite group such that $\mathbb{F}[G]$ is indecomposable, then $|G| = p^n$ for some positive integer n.

4.15. In the following, G is a finite group, $\mathbb{F}$ is a field, and $A = \mathbb{F}[G]$. No assumption is made about the characteristic of $\mathbb{F}$. An A-module is said to be *indecomposable* if it is not 0 and is not the direct sum of two nonzero submodules.

(a) Show that if e and f_1 are idempotents in A with $f_1 e = f_1$, then $e_1 \stackrel{\text{def}}{=} e f_1 e$ and $e_2 \stackrel{\text{def}}{=} e - e_1$ are orthogonal idempotents, with $e = e_1 + e_2$, with $e_1 e = e_1$ and $e_2 e = e_2$.

(b) Show that if y is an indecomposable idempotent in A, then the left ideal Ay cannot be written as a direct sum of two distinct nonzero left ideals.

(c) Suppose L is a left ideal in A that has a complementary ideal L_c, such that A is the direct sum of L and L_c. Show that there is an idempotent $y \in L$ such that $L = Ay$.

(d) Prove that there is a largest positive integer n such that there exist nonzero orthogonal idempotents $y_1, \ldots, y_n \in A$ whose sum is 1. Show that each y_i is indecomposable.

(e) Prove that there is a largest positive integer s such that there exist nonzero central idempotents $u_1, \ldots, u_s$, orthogonal to each other, for which $u_1 + \cdots + u_s = 1$.

(f) Prove that any central idempotent u is a sum of some of the u_i in (4.15e). Then show that the set $\{u_1, \ldots, u_s\}$ is uniquely specified as the largest set of nonzero central idempotents adding up to 1.

(g) With $u_1, \ldots, u_s$ as above, show that each u_i is a sum of some of the idempotents $e_1, \ldots, e_n$ in (4.15d). If e_i appears in the sum for u_r. then $e_i u_r = e_i$ and $e_i u_t = 0$ for $t \neq r$.

(h) Show that Au_i is indecomposable in the sense that it is not 0 and is not the direct sum of two nonzero left ideals, and that the map

$$\prod_{i=1}^{s} Au_i \to A : (a_1, \ldots, a_s) \mapsto a_1 + \cdots + a_s$$

is an isomorphism of algebras.

(i) Show that A is the direct sum of indecomposable submodules $V_1, \ldots, V_n$.

(j) Let E be a finite-dimensional indecomposable A-module. Prove that there is a submodule $E_0 \subset E$ that is maximal in the sense that E is the only submodule of E which contains E_0 as a proper subset. Then show that E/E_0 is a simple A-module.

(k) Let $\phi : \mathbb{F} \to \mathbb{F}$ be an automorphism of the field $\mathbb{F}$ (e.g., ϕ could be simply the identity or, in the case of the complex field, ϕ could be conjugation). Suppose $\Phi : A \to A$ is a bijection which is additive, ϕ-linear,

$$\Phi(kx) = \phi(k)\Phi(x) \qquad \text{for all } k \in \mathbb{F} \text{ and } x \in \mathbb{F}[G],$$

and for which either $\Phi(ab) = \Phi(a)\Phi(b)$ for all $a, b \in A$ or $\Phi(ab) = \Phi(b)\Phi(a)$ for all $a, b \in A$. Show that

$$\{\Phi(u_1), \ldots, \Phi(u_s)\} = \{u_1, \ldots, u_s\}.$$

Thus, for each i there is a unique $\Phi(i)$ such that $\Phi(u_i) = u_{\Phi(i)}$.

(l) Let
$$\mathrm{Tr}_e : \mathbb{F}[G] \to \mathbb{F} : x \mapsto x_e.$$

Show that
$$\mathrm{Tr}_e(xy) = \mathrm{Tr}_e(yx).$$

Assuming that Φ maps G into itself, show that

$$\mathrm{Tr}_e \Phi(x) = \phi(\mathrm{Tr}_e x).$$

(m) Consider the pairing

$$(\cdot, \cdot)_\Phi : A \times A \to \mathbb{F} : (x, y) \mapsto \mathrm{Tr}_e\big(x\Phi(y)\big),$$

which is linear in x and ϕ-linear in y. Prove that this pairing is nondegenerate in the sense that (i) if $(x, y)_\Phi = 0$ for all $y \in A$, then x is 0, and (ii) if $(x, y)_\Phi = 0$ for all $x \in A$, then y is 0. Check that this means that the map $y \mapsto y'$ of A to its dual vector space A' specified by

$$y'(x) = (x, y)_\Phi$$

is an isomorphism of vector spaces over $\mathbb{F}$, where for the vector space structure on A' multiplication by scalars is specified by

$$(cf)(x) = \phi(c)f(x)$$

for all $c \in \mathbb{F}$, all $f \in A'$, and all $x \in A$. Assuming that Φ maps $G \subset \mathbb{F}[G]$ into itself, show that

$$(\Phi(x), \Phi(y))_\Phi = \phi\left((x, y)_\Phi\right).$$

(n) Show that for each $i \in [s]$ the pairing

$$Au_i \times Au_j \to A : (x, y) \mapsto (x, y)_\Phi$$

is nondegenerate if $j = \Phi^{-1}(i)$, and is 0 otherwise.

(o) Take the special case Ψ for Φ given by

$$\Psi(x) = \check{x} = \sum_{g \in G} x_g g^{-1}.$$

Show that the pairing $(\cdot, \cdot)_\Psi$ is G-invariant in the sense that

$$(gx, gy)_\Psi = (x, y)_\Psi$$

for all $x, y \in \mathbb{F}[G]$ and $g \in G$. Then show that the induced map $A \to A' : y \mapsto y'$ is an isomorphism of left $\mathbb{F}[G]$-modules, where the dual space A' is a left $\mathbb{F}[G]$-module through the dual representation of G on A' given by

$$\rho'_{\text{reg}}(g)f \stackrel{\text{def}}{=} f \circ \rho_{\text{reg}}(g)^{-1}.$$

(p) Let $L_k = Ay_k$, where y_k is one of the idempotents in a maximal string of orthogonal indecomposable idempotents $y_1,..., y_n$ adding up to 1. Prove that the dual vector space L'_k, with the left $\mathbb{F}[G]$-module structure given by the dual representation $(\rho_{\text{reg}}|L_k)'$, is isomorphic to L_j for some $j \in [n]$. (We saw a version of this in Theorem 1.1.) Moreover, $L_k \simeq L'_j$.

(q) Let E be an indecomposable left A-module and let $y_1,..., y_n$ be a string of indecomposable orthogonal idempotents in A adding up to 1. Show that $y_j E \neq 0$ for some $j \in [n]$.

(r) Let F be a simple left A-module and suppose $y_j F \neq 0$, as above. Let $W = \{x \in Ay_j : xF = 0\}$, which is a left ideal of A contained inside Ay_j. Show that $Ay_j/W \simeq F$, isomorphic as A-modules, and conclude that W is a maximal proper submodule of Ay_j.

(s) Let E be a simple left A-module and apply the previous step with $F = E'$, where E' is the dual vector space with the usual dual representation/A-module structure to obtain $j \in [n]$ with $y_j E' \neq 0$ and a maximal proper submodule W in Ay_j. Continuing with the notation from above, $Ay_j \simeq (Ay_k)'$ (we use $\simeq$ to denote isomorphism of A-modules) for some $k \in [n]$. Let $\tilde{W}$ be the image of W in $(Ay_k)' \simeq Ay_j$. Then

$$(Ay_j)/W \simeq (Ay_k)'/\tilde{W} \simeq \tilde{W}_0', \qquad (4.41)$$

where we used Lemma 1.1 with $\tilde{W}_0$ being the annihilator,

$$\tilde{W}_0 \stackrel{\text{def}}{=} \{x \in Ay_k : f(x) = 0 \text{ for all } f \in \tilde{W}\}, \qquad (4.42)$$

as A-modules. Using Lemma 1.1, show that $\tilde{W}_0$ is a simple submodule of Ay_k. Conclude (by Exercise 1.11) that

$$E' \simeq \tilde{W}_0', \qquad (4.43)$$

and then $E \simeq W_0$, as A-modules (see Exercise 1.11). Thus, every simple A-module is isomorphic to a submodule of one of the indecomposable A-modules Ay_k.

4.16. A *Frobenius algebra* is a finite-dimensional algebra A over a field $\mathbb{F}$, containing a multiplicative identity element, along with a linear map $\epsilon : A \to \mathbb{F}$ for which the bilinear form on A given by $A \times A \to \mathbb{F}$: $(a, b) \mapsto \epsilon(ab)$ is nondegenerate. We work now with the motivating example, where $A = \mathbb{F}[G]$, the group algebra of a finite group G, and ϵ is the trace $\mathrm{Tr}_e : \mathbb{F}[G] \to \mathbb{F} : x \mapsto x_e$. Let B_0 be the bilinear form on A given by $B_0(x, y) = \mathrm{Tr}_e(xy)$. Note that the bilinear form B_0 is non-degenerate as well as symmetric. Let $A' = \mathrm{Hom}_{\mathbb{F}}(A, \mathbb{F})$ be the dual space consisting of all $\mathbb{F}$-linear maps $A \to \mathbb{F}$ and, for each $g \in G$, let $\delta_g \in A'$ be given by $\delta_g(x) = x_g$ for all $x \in A$. Consider also the multiplication map

$$m : A \otimes A \to A : x \otimes y \mapsto xy.$$

(a) Let $\alpha_0 : A \to A'$ be the linear map induced by B_0: $\alpha_0(x)(y) = B_0(x, y)$. Work out $\alpha_0(g)$ for $g \in G$.

(b) The multiplication m induces dually a linear map

$$m' : A' \to (A \otimes A)'$$

by $m'(f)(w) = f(m(w))$ for all $f \in A'$ and $w \in A \otimes A$. Identify A' with A and identify $(A \otimes A)'$ with $A \otimes A$ using α_0, and let the *comultiplication*

$$\hat{m} : A \to A \otimes A$$

be obtained from m' by identifying A with A' using α_0. Work out the value of $\hat{m}(g)$ explicitly for each $g \in G$.

(c) The unit element (multiplicative identity) $1_A \in A$ may be viewed as specified through the linear map $\mathbb{F} \to A : c \mapsto c1_A$. Dually there is a linear map $A' \to \mathbb{F}$ which, using the identification of A' with A via α_0, yields a linear map $A \to \mathbb{F}$. Show that this is just ϵ, which could therefore be called a *counit*.

(d) Check the *coassociativity* condition: $(\mathrm{id} \otimes \hat{m})\hat{m} = (\hat{m} \otimes \mathrm{id})\hat{m}$.

4.17. Let G be a finite group, $\mathbb{F}$ be a field, and ZG be the center of the group algebra $\mathbb{F}[G]$. Let $\epsilon : ZG \to \mathbb{F} : x \mapsto x_e$, the restriction of Tr_e to ZG. Consider the bilinear form B_ϵ on ZG given by $B_\epsilon(x, y) = \epsilon(xy)$ for all $x, y \in ZG$. It is clearly symmetric. In the following it will be useful to recall that a basis of ZG is given by the conjugacy class sums $z_C = \sum_{g \in C} g$, with C running over the set $\mathcal{C}_G$ of all conjugacy classes in G.

(a) Show that B_ϵ is nondegenerate. Thus, ZG is a commutative Frobenius algebra.

(b) Let $\alpha : ZG \to (ZG)'$ be the linear map specified by requiring that $\alpha(x)(y) = B_\epsilon(x, y)$ for all $x, y \in ZG$. Using the basis of ZG given by the elements z_C, and the corresponding dual basis in $(ZG)'$, describe the map α in terms of matrices.

(c) Let $m : ZG \otimes ZG \to ZG : (x, y) \mapsto xy$ be the multiplication map. Dually there is a map $m' : (ZG)' \to (ZG \otimes ZG)'$, and identifying $(ZG)'$ with ZG by means of α gives a linear map $\hat{m} : ZG \to ZG \otimes ZG$. Express $\hat{m}(z_C)$ explicitly for $C \in \mathcal{C}_G$. Compare this with the Frobenius group matrix in Exercise 4.4.

For more on Frobenius algebras, and the relationship between commutative Frobenius algebras and two-dimensional topological quantum field theories, see the book by Kock [51].

Chapter 5

Simply Semisimple

We have seen that the group algebra $\mathbb{F}[G]$ is especially rich and easy to explore when $|G|$, the number of elements in the group G, is not divisible by the characteristic of the field $\mathbb{F}$. What makes everything flow so well in this case is that the algebra $\mathbb{F}[G]$ is semisimple. In this chapter we are going to fly over largely the same terrain as we have already, but this time we will replace $\mathbb{F}[G]$ by a more general ring, and look at everything directly through semisimplicity. This chapter can be read independently of the previous ones, although occasionally looking back at previous chapters would be useful.

We will be working with modules over a ring A; for us a ring always contains a multiplicative identity $1 \neq 0$. So, all through this chapter A denotes such a ring. Note that A need not be commutative. Occasionally, we will comment on the case where the ring A is an *algebra* over a field $\mathbb{F}$.

By definition, a module E over the ring A is *semisimple* if for any submodule F in E there is a submodule F_c in E such that E is the direct sum of F and F_c.

A ring is said to be *semisimple* if it is semisimple as a left module over itself.

A module is said to be *simple* if it is not 0 and contains no submodule other than 0 and itself.

A (termino)logical pitfall to note: the zero module 0 is semisimple but not simple.

Aside from the group ring $\mathbb{F}[G]$, the algebra $\mathrm{End}_{\mathbb{F}}(V)$ of all endomorphisms of a finite-dimensional vector space V over a field $\mathbb{F}$ is a semisimple algebra (a matrix formalism verification is traced out in Exercise 5.5).

A.N. Sengupta, *Representing Finite Groups: A Semisimple Introduction*,
DOI 10.1007/978-1-4614-1231-1_5, © Springer Science+Business Media, LLC 2012

5.1   Schur's Lemma

We will work with a ring A. Suppose

$$f : E \to F$$

is linear, where E is a simple A-module and F is an A-module. The kernel

$$\ker f = f^{-1}(0)$$

is a submodule of E and hence is either $\{0\}$ or E itself. If, moreover, F is also simple, then $f(E)$, being a submodule of F, is either $\{0\}$ or F. This is *Schur's lemma*:

Theorem 5.1 *If E and F are simple modules over a ring A, then every nonzero element in*

$$\mathrm{Hom}_A(E, F)$$

is an isomorphism of E onto F.

For a simple A-module $E \neq 0$, this implies that every nonzero element in the ring

$$\mathrm{End}_A(E)$$

has a multiplicative inverse. Such a ring is called a *division ring*; it falls short of being a field only in that multiplication (which is composition in this case) is not necessarily commutative.

We can now specialize to a case of interest, where A is a finite-dimensional algebra over an algebraically closed field $\mathbb{F}$. We can view $\mathbb{F}$ as a subring of $\mathrm{End}_A(E)$:

$$\mathbb{F} \simeq \mathbb{F}1 \subset \mathrm{End}_A(E),$$

where 1 is the identity element in $\mathrm{End}_A(E)$. The assumption that $\mathbb{F}$ is algebraically closed implies that $\mathbb{F}$ has no proper finite extension, and this leads to the following consequence:

Theorem 5.2 *Suppose A is a finite-dimensional algebra over an algebraically closed field $\mathbb{F}$. Then for any simple A-module E that is finite-dimensional as a vector space over $\mathbb{F}$,*

$$\mathrm{End}_A(E) = \mathbb{F},$$

upon identifying $\mathbb{F}$ with $\mathbb{F}1 \subset \mathrm{End}_A(E)$. Moreover, if E and F are simple A-modules, then $\mathrm{Hom}_A(E, F)$ is either $\{0\}$ or a one-dimensional vector space over $\mathbb{F}$.

Proof. Let $x \in \text{End}_A(E)$ and suppose $x \notin \mathbb{F}1$. Note that x commutes with all elements of $\mathbb{F}1$. Since $\text{End}_A(E) \subset \text{End}_\mathbb{F}(E)$ is a finite-dimensional vector space over $\mathbb{F}$, there is a smallest natural number $n \in \{1, 2, ...\}$ such that $1, x, ..., x^n$ are linearly dependent over $\mathbb{F}$; put another way, there is a polynomial $p(X) \in \mathbb{F}[X]$, of lowest degree, with $\deg p(X) = n \geq 1$, such that

$$p(x) = 0.$$

Since $\mathbb{F}$ is algebraically closed, $p(X)$ factorizes over $\mathbb{F}$ as

$$p(X) = (X - \lambda)q(X)$$

for some $\lambda \in \mathbb{F}$. Consequently, $x - \lambda 1$ is not invertible, because otherwise $q(x)$, of lower degree, would be 0. Thus, by Schur's lemma (Theorem 5.1), $x = \lambda 1 \in \mathbb{F}1$.

Now suppose E and F are simple A-modules and there is a nonzero element $f \in \text{Hom}_A(E, F)$. By Theorem 5.1, f is an isomorphism. If g is also an element of $\text{Hom}_A(E, F)$, then $f^{-1}g$ is in $\text{End}_A(E, E)$, and so, by the first part, is an $\mathbb{F}$-multiple of the identity element in $\text{End}_A(E)$. Consequently, g is an $\mathbb{F}$-multiple of f. $\boxed{\text{QED}}$

The preceding proof can be shortened by appeal to Wedderburn's result (Theorem 1.3) that every finite-dimensional division algebra $\mathbb{D}$ over any algebraically closed field $\mathbb{F}$ is $\mathbb{F}$ itself, viewed as a subset of $\mathbb{D}$ (alternatively, the first part of the preceding proof can be viewed as a fully worked out version of a proof of Wedderburn's theorem).

5.2 Semisimple Modules

We will work with modules over a ring A with unit element $1 \neq 0$.

Proposition 5.1 *Submodules and quotient modules of semisimple modules are semisimple.*

Proof. Let F be a submodule of a semisimple module E. We will show that F is also semisimple. To this end, let L be a submodule of F. Then, by semisimplicity of E, the submodule L has a complement L_c in E:

$$E = L \oplus L_c.$$

If $f \in F$, we can decompose it uniquely as

$$f = \underbrace{a}_{\in L} + \underbrace{a_c}_{\in L_c}.$$

Then

$$a_c = f - a \in F,$$

and so, in the decomposition of $f \in F$ as $a + a_c$, both $a \in L \subset F$ and a_c are in F. Hence,

$$F = L \oplus (L_c \cap F).$$

Having found a complement of any submodule inside F, we have semisimplicity of F.

If F_c is the complementary submodule to F in E, then the map

$$F_c \to E/F : x \mapsto x + F$$

is an isomorphism of modules.

So E/F, being isomorphic to the submodule F_c in E, is semisimple.
$\boxed{\text{QED}}$

For another perspective on the preceding result, see Exercise 5.19.

Complements are not unique, but something can be said about different choices of complements:

Proposition 5.2 *Let L be a submodule of a module E over a ring. Then a submodule L_c of E is a complement of L if and only if the quotient map $q : E \to E/L$ maps L_c isomorphically onto E/L.*

Proof. Suppose E is the direct sum of L and L_c. Then $q(L_c) = q(E) = E/L$ and $\ker(q|L_c) = L_c \cap L = 0$, and so $q|L_c$ is an isomorphism onto E/L.

Conversely, suppose $q|L_c$ is an isomorphism onto E/L. Then $L_c \cap L = \ker(q|L_c) = 0$. Moreover, for any $e \in E$ there is an element $e_c \in L_c$ with $q(e_c) = q(e)$, and so

$$e = \underbrace{(e - e_c)}_{\in \ker q = L} + e_c,$$

showing that $E = L + L_c$. $\boxed{\text{QED}}$

Our goal is to decompose a module over a semisimple ring into a direct sum of simple submodules. The first obstacle in reaching this goal is a strange one: how do we even know there is a simple submodule? If the module happens to come automatically equipped with a vector space structure, then we can use the dimension to form the steps of a ladder to climb down all the way to a minimal dimensional submodule. Without a vector space structure, it seems we are looking down an endless abyss of uncountable descent. Fortunately, this transfinite abyss can be plumbed using Zorn's lemma.

Proposition 5.3 *Let E be a nonzero semisimple module over a ring A. Then E contains a simple submodule.*

Proof. Pick a nonzero $v \in E$, and consider Av. A convenient feature of Av is that a submodule of Av is proper if and only if it does not contain v. We will produce a simple submodule inside Av, as a complement of a maximal proper submodule. A maximal proper submodule is produced using Zorn's lemma. Let $\mathcal{F}$ be the set of all proper submodules of Av. If $\mathcal{G}$ is a nonempty subset of $\mathcal{F}$ that is a chain in the sense that if $H, K \in \mathcal{G}$ then $H \subset K$ or $K \subset H$, then $\cup \mathcal{G}$ is a submodule of Av that does not contain v. Hence, Zorn's lemma is applicable to $\mathcal{F}$ and implies that there is a maximal element M in $\mathcal{F}$. This means that a submodule of Av that contains M is Av or M itself. Since E is semisimple, so is the submodule Av. Then there is a submodule $M_c \subset Av$ such that Av is the direct sum of M and M_c. We claim that M_c is simple. First, $M_c \neq 0$ because otherwise M would be all of Av, which it is not since it is missing v. Next, if L is a nonzero submodule of M_c, then $M + L$ is a submodule of Av properly containing M and hence is all of Av, and this implies $L = M_c$. Thus, M_c is a simple module. $\boxed{\text{QED}}$

Now we will prove some convenient equivalent forms of semisimplicity. The idea of producing a minimal module as a complement of a maximal one will come in useful. The argument, at one point, will also use the reasoning that leads to a basic fact about vector spaces: if T is a linearly independent subset of a vector space, and S is a subset that spans the whole space, then a basis B of the vector space is formed by adjoining to T a maximal subset S_0 of S for which $B = T \cup S_0$ is linearly independent.

Theorem 5.3 *The following conditions are equivalent for an E over any ring:*

1. *E is semisimple.*

2. *E is a sum of simple submodules.*

3. *E is a direct sum of simple submodules.*

If $E = \{0\}$ then the sums in statements 2 and 3 are empty sums. The proof also shows that if E is the sum of a set of simple submodules, then E is a direct sum of a subset of this collection of submodules.

Proof. Assume that statement 1 holds. Let F be the sum of a maximal collection of simple submodules of E; such a collection exists, by Zorn's lemma. Then $E = F \oplus F_c$ for a submodule F_c of E. We will show that $F_c = 0$. Suppose $F_c \neq 0$. Then, by Proposition 5.3, F_c has a simple submodule, and this contradicts the maximality of F. Thus, E is a sum of simple submodules. We have thus shown that statement 1 implies statement 2.

Suppose F is a submodule of E, contained in the sum of a family $\{F_j\}_{j \subset J}$ of simple submodules of E:

$$F \subset \sum_{j \in J} E_j.$$

Zorn's lemma extracts a maximal subset K (possibly empty) of J such that the sum

$$H = F + \sum_{k \in K} E_k$$

is a direct sum of the family $\{F\} \cup \{E_k : k \in K\}$. For any $j \in J$, the intersection $E_j \cap H$ is a submodule of E_j and so is either 0 or E_j. It cannot be 0 by maximality of K. Thus, $E_j \subset H$ for all $j \in J$, and so $\sum_{j \in J} E_j \subset H$. Hence,

$$\sum_{j \in J} E_j = F + \sum_{k \in K} E_k, \tag{5.1}$$

which is a direct sum of the family $\{F\} \cup \{E_k : k \in K\}$.

Now suppose statement 2 holds: E is the sum of a family of simple submodules E_j with j running over a set J. Then by (5.1), with $F = 0$, we see that E is a direct sum of some of the simple submodules E_k. This proves that statement 2 implies statement 3.

Now we show that statement 3 implies statement 1. Applying our observations in (5.1) to a family $\{E_j\}_{j \in J}$ that gives a direct sum decomposition of E, and taking F to be any submodule of E, we find that

$$E = F \oplus F_c,$$

where F_c is a direct sum of some of the simple submodules E_k. Thus, statement 3 implies statement 1. $\boxed{\text{QED}}$

5.3 Deconstructing Semisimple Modules

In Theorem 5.3 we saw that a semisimple module is a sum of simple sub-modules. In this section we will use this to obtain a full structure theorem for semisimple modules.

We begin with an observation about simple modules that is analogous to the situation for vector spaces. Indeed, the proof is accomplished by viewing a module as a vector space (for more logical handwringing, see Theorem 5.6).

Theorem 5.4 *If E is a simple A-module, then E is a vector space over the division ring $\operatorname{End}_A(E)$. If $E^n \simeq E^m$ as A-modules, then $n = m$.*

Proof. If E is a simple A-module, then, by Schur's lemma,

$$\mathbb{D} \overset{\text{def}}{=} \operatorname{End}_A(E)$$

is a division ring. Thus, E is a vector space over $\mathbb{D}$. Then E^n is the product vector space over $\mathbb{D}$. If $\dim_{\mathbb{D}} E$ were finite, then we would be finished. In the absence of this, there is a clever alternative route. Look at $\operatorname{End}_A(E^n)$. This is a vector space over $\mathbb{D}$, because for any $\lambda \in \mathbb{D}$ and A-linear $f : E^n \to E^n$, the map λf is also A-linear. In fact, each element of $\operatorname{End}_A(E^n)$ can be displayed, as usual, as an $n \times n$ matrix with entries in $\mathbb{D}$. Moreover, this provides a basis of the $\mathbb{D}$-vector space $\operatorname{End}_A(E^n)$ consisting of n^2 elements. Thus, $E^n \simeq E^m$ implies $n = m$. $\boxed{\text{QED}}$

Now we can turn to the uniqueness of the structure of semisimple modules of finite type:

Theorem 5.5 *Suppose a module E over a ring A can be expressed as*

$$E \simeq E_1^{k_1} \oplus \cdots \oplus E_n^{k_n}, \tag{5.2}$$

where $E_1, ..., E_n$ are nonisomorphic simple modules and each k_i is a positive integer. Suppose E can also be expressed as

$$E \simeq F_1^{j_1} \oplus \cdots \oplus F_m^{j_m},$$

where $F_1, ..., F_m$ are nonisomorphic simple modules and each j_i is a positive integer. Then $m = n$, and each E_a is isomorphic to one and only one F_b, and then $k_a = j_b$. Every simple submodule of E is isomorphic to E_l for exactly one $l \in [n]$.

Proof. Let H be any simple module isomorphic to a submodule of E. Then composing an isomorphism $H \to E$ with the projection $E \to E_r$, we see that there exists an a for which the composite $H \to E_a$ is not zero, and hence $H \simeq E_a$. Similarly, there is a b such that $H \simeq F_b$. Thus, each E_a is isomorphic to some F_b. The rest follows by Theorem 5.4. $\boxed{\text{QED}}$

The preceding results, or variations on them, are generally called, in combination, the Krull–Schmidt theorem. There is a way to understand them without peering too far into the internal structure or elements of a module; instead we can look at the partially ordered set, or lattice, of submodules of a module. Exercises 5.18 and 5.19 provide a glimpse into this approach, and we include it as a token tribute to Dedekind's much-maligned foundation of lattice theory [17, 18] (see the ever-readable Rota [68] for the historical context).

The arguments proving the preceding results rely on the uniqueness of the dimension of a vector space over a division ring. The proof of this is identical to the case of vector spaces over fields, and is elementary in the finite-dimensional case. The proof of the uniqueness of dimension for infinite-dimensional spaces is an unpleasant application of Zorn's lemma (see Hungerford [47]). Alternatively, the tables can be turned and the decomposition theory for semisimple modules, specialized all the way down to the case of division rings, can be used as proof for the existence of a basis and the uniqueness of the dimension of a vector space over a division ring. With this perspective, we have the following (adapted from Chevalley [12]):

Theorem 5.6 *Let E and F be modules over a ring A such that E and F are both sums of simple submodules. Assume that every simple submodule of E is isomorphic to every simple submodule of F. Then the following are equivalent: (1) E and F are isomorphic; (2) any set of simple submodules of E whose direct sum is all of E has the same cardinality as any set of simple submodules of F whose direct sum is F. In particular, if A is a division ring, then any two bases of a vector space over A have the same cardinality.*

Proof. By Theorem 5.3, if a module is the sum of simple submodules, then it is also a direct sum of a family of simple submodules. Let E be the direct sum of simple submodules E_i, with i running over a set I, and F be the direct sum of simple submodules F_j with j running over a set J. Suppose each E_i is isomorphic to each F_j; if $|I| = |J|$, then we clearly obtain an isomorphism $E \to F$.

Now assume, for the converse, that $f : E \to F$ is an isomorphism. First we work with the case where I is a finite set. The argument is by induction on $|I|$. If $I = \emptyset$, then $E = 0$ and so $F = 0$ and $J = \emptyset$. Now suppose $I \neq \emptyset$, assume the claimed result for smaller values of $|I|$, and pick $a \in I$. Then, by Theorem 5.3, a complement H of $f(E_a)$ in F is formed by adding up a suitable set of F_j's:

$$F = f(E_a) +_d H,$$

where $+_d$ signifies (internal) direct sum, with

$$H = \sum_{j \in S} F_j,$$

and S is a subset of J. Now choose $b \in J$ such that F_b is not contained inside H; such a b exists because $f(E_a)$, being an isomorphic copy of the simple module E_a, is not 0. Then the quotient map $q : F \to F/H$ is not 0 when it is restricted to F_b and so, by Schur's lemma used on the simplicity of F_b and of $F/H \simeq f(E_a) \simeq E_a$, the restriction $q|F_b : F_b \to F/H$ is an isomorphism. Then by Proposition 5.2, F_b is also a complement of H. But then

$$F_b +_d \sum_{j \in S} F_j = F_b +_d H = F = F_b +_d \sum_{j \in J - \{b\}} F_j,$$

and, these being direct sums, we conclude that $S = J - \{b\}$. Combining the various isomorphisms, we have

$$E/E_a \simeq F/f(E_a) \simeq H \simeq F/F_b.$$

This implies that the direct sum of the simple modules E_i, with $i \in I - \{a\}$, is isomorphic to the direct sum of the simple modules F_j, with $j \in J - \{b\}$. Then by the induction hypothesis, $|I - \{a\}| = |J - \{b\}|$, whence $|I| = |J|$.

Consider now the case of infinite I. For any $i \in I$, pick nonzero $x_i \in E_i$, and observe that there is a *finite* set $S_i \subset J$ such that $f(x_i) \in \sum_{j \in S_i} F_i$, and so

$$f(E_i) \subset \sum_{j \in S_i} F_j,$$

because $E_i = A x_i$. Let S_* be the union of all the S_i; then

$$f(E) \subset \sum_{j \in S_*} F_j.$$

But $f(E) = F$, and so $S_* = J$. The cardinality of S_* is the same as that of I because I is infinite (this is a little set theory observation courtesy of Zorn's lemma). Hence, $|I| = |J|$.

Lastly, suppose A is a division ring. Observe that an A-module is simple if and only if it is of the form Av for a nonzero element v in the module. Thus, every decomposition $\{E_i\}_{i \in I}$ of an A-module E into a direct sum of simple modules gives rise to a choice of a basis $\{v_i\}_{i \in I}$ for E of the same cardinality $|I|$ and, conversely, every choice of a basis of E gives rise to a decomposition into a direct sum of simple submodules. Hence, by statement 2 any two bases of E have the same cardinality. $\boxed{\text{QED}}$

5.4 Simple Modules for Semisimple Rings

An element y in a ring A is an *idempotent* if $y^2 = y$. Idempotents v and w are *orthogonal* if $vw = wv = 0$. An idempotent y is *indecomposable* if it is not 0 and is not the sum of two nonzero, orthogonal idempotents. A *central* idempotent is one which lies in the center of A.

Here is an ambidextrous upgrade of Proposition 4.1, formulated without using semisimplicity.

Proposition 5.4 *If y is an idempotent in a ring A, then the following are equivalent:*

   1. y is an indecomposable idempotent.

   2. Ay is not the direct sum of two nonzero left ideals in A.

   3. yA is not the direct sum of two nonzero right ideals in A.

We omit the proof, which you can read by replacing $\mathbb{F}[G]$ with A in the proof of Proposition 4.1, and then going through a second run with "left" replaced by "right."

If a left ideal L can be expressed as Ay, we say that y is a *generator* of L. Similarly, if a right ideal has the form yA, we call y a generator of the right ideal.

Theorem 5.7 *Let L be a left ideal in a ring A. The following are equivalent:*

1. *There is a left ideal L_c such that A is the direct sum of L and L_c.*

2. *There is an idempotent $y_L \in L$ such that $L = Ay_L$.*

If statements 1 and 2 hold, then

$$LL = L. \tag{5.3}$$

In particular, in a semisimple ring every left ideal has an idempotent generator.

Proof. Suppose
$$A = L \oplus L_c,$$
where L_c is also a left ideal in A. Then the multiplicative unit $1 \in A$ decomposes as
$$1 = y_L + y_c,$$
where $y_L \in L$ and $y_c \in L_c$. For any $a \in A$ we then have
$$a = a1 = \underbrace{ay_L}_{\in L} + \underbrace{ay_c}_{\in L_c}.$$

This shows that a belongs to L if and only if it is equal to ay_L. In particular, y_L^2 equals y_L, and $L = Ay_L$. Moreover,
$$L = Ay_L = Ay_L y_L \subset LL.$$

Of course, L being a left ideal, we also have $LL \subset L$. Thus, LL equals L.

Conversely, suppose $L = Ay_L$, where $y_L \in L$ is an idempotent. Then A is the direct sum of $L = Ay_L$, and $L_c = A(1 - y_L)$. $\boxed{\text{QED}}$

Next we see why simple modules are isomorphic to simple left ideals. The criteria obtained here for simple modules to be isomorphic will prove useful later.

Theorem 5.8 *Let L be a left ideal in a ring A and E be a simple A-module. Then exactly one of the following holds:*

1. $LE = 0$.

2. $LE = E$ and L is isomorphic to E.

If, moreover, the ring A is semisimple and $LE = 0$, then E is not isomorphic to L as A-modules.

Proof. Since LE is a submodule of E, it is either $\{0\}$ or E. Suppose $LE = E$. Then take a $y \in E$ with $Ly \neq 0$. By simplicity of E, then $Ly = E$. The map

$$L \mapsto E = Ly : a \mapsto ay$$

is an A-linear surjection, and it is injective because its kernel, being a submodule of the simple module L, is $\{0\}$. Thus, if $LE = E$, then L is isomorphic to E.

Now assume that A is semisimple. If $f : L \to E$ is A-linear, then

$$f(L) = f(LL) = Lf(L).$$

Hence, if f is an isomorphism, so that $f(L) = E$, then $E = LE$. $\boxed{\text{QED}}$

Finally, we have a curious but convenient fact about left ideals that are isomorphic as A-modules:

Proposition 5.5 *If L and M are isomorphic left ideals in a semisimple ring A, then*

$$L = Mx,$$

for some $x \in A$.

Proof. We know by Theorem 5.7 that $M = Ay_M$ for some idempotent y_M. Let $f : M \to L$ be an isomorphism of A-modules. Then

$$L = f(M) = f(Ay_M y_M) = Ay_M f(y_M) = Mx,$$

where $x = f(y_M)$. $\boxed{\text{QED}}$

5.5   Deconstructing Semisimple Rings

We will work with a semisimple ring A. Recall that this means that A is semisimple as a left module over itself.

Semisimplicity decomposes A as a direct sum of simple submodules. A submodule in A is just a left ideal. Thus, we have a decomposition

$$A = \sum \{\text{all simple left ideals of } A\}. \tag{5.4}$$

Let

$$\{L_i\}_{i \in \mathcal{R}}$$

be a maximal family of nonisomorphic simple left ideals in A; such a family exists by Zorn's lemma. Let

$$A_i = \sum \{L : L \text{ is a left ideal isomorphic to } L_i\}. \tag{5.5}$$

Another convenient way to express A_i is as L_iA:

$$A_i = L_iA,$$

which makes it especially clear that A_i is a two-sided ideal.

By Theorem 5.8, we have

$$LL' = 0 \qquad \text{if } L \text{ is not isomorphic to } L'.$$

So

$$A_iA_j = 0 \qquad \text{if } i \neq j. \tag{5.6}$$

Since A is semisimple, it is the sum of all its simple left ideals, and so

$$A = \sum_{i \in \mathcal{R}} A_i. \tag{5.7}$$

Thus, A is a sum of two-sided ideals A_i. As it stands there seems to be no reason why $\mathcal{R}$ should be a finite set; yet, remarkably, it is finite!

The finiteness of $\mathcal{R}$ becomes visible when we look at the decomposition of the unit element $1 \in A$:

$$1 = \sum_{i \in \mathcal{R}} \underbrace{u_i}_{\in A_i}. \tag{5.8}$$

The sum here, of course, is finite; that is, *all but finitely many u_i are 0*. For any $a \in A$ we can write

$$a = \sum_{i \in \mathcal{R}} a_i \qquad \text{with each } a_i \text{ in } A_i.$$

Then, on using (5.6),

$$a_j = a_j1 = a_ju_j = au_j.$$

Thus, a determines the "components" a_j uniquely, and so

$$\text{the sum } A = \sum_{i \in \mathcal{R}} A_i \text{ is a direct sum.} \tag{5.9}$$

If some u_j were 0, then all the corresponding a_j would be 0, which cannot be since each A_j is nonzero. Consequently,

$$\text{the index set } \mathcal{R} \text{ is finite.} \tag{5.10}$$

Since we also have, for any $a \in A$,

$$a = 1a = \sum_{i \in \mathcal{R}} u_i a,$$

we have from the fact that the sum $A - \sum_i A_i$ is direct,

$$u_i a = a_i = a u_i.$$

Hence, u_i is the multiplicative identity in A_i.

We have arrived at a first view of the structure of semisimple rings:

Theorem 5.9 *Suppose A is a semisimple ring. Then there are finitely many left ideals $L_1, ..., L_r$ in A such that every left ideal of A is isomorphic, as a left A-module, to L_j for exactly one $j \in [r]$. Furthermore,*

$$A_j = L_j A = \text{sum of all left ideals isomorphic to } L_j$$

is a two-sided ideal, with a nonzero unit element u_j, and A is the product of the rings A_j, in the sense that the map

$$\prod_{i=j}^{r} A_i \to A : (a_1, \dots, a_r) \mapsto a_1 + \cdots + a_r \tag{5.11}$$

is an isomorphism of rings. Any simple left ideal in A_j is isomorphic to L_j. Moreover,

$$\begin{aligned} 1 &= u_1 + \cdots + u_r, \\ A_j &= A u_j, \\ A_i A_j &= 0 \qquad \text{for } i \neq j. \end{aligned} \tag{5.12}$$

Here is a summary of the properties of the elements u_i:

Proposition 5.6 *Let $L_1, \ldots, L_r$ be simple left ideals in a semisimple ring A such that every left ideal of A is isomorphic, as a left A-module, to exactly one of the L_j. Let $A_j = L_j A$ and u_j be an idempotent generator of A_j. Then $u_1, \ldots, u_r$ are nonzero, lie in the center of the algebra, and satisfy*

$$u_i^2 = u_i, \quad u_i u_j = 0 \quad if\ i \neq j,$$
$$u_1 + \cdots + u_r = 1.$$
(5.13)

Moreover, $u_1, \ldots, u_r$ is a longest set of nonzero central idempotents satisfying (5.13). Multiplication by u_i in A is the identity on A_i and is 0 on all A_j for $j \neq i$.

The two-sided ideals A_j are, it turns out, minimal two-sided ideals, and every two-sided ideal in A is a sum of certain A_j.

Theorem 5.10 *Let $A_j = L_j A$, where $L_1, \ldots, L_r$ are simple left ideals in a semisimple ring A such that every simple left ideal is isomorphic, as a left A-module, to L_j for exactly one $j \in [r]$. Then A_j is a ring in which the only two-sided ideals are 0 and A_j. Every two-sided ideal in A is a sum of some of the A_j.*

Proof. Suppose $J \neq 0$ is a two-sided ideal of A_j. Since $A_i A_k = 0$ for $i = k$, it follows that J is also a two-sided ideal in A. Since A is semisimple, so is J as a left submodule of A. Then J is a sum of simple left ideals of A. Let L be a simple left ideal of A contained inside J. Now recall that A_j is the sum of all left ideals isomorphic to a certain simple left ideal L_j, and that all such left ideals are of the form $L_j x$ for $x \in A$. Then, since J is also a right ideal, each such $L_j x$ is inside J and so $A_j \subset J$. Thus, the only nonzero two-sided ideals of A_j are 0 and itself.

Now consider any two-sided ideal I in A. Then $AI \subset I$, but also $I \subset AI$ since $1 \in A$. Hence,

$$I = AI = A_1 I + \cdots + A_r I.$$

Note that $A_j I$ is a two-sided ideal, and $A_j I \subset A_j$. By the property we have already proved, it follows that $A_j I$ is either 0 or A_j. Consequently,

$$I = \sum_{j: A_j I \neq 0} A_j. \quad \boxed{\text{QED}}$$

5.6   Simply Simple

Let A be a semisimple ring; as we have seen, A is the product of minimal two-sided ideals $A_1,...,\ A_r$, where each A_j is the sum of all left ideals isomorphic, as left A-modules, to a specific simple left ideal L_j. Each subring A_j is *isotypical*: it is the sum of simple left ideals that are all isomorphic to one common left ideal.

A ring B is *simple* if it is a sum of simple left ideals that are all isomorphic to each other as left B-modules. Note that a simple ring is also semisimple.

By Proposition 5.5, all isomorphic left ideals are right translates of one another, a simple ring B is a sum of right translates of any given simple left ideal L. Consequently,

$$B = LB \qquad \text{if } B \text{ is a simple ring and } L \text{ is any simple left ideal.} \qquad (5.14)$$

As a consequence, we have:

Proposition 5.7 *The only two-sided ideals in a simple ring are 0 and the whole ring itself.*

Proof. Let I be a two-sided ideal in a simple ring B and suppose $I \neq 0$. By semisimplicity of B, the left ideal I is a sum of simple left ideals, and so, in particular, contains a simple left ideal L. Then by (5.14), we see that $LB = B$. But $LB \subset I$, because I is also a right ideal. Hence, $I = B$. $\boxed{\text{QED}}$

For a ring B, any B-linear map $f : B \to B$ is completely specified by the value $f(1)$, because

$$f(b) = f(b1) = bf(1).$$

Moreover, if $f, g \in \operatorname{End}_B(B)$, then

$$(fg)(1) = f(g(1)) = g(1)f(1),$$

and so we have a ring isomorphism

$$\operatorname{End}_B(B) \to B^{\text{opp}} : f \mapsto f(1), \qquad (5.15)$$

where B^{opp}, the *opposite ring*, is the ring B with multiplication in "opposite" order:

$$(a, b) \mapsto ba.$$

We then have

Theorem 5.11 *If B is a simple ring, then B is isomorphic to a ring of matrices*

$$B \simeq \mathrm{Matr}_n(\mathbb{D}^{\mathrm{opp}}), \tag{5.16}$$

where n is a positive integer, and $\mathbb{D}$ is the division ring $\mathrm{End}_B(M)$ for any simple left ideal M in B.

Proof. We know that B is the sum of a finite number of simple left ideals, each of which is isomorphic, as a left B-module, to any one simple left ideal M. Then $B \simeq M^n$, as left B-modules, for some positive integer n. We also know that there are ring isomorphisms

$$B^{\mathrm{opp}} \simeq \mathrm{End}_B(B) = \mathrm{End}_B(M^n) \simeq \mathrm{Matr}_n(\mathbb{D}).$$

Taking the opposite ring, we obtain an isomorphism of B with $\mathrm{Matr}_n(\mathbb{D})^{\mathrm{opp}}$. But now consider the transpose of $n \times n$ matrices:

$$\mathrm{Matr}_n(\mathbb{D})^{\mathrm{opp}} \to \mathrm{Matr}_n(\mathbb{D}^{\mathrm{opp}}) : A \mapsto A^{\mathrm{tr}}.$$

Then, working in components of the matrices, and denoting "opposite" multiplication by $*$ (and by S^{tr} the transpose of S), we have

$$(A * B)^{\mathrm{tr}}_{ik} = (BA)_{ki} = \sum_{j=1}^{n} B_{kj} A_{ji} = \sum_{j=1}^{n} A_{ji} * B_{kj},$$

which is the ordinary matrix product $A^{\mathrm{tr}} B^{\mathrm{tr}}$ in $\mathrm{Matr}_n(\mathbb{D}^{\mathrm{opp}})$. Thus, the transpose gives an isomorphism $\mathrm{Matr}_n(\mathbb{D})^{\mathrm{opp}} \simeq \mathrm{Matr}_n(\mathbb{D}^{\mathrm{opp}})$. $\boxed{\text{QED}}$

The opposite ring often arises in matrix representations of endomorphisms. If M is a one-dimensional vector space over a division ring $\mathbb{D}$, with a basis element v, then with each $T \in \mathrm{End}_{\mathbb{D}}(M)$ we can associate the "matrix" $\hat{T} \in \mathbb{D}$ specified through $T(v) = \hat{T}v$. But then for any $S, T \in \mathrm{End}_{\mathbb{D}}(M)$ we have

$$\widehat{ST} = \hat{T}\hat{S}.$$

Thus, $\mathrm{End}_{\mathbb{D}}(M)$ is isomorphic to $\mathbb{D}^{\mathrm{opp}}$, via its matrix representation.

5.7 Commutants and Double Commutants

There is a more abstract, "coordinate-free" version of Theorem 5.11. First let us observe that for a module M over a ring A, the endomorphism ring

$$A_{\mathrm{c}} = \mathrm{End}_A(M)$$

is the *commutant* for A, consisting of all additive maps $M \to M$ that commute with the action of A. The double commutant is

$$A_{dc} = \text{End}_{A_c}(M).$$

Since for any $a \in A$ the multiplication

$$l_a : M \to M : x \mapsto ax \tag{5.17}$$

commutes with the action of every element of A_c, it follows that

$$l_a \in A_{dc}.$$

Note that

$$l_{ab} = l_a l_b,$$

and l maps the identity element in A to that in A_{dc}. Thus, l is a ring homomorphism. The following result is due to Rieffel (see Lang [55]):

Theorem 5.12 *Let B be a simple ring, L be a nonzero left ideal in B,*

$$B_c = \text{End}_B(L), \qquad B_{dc} = \text{End}_{B_c}(L),$$

and

$$l : B \to B_{dc} : a \mapsto l_a,$$

the natural ring homomorphism given by (5.17). Then l is an isomorphism of B onto $\text{End}_{B_c}(L)$. In particular, every simple ring is isomorphic to the ring of endomorphisms on a module.

Proof. To avoid confusion, it is useful to keep in mind that elements of B_c and B_{dc} are $\mathbb{Z}$-linear maps $L \to L$.

The ring morphism

$$l : B \to B_{dc} : b \mapsto l_b$$

is given explicitly by

$$l_b x = bx \qquad \text{for all } b \in B, \text{ and all } x \in L.$$

It maps the unit element in B to the unit element in B_{dc}, and so is not 0. The kernel of $l \neq 0$ is a two-sided ideal in a simple ring, and hence is 0. Thus, l is injective.

We will show that $l(B)$ is B_{dc}. Since $1 \in l(B)$, it will suffice to prove that $l(B)$ is a left ideal in B_{dc}.

Since LB contains L as a subset, and is thus not $\{0\}$, and is clearly a two-sided ideal in B, it is equal to B:

$$LB = B.$$

Hence,

$$l(L)l(B) = l(B).$$

Thus, it will suffice to prove that $l(L)$ is a left ideal in B_{dc}. We can check this as follows: if $f \in B_{\mathrm{dc}}$ and $b, y \in L$, then

$$
\begin{aligned}
\left(f l_b\right)(y) &= f(by) \\
&= f(b)y \quad \text{because } L \to L : x \mapsto xy \text{ is in } B_c = \mathrm{End}_B(L) \\
&= l_{f(b)}(y),
\end{aligned}
$$

thus showing that

$$f \cdot l_b = l_{f(b)},$$

and hence $l(L)$ is a left ideal in B_{dc}. $\boxed{\text{QED}}$

Lastly, let us make an observation about the center of a simple ring:

Proposition 5.8 *If B is a simple ring, then its center $Z(B)$ is a field. If B is a finite-dimensional simple algebra over an algebraically closed field $\mathbb{F}$, then $Z(B) = \mathbb{F}1$.*

Proof. For each $z \in Z(B)$ the map

$$l_z : B \to B : b \mapsto zb$$

is both left and right B-linear. As we have seen before, $l_z \in B_{\mathrm{dc}}$. Assume now that $z \neq 0$. We need to produce z^{-1}. We have the ring isomorphism

$$B \to B_{\mathrm{dc}} : x \mapsto l_x,$$

so we need only produce l_z^{-1}. Since $l_z : B \to B : a \mapsto za$ is left and right B-linear, $\ker l_z$ is a two-sided ideal. This ideal is not B because $z \neq 0$; so $\ker l_z = 0$. Hence, the two-sided ideal $l_z(B)$ in B is not 0 and hence is all of B. So l_z is invertible as an element of B_{dc}, and so z is invertible. Thus, every nonzero element in $Z(B)$ is invertible. Since $Z(B)$ is also commutative and contains $1 \neq 0$, it is a field.

Suppose now that B is a finite-dimensional $\mathbb{F}$-algebra and $\mathbb{F}$ is algebraically closed. Then any $z \in Z(B)$ not in $\mathbb{F}$ would give rise to a proper finite extension of $\mathbb{F}$, but this is impossible (see the proof of Theorem 5.2).
⊡ QED

5.8   Artin–Wedderburn Structure

We need only bring together the understanding we have gained of the structure of semisimple rings to formulate the full structure theorem for semisimple rings:

Theorem 5.13 *If A is a semisimple ring, then there are positive integers s, $d_1, ..., d_s$, and division rings $\mathbb{D}_1, ..., \mathbb{D}_s$, and an isomorphism of rings*

$$A \to \prod_{j=1}^{s} \mathrm{Matr}_{d_j}(\mathbb{D}_j), \qquad (5.18)$$

where $\mathrm{Matr}_{d_j}(\mathbb{D}_j)$ is the ring of $d_j \times d_j$ matrices with entries in $\mathbb{D}_j$. Conversely, the ring $\mathrm{Matr}_d(\mathbb{D})$ for any positive integer d and division ring $\mathbb{D}$ is simple and every finite product of such rings is semisimple. If a semisimple ring A is a finite-dimensional algebra over an algebraically closed field $\mathbb{F}$, then each $\mathbb{D}_j$ is the field $\mathbb{F}$.

The decomposition of a semisimple ring into a product of matrix rings is generally called the Artin–Wedderburn theorem.

Proof. In Theorem 5.9 we proved that every semisimple ring is a product of simple rings. Then in Theorem 5.11 we proved that every simple ring is isomorphic to a matrix ring over a division ring. For the converse direction, work out Exercise 5.5a. By Theorem 5.11, the division ring $\mathbb{D}_j$ is the opposite ring of $\mathrm{End}_A(L_j)$ for a suitable simple left ideal L_j in A, and then by Schur's lemma (in the form of Theorem 5.2) $\mathbb{D}_j = \mathbb{F}$ if $\mathbb{F}$ is algebraically closed.
⊡ QED

Note that for the second part of the conclusion in the preceding result all we need is for $\mathbb{F}$ to be a splitting field for the algebra A.

5.9   A Module as the Sum of Its Parts

We will now see how the decomposition of a semisimple ring A yields a decomposition of any A-module E.

Let A be a semisimple ring. Recall that there is a finite collection of simple left ideals

$$L_1, \ldots, L_r \subset A$$

such that every simple left ideal is isomorphic to L_i for exactly one $i \in [r]$. Moreover,

$$A_i \overset{\text{def}}{=} \text{the sum of all left ideals isomorphic to } L_i$$

is a two-sided ideal in A, and the map

$$I : \prod_{i=1}^{r} A_i \to A : (a_1, \ldots, a_r) \mapsto a_1 + \cdots + a_r$$

is an isomorphism of rings. Recall that each A_i has a unit element u_i, and

$$u_1 + \cdots + u_r = 1.$$

Every $a \in A$ decomposes uniquely as

$$a = \sum_{i=1}^{r} a_i,$$

where

$$a u_i = a_i = u_i a \in A_i.$$

Consider now any A-module E. Any element $x \in E$ can then be decomposed as

$$x = 1x = \sum_{j=1}^{r} u_j x.$$

Note that

$$u_j x \in E_j \overset{\text{def}}{=} A_j E, \tag{5.19}$$

and E_j is a submodule of E. Observe also that since

$$A_j = u_j A,$$

we have

$$E_j = u_j E. \tag{5.20}$$

Moreover,

$$E_j = A_j E = \sum_{\text{left ideal } L \simeq L_j} LE. \tag{5.21}$$

Proposition 5.9 *If A is a semisimple ring and $E \neq \{0\}$ is an A-module, then E has a submodule isomorphic to some simple left ideal in A. In particular, every simple A-module is isomorphic to a simple left ideal in A.*

Proof. Observe that $E = AE \neq \{0\}$. Now A is the sum of its simple left ideals. Thus, there is a simple left ideal L in A and an element $v \in E$ such that $Lv \neq \{0\}$. The map

$$L \to Lv : x \mapsto xv$$

is surjective and, by simplicity of L, is also injective. Thus, $L \simeq Lv$, and Lv is therefore a simple submodule of E. $\boxed{\text{QED}}$

Theorem 5.14 *Suppose A is a semisimple ring. Let $L_1, \ldots, L_r$ be left ideals of A such that every simple left ideal of A is isomorphic, as a left A-module, to L_i for exactly one $i \in [r]$, and let A_j be the sum of all left ideals of A isomorphic to L_j. Let u_i be a central idempotent for which $A_i = Au_i$ for each $i \in [r]$. If E is a left A-module, then*

$$E = E_1 \oplus \cdots \oplus E_r,$$

where

$$E_i = A_i E = u_i E$$

is the sum of all simple left submodules of E isomorphic to L_i, this sum being taken to be $\{0\}$ when there is no such submodule.

Proof. Let F be a simple submodule of E. We know that it must be isomorphic to one of the simple ideals L_j in A. Then, since $LF = 0$ whenever L is a simple ideal not isomorphic to L_j, we have

$$F = AF = A_j F \subset E_j.$$

Thus, every submodule isomorphic to L_j is contained in E_j. On the other hand, A_j is the sum of simple left ideals isomorphic to L_j, and so $E_j = A_j E$ is a sum of simple submodules isomorphic to L_j. The module E is the direct sum of simple submodules, and each such submodule is isomorphic to some L_j. Summing up the submodules isomorphic to L_j yields E_j. $\boxed{\text{QED}}$

5.10 Readings on Rings

The general subject of which we have seen a special sample in this chapter is the theory of noncommutative rings. Books on noncommutative rings and algebras generally subscribe to the "beatings shall continue until morale improves" school of exposition. A delightful exception is the page-turner account in the book by Lam [52]. The accessible book by Farb and Dennis [26] also includes a slim yet substantive chapter on representations of finite groups. Lang's *Algebra* [55] is a very convenient and readable reference for the basic major results.

5.11 Afterthoughts: Clifford Algebras

Clifford algebras are algebras of great use and interest that lie just at the borders of our exploration. Here we take a very quick look at this family of algebras.

A *quadratic form* Q on a vector space V, over a field $\mathbb{F}$, is a mapping $Q : V \to \mathbb{F}$ for which

$$Q(cv) = c^2 Q(v) \qquad \text{for all } c \in \mathbb{F} \text{ and } v \in V, \tag{5.22}$$

and such that the map

$$V \times V \to \mathbb{F} : (u, v) \mapsto B_Q(u, v) \stackrel{\text{def}}{=} Q(u + v) - Q(u) - Q(v) \tag{5.23}$$

is bilinear.

If $w \in V$ has $Q(w) \neq 0$, then the mapping

$$r_w : V \to V : v \mapsto v - \frac{B_Q(v, w)}{Q(w)} w \tag{5.24}$$

fixes each point on the subspace

$$w^\perp = \{v \in V : B_Q(v, w) = 0\},$$

and maps w to $-w$. If we assume that the characteristic of $\mathbb{F}$ is not 2, it follows that $r_w \neq I_V$ (where I_V is the identity map on V), $r_w^2 = I_V$, and r_w fixes each point on the hyperplane H. Thus, we can define r_w to be the *reflection* across the hyperplane H.

If the characteristic of $\mathbb{F}$ is 2, we can construct reflections "by hand": for a hyperplane H in V, pick a vector w outside H and a vector v_0 in H; define $r_w : V \to V$ to be the linear map which is the identity on H and maps w to $w + v_0$. It is readily checked that $r_w \neq I_V$ (where I_V is the identity map on V), $r_w^2 = I_V$ and, of course, r_w fixes each point on the hyperplane H.

The *Clifford algebra* C_Q for a quadratic form Q on a vector space V is the quotient algebra

$$C_Q = T(V)/J_Q, \tag{5.25}$$

where $T(V)$ is the tensor algebra

$$T(V) = \mathbb{F} \oplus V \oplus V^{\otimes 2} \oplus \cdots$$

and J_Q is the two-sided ideal in $T(V)$ generated by all elements of the form

$$v \otimes v + Q(v)1 \qquad \text{for all } v \in V.$$

The natural injection $V \to T(V)$ induces, by composition with the projection down to the quotient $C_Q(V)$, a linear map

$$j_Q : V \to C_Q(V) \tag{5.26}$$

which satisfies

$$j_Q(v)^2 + Q(v)1 = 0 \qquad \text{for all } v \in V. \tag{5.27}$$

The map $j_Q : V \to C_Q(V)$ specifies $C_Q(V)$ as the "minimal" such algebra in the sense that it has the *universal property* that if $f : V \to A$ is any linear map from V to an $\mathbb{F}$-algebra A with multiplicative identity 1_A for which $f(v)^2 + Q(v)1_A = 0$ for all $v \in V$, then there is a unique algebra morphism $f_Q : C_Q(V) \to A$, mapping 1 to 1_A, such that

$$f = f_Q \circ j_Q.$$

For our discussion, let us focus on a complex vector space V of finite dimension d, and where the bilinear form B_Q is specified by the matrix

$$B_Q(e_a, e_b) = -2\delta_{ab} \qquad \text{for all } a, b \in [d],$$

where $e_1, \ldots, e_d$ is some basis of V. The corresponding Clifford algebra, which we denote by C_d, can be taken to be the complex algebra generated by the $e_1, \ldots, e_d$, subject to the relations

$$\{e_a, e_b\} \overset{\text{def}}{=} e_b e_b + e_b e_a = -2\delta_{ab}1 \qquad \text{for all } a, b \in [d]. \tag{5.28}$$

A basis of the algebra is given by all products of the form

$$e_{s_1} \ldots e_{s_m},$$

where $m \geq 0$, and $1 \leq s_1 < s_2 < \cdots \leq s_m \leq d$. Writing S for such a set $\{s_1, \ldots, s_m\} \subset [d]$, with the elements s_i always in increasing order, we see that C_d, as a vector space, has a basis consisting of one element e_S for each subset S of $[d]$. Notice also that condition (5.28) implies that every time a term $e_s e_t$, with $s > t$, is replaced by $e_t e_s$, one picks up a minus sign:

$$e_t e_s = -e_s e_t \qquad \text{if } s \neq t. \tag{5.29}$$

Keeping in mind also the condition $e_s^2 = 1$ for all $s \in [d]$, we have

$$e_S e_T = \epsilon_{ST} e_{S \Delta T}, \tag{5.30}$$

where $S \Delta T$ is the symmetric difference of the sets S and T, and

$$\epsilon_{ST} = \prod_{s \in S, t \in T} \epsilon_{st},$$

$$\epsilon_{st} = \begin{cases} +1 & \text{if } s < t, \\ +1 & \text{if } s = t, \\ -1 & \text{if } s > t, \end{cases} \tag{5.31}$$

and the empty product, which occurs if S or T is $\emptyset$, is taken to be 1. The algebra C_d can be reconstructed more officially as the 2^d-dimensional free vector space over the set of formal variables e_S, and then by specifying multiplication by (5.30). (For more details, see the book by Artin [2].)

Each basis vector e_a gives rise to idempotents

$$\frac{1}{2}(1 + e_a) \qquad \text{and} \qquad \frac{1}{2}(1 - e_a).$$

In fact, the relation

$$(e_{s_1} \ldots e_{s_m})^2 = (-1)^{m(m-1)} \tag{5.32}$$

shows that any basis element e_S in C_d, where $S = \{s_1, \ldots, s_m\}$ contains m elements, produces orthogonal idempotents

$$y_{+,S} = \frac{1}{2}(1 - (-1)^{m(m-1)/2} e_S) \qquad \text{and} \qquad y_{-,S} = \frac{1}{2}(1 + (-1)^{m(m-1)/2} e_S).$$

If d is *odd*, then the full product $e_{[d]} = e_1 \ldots e_d$ is in the center of the algebra C_d, and the idempotents $y_{\pm,[d]}$ are central idempotents. Thus, for d is odd, C_d is the product of the two-sided ideals $C_d y_{+,[d]}$ and $C_d y_{-,[d]}$; more precisely, the map

$$C_d y_{+,[d]} \times C_d y_{-,[d]} \to C_d : (a, b) \mapsto a + b$$

is an isomorphism algebra.

Particularly useful are the orthogonal idempotents arising from pairs $\{a, b\} \subset [d]$:

$$y_{+,\{a,b\}} = \frac{1}{2}(1 + e_a e_b) \quad \text{and} \quad y_{-,\{a,b\}} = \frac{1}{2}(1 - e_a e_b),$$

where $a < b$. Could this be an indecomposable idempotent? Recall the criterion for indecomposability from Proposition 4.7 for a nonzero idempotent y:

y is indecomposable if yxy is a scalar multiple of y for every $x \in C_d$.

$$(5.33)$$

A simple calculation shows that

$$y_{\pm,\{a,b\}} e_c = \begin{cases} e_c y_{\pm,\{a,b\}} & \text{if } c \notin \{a, b\}, \\ e_c y_{\mp,\{a,b\}} & \text{if } c \in \{a, b\}. \end{cases} \qquad (5.34)$$

Thus, to construct an indecomposable idempotent we can take a product of the idempotents $y_{\pm,\{a,b\}}$. Suppose first that d is even and let π_d be the partition of $[d]$ into pairs of consecutive integers:

$$\pi_d = \{\{1, 2\}, \ldots, \{d-1, d\}\}.$$

Let ϵ be any mapping of π_d to $\{+1, -1\}$, giving a choice of sign for each pair $\{j, j+1\}$ in π_d. Then we have the idempotent

$$y_\epsilon = \prod_{B \in \pi_d} y_{\epsilon(B),B}, \qquad (5.35)$$

where, observe, the terms $y_{\epsilon(B),B}$ commute with each other since the distinct B's are disjoint. An example of such an idempotent, for $d = 4$, is

$$\frac{1}{2}(1 + e_1 e_2)\frac{1}{2}(1 - e_3 e_4).$$

If we apply criterion (5.33) with $x = e_c$, and use (5.34), it follows that the idempotent y_ϵ is indecomposable. Thus, we have the full decomposition of C_d, for even d, into a sum of simple left ideals:

$$C_d = \bigoplus_{\epsilon \in \{+1, -1\}^{\pi_d}} C_d y_\epsilon. \qquad (5.36)$$

This explicitly exhibits the semisimple structure of C_d for even d. A straightforward extension produces the semisimple structure of C_d for odd d on using the central idempotents $y_{\pm, [d]}$.

If one thinks of $e_1, \ldots, e_d$ as forming an orthonormal basis for a real vector space V_0 sitting inside V, the relation $e_a^2 = 1$ is suggestive of reflection across the hyperplane $e_a^\perp$. More precisely, for any nonzero vector $w \in V_0$, the map

$$V_0 \to V_0 : v \mapsto -wvw^{-1}$$

takes w to $-w$ and takes any $v \in w^\perp$ to

$$-wvw^{-1} = vww^{-1} = v,$$

and is thus just the reflection map r_w across the hyperplane $w^\perp$. A linear map $T : V_0 \to V_0$ is an *orthogonal* transformation, relative to Q, if $Q(Tv) = Q(v)$ for all $v \in V_0$. A general orthogonal transformation is a composition of reflections, and the Clifford algebra is a crucial structure in the study of representations of the group of orthogonal transformations.

Exercises

5.1. Sanity check:

    (a) Is $\mathbb{Z}$ a semisimple ring?

    (b) Is $\mathbb{Q}$ a semisimple ring?

    (c) Is a subring of any semisimple ring also semisimple?

5.2. Show that in the ring of all matrices $M_{a,b} = \begin{pmatrix} a & b \\ 0 & a \end{pmatrix}$, with a, b running over $\mathbb{C}$, the left ideal $\{M_{0,b} : b \in \mathbb{C}\}$ has no complement.

5.3. Show that a commutative simple ring is a field.

5.4. Let A be a finite-dimensional semisimple algebra over a field $\mathbb{F}$, and define $\chi_{\text{reg}} : A \to \mathbb{F}$ by

$$\chi_{\text{reg}}(a) = \text{Tr}\left(\rho_{\text{reg}}(a)\right), \quad \text{where } \rho_{\text{reg}}(a) : A \to A : x \mapsto ax. \quad (5.37)$$

Let $L_1, \ldots, L_s$ be a maximal collection of nonisomorphic simple left ideals in A, so that $A \simeq \prod_{i=1}^{s} A_i$, where A_i is the two-sided ideal formed by the sum of all left ideals isomorphic to L_i. As usual, let $1 = u_1 + \cdots + u_s$ be the decomposition of 1 into idempotents $u_i \in A_i = Au_i$. Viewing L_i as a vector space over $\mathbb{F}$, define

$$\chi_i(a) = \text{Tr}(\rho_{\text{reg}}(a)|L_i). \quad (5.38)$$

Note that since L_i is a left ideal, $\rho_{\text{reg}}(a)(L_i) \subset L_i$. Do the following:

(a) Show that $\chi_{\text{reg}} = \sum_{i=1}^{s} d_i \chi_i$, where d_i is the integer for which $A_i \simeq L_i^{d_i}$ as A-modules.

(b) Show that $\chi_i(u_j) = \delta_{ij} \dim_{\mathbb{F}} L_i$.

(c) Assume that the characteristic of $\mathbb{F}$ does not divide any of the numbers $\dim_{\mathbb{F}} L_i$ (there is an important case of this in Exercise 3.7). Use (b) to show that the functions $\chi_1, \ldots, \chi_s$ are linearly independent over $\mathbb{F}$.

(d) Let E be an A-module, and define $\chi_E : A \to \mathbb{F}$ by

$$\chi_E(a) = \text{Tr}\left(\rho_E(a)\right), \quad \text{where } \rho_E(a) : E \to E : x \mapsto ax. \quad (5.39)$$

Show that χ_E is a linear combination of the functions χ_i with nonnegative integer coefficients:

$$\chi_E = \sum_{i=1}^{s} n_i \chi_i,$$

where n_i is the number of copies of L_i in a decomposition of E into simple A-modules.

(e) Under the assumption made in (c), show that if E and F are A-modules with $\chi_E = \chi_F$, then $E \simeq F$.

5.5. Let $B = \text{Matr}_n(\mathbb{D})$ be the algebra of $n \times n$ matrices over a division ring $\mathbb{D}$.

(a) Show that for each $j \in [n]$, the set L_j of all matrices in B that have all entries 0 except possibly those in column j is a simple left ideal. Since $B = L_1 + \cdots + L_n$, this implies that B is a semisimple ring.

(b) Show that if L is a simple left ideal in B, then there is a basis $b_1, ..., b_n$ of $\mathbb{D}^n$, treated as a right $\mathbb{D}$-module, such that L consists exactly of those matrices T for which $Tb_i = 0$ whenever $i \neq 1$.

(c) With notation as in (a), produce orthogonal idempotent generators in $L_1, ..., L_n$.

5.6. Prove that if a module N over a ring is the direct sum of simple submodules, no two of which are isomorphic to each other, then every simple submodule of N *is* one of these submodules.

5.7. Suppose L_1 and L_2 are simple left ideals in a semisimple ring A. Show that the following are equivalent: (a) $L_1 L_2 = 0$; (b) L_1 and L_2 are not isomorphic as A-modules; (c) $L_2 L_1 = 0$.

5.8. Suppose N_1 and N_2 are left ideals in a semisimple ring A. Show that the following are equivalent: (a) $N_1 N_2 = 0$; (b) there is no nonzero A-linear map $N_1 \to N_2$; (c) there is a simple submodule of N_1 that is isomorphic to a submodule of N_2; (d) $N_2 N_1 = 0$.

5.9. Let u and v be indecomposable idempotents in a semisimple ring A for which $uA = vA$. Show that Au is isomorphic to Av as left A-modules.

5.10. Prove the results of Sect. 4.10 for semisimple algebras, assuming where needed that the algebra is finite-dimensional over an algebraically closed field.

5.11. Suppose y is an idempotent in a ring A such that the left ideal Ay is simple. Show that $\mathbb{D}_y = \{yxy : x \in A\}$ is a division ring under the addition and multiplication operations inherited from A.

5.12. Let I be a nonempty finite set of commuting nonzero idempotents in a ring A. Show that there is a set P of orthogonal nonzero idempotents in A that add up to 1 such that every element of I is the sum of a unique subset of P.

5.13. For an algebra A over a field $\mathbb{F}$, define an element $s \in A$ to be *semisimple* if $s = c_1 e_1 + \cdots + c_m e_m$ for some orthogonal nonzero idempotents e_j and $c_1, \ldots, c_m \in \mathbb{F}$. For such s, show that each e_j is equal to $p_j(s)$ for some polynomial $p_j(X) \in \mathbb{F}[X]$. Show also that the elements in A that are polynomials in s form a semisimple subalgebra of A.

5.14. Let C be a finite nonempty set of commuting semisimple elements in an algebra A over a field $\mathbb{F}$. Show that there are orthogonal nonzero idempotents $e_1, \ldots, e_n$ such that every element of C is an $\mathbb{F}$-linear combination of the e_j.

5.15. Let A be a semisimple algebra over an algebraically closed field $\mathbb{F}$, $\{L_i\}_{i \in \mathcal{R}}$ be a maximal collection of nonisomorphic simple left ideals in A, and A_i be the sum of all left ideals isomorphic to L_i. We know that $A_i \simeq \operatorname{End}_{\mathbb{F}}(L_i)$ and $A \simeq \prod_{i \in \mathcal{R}} A_i$, as algebras (where $\simeq$ denotes algebra isomorphism). Show that an element $a \in A$ is an idempotent if and only if its representative block diagonal matrix in $\prod_{i \in \mathcal{R}} \operatorname{End}_{\mathbb{F}}(L_i)$ is a projection matrix, and that it is an indecomposable idempotent if and only if the matrix is a projection matrix of rank 1.

5.16. Let A be a finite-dimensional semisimple algebra over an algebraically closed field $\mathbb{F}$. Let $L_1, \ldots, L_s$ be simple left ideals in A such that every simple A-module is isomorphic to L_i for exactly one $i \in [s]$. For every $a \in A$ let $\rho_i(a)$ be the $d_i \times d_i$ matrix for the map $L_i \to L_i$: $x \mapsto ax$ relative to a fixed basis $|b_1(i)\rangle, \ldots, |b_{d_i}(i)\rangle$ of L_i. Denote by $\langle b_1(i)|, \ldots, \langle b_{d_i}(i)|$ the corresponding dual basis in L_i'. Prove that the matrix-entry functions $\rho_{i,jk} : a \mapsto \langle b_j(i)|a\, b_k(i)\rangle$, with $j, k \in [d]$ and $i \in [r]$, are linearly independent over $\mathbb{F}$. Using this, conclude that the characters $\chi_i = \operatorname{Tr}\rho_i$ are linearly independent.

5.17. Show that if u and v are indecomposable idempotents in a semisimple $\mathbb{F}$-algebra A, where $\mathbb{F}$ is algebraically closed, then uv is either 0, or has square equal to 0, or is an $\mathbb{F}$-multiple of an indecomposable idempotent. What can be said if u and v are commuting indecomposable idempotents?

5.18. Let $(S, \leq)$ be a partially ordered set. For $a, b \in S$, we denote by $a \vee b$ the least element $\geq a, b$, and by $a \wedge b$ the largest element $\leq a, b$. If $a \vee b$ and $a \wedge b$ exist for all $a, b \in S$, then S is said to be a *lattice*. More generally, the least upper bound of a subset $T \subset S$ is the *supremum* of S, denoted sup S, and the greatest lower bound is the *infimum* of S, denoted inf S. If every subset of S has both a supremum and an infimum, then the lattice S is said to be *complete*. If S has a least element, it is denoted 0, and the greatest element, if it exists, is denoted 1. An *atom* in S is an element $a \in S$ such that $a \neq 0$ and if $b \leq a$ then $b \in \{0, a\}$. If S is a subset of a partially ordered set, a *maximal element* of S is an element $a \in S$ such that if $b \in S$ with $a \leq b$ then $b = a$; a *minimal element* of S is an element $a \in S$ such that if $b \in S$ with $b \leq a$ then $b = a$. A partially ordered set $(S, \leq)$ satisfies the *ascending chain condition* if every nonempty subset of S contains a maximal element; it satisfies the *descending chain condition* if every nonempty subset of S contains a minimal element (other equivalent forms of these notions are more commonly used). Now let $\mathbb{L}_M$ be the set of all submodules of a module M over a ring A, and take the inclusion relation $L_1 \subset L_2$ as a partial order on $\mathbb{L}_M$. Thus, an atom in $\mathbb{L}_M$ is a simple submodule. Prove the following:

(a) $\mathbb{L}_M$ is a complete lattice.

(b) The lattice $\mathbb{L}_M$ is *modular*:

If $p, m, b \in \mathbb{L}_M$ and $m \subset b$ then $(p + m) \cap b = (p \cap b) + m$.

$$(5.40)$$

(The significance of modularity in a lattice was underlined by Dedekind [17, Sect. 4, (M)], [18, Sect. II.8].)

(c) Prove that if A is a finite-dimensional algebra over a field, then A is *left-Artinian* in the sense that the lattice of left ideals in A satisfies the descending chain condition.

(d) If A is a semisimple ring, then A is *left-Noetherian* in the sense that the lattice of left ideals in A satisfies the ascending chain condition.

(e) If A is a semisimple ring and I and J are two-sided ideals in A, then $I \cap J = IJ$.

(f) If A is a semisimple ring, then the lattice of two-sided ideals in A is *distributive*:

$$I \cap (J + K) = (I \cap J) + (I \cap K),$$
$$I + (J \cap K) = (I + J) \cap (I + K)$$

$$(5.41)$$

for all two-sided ideals I, J, K in A.

5.19. Let $(\mathbb{L}, \leq)$ be a modular lattice with 0 and 1 (these and other related terms are as defined in Exercise 5.18). Let $\mathcal{A}$ be the set of atoms in $\mathbb{L}$. Denote by $a + b$ the supremum of $\{a, b\}$, and by $a \cap b$ the infimum of $\{a, b\}$, and, more generally, denote the supremum of a subset $S \subset \mathbb{L}$ by $\sum S$. Elements $a, b \in \mathbb{L}$ are *complements* of each other if $a + b = 1$ and $a \cap b = 0$. Say that a subset $S \subset \mathcal{A}$ is *linearly independent* if $\sum T_1 = \sum T_2$ for some finite subsets $T_1 \subset T_2 \subset S$ implies $T_1 = T_2$.

  (a) Suppose every element of $\mathbb{L}$ has a complement. Show that if $t \leq s$ in $\mathbb{L}$, then there exists $v \in \mathbb{L}$ such that $t + v = s$ and $t \cap v = 0$.

  (b) $S \subset \mathcal{A}$ is independent if and only if $a \cap \sum T = 0$ for every finite $T \subset S$ and all $a \in S - T$.

  (c) Suppose every $s \in \mathbb{L}$ has a complement and $\mathbb{L}$ satisfies the ascending chain condition. Show that for every nonzero $m \in \mathbb{L}$ there is an $a \in \mathcal{A}$ with $a \leq m$.

  (d) Here is a primitive (in the logical, not historical) form of the Chinese remainder theorem: For any elements A, B, I, and J in a modular lattice for which $I + J = 1$, show that there is an element C such that $C + I = A + I$ and $C + J = B + J$. Next, working with the lattice $\mathbb{L}_R$ of two-sided ideals in a ring R, show that if $I_1, \ldots, I_m \in \mathbb{L}_R$ for which $I_a + I_b = R$ for $a \neq b$, and if $K_1, \ldots, K_m \in \mathbb{L}_R$, then there exists $C \in \mathbb{L}_R$ such that $C + I_a = K_a + I_a$ for all $a \in \{1, \ldots, m\}$.

5.20. We return to the theme of Frobenius algebras, which we saw earlier in Exercise 4.16. Let A be a finite-dimensional algebra over a field $\mathbb{F}$, containing a multiplicative identity. Show that the following are equivalent:

  (a) There is a linear functional $\epsilon : A \to \mathbb{F}$ such that $\ker \epsilon$ contains no nonzero left or right ideal of A.

  (b) There is a nondegenerate bilinear map $B : A \times A \to \mathbb{F}$ that is *associative* in the sense that $B(ac, b) = B(a, cb)$ for all $a, b, c \in A$.

Chapter 6

Representations of S_n

Having survived the long exploration of semisimple structure, we may think that midway on our journey we find ourselves in deep woods, the right path lost [16]. But this is not the time to abandon hope; instead, we plunge right into untangling the structure of representations of an important family of groups, the permutation groups S_n. This will be the only important class of finite groups to which we will apply all the machinery we have manufactured. A natural pathway beyond this is the study of representations of reflection groups.

There are several highly efficient ways to speed through the basics of the representations of S_n. We choose a more leisurely path, beginning with a look at permutations of $[n] = \{1, ..., n\}$ and partitions of $[n]$. This will lead us naturally to a magically powerful device: Young tableaux, which package special pairs of partitions of $[n]$. We will then proceed to Frobenius's construction of indecomposable idempotents, or, equivalently, irreducible representations of S_n, by using symmetries of Young tableaux.

6.1 Permutations and Partitions

To set the strategy for constructing the irreducible representations of S_n in its natural context, let us begin by looking briefly at the relationship between subgroups of S_n and partitions of $[n] = \{1, ..., n\}$.

A *partition* π of $[n]$ is a set of disjoint nonempty subsets of $[n]$ whose union is $[n]$; we will call the elements of π the *blocks* of π. For example, the set

$$\{\{2, 5, 3\}, \{1\}, \{4, 6\}\}$$

A.N. Sengupta, *Representing Finite Groups: A Semisimple Introduction*,
DOI 10.1007/978-1-4614-1231-1_6, © Springer Science+Business Media, LLC 2012

is a partition of [6] consisting of the blocks $\{2, 3, 5\}$, $\{1\}$, $\{4, 6\}$. Let

$$\mathbb{P}_n = \text{the set of all partitions of } [n]. \tag{6.1}$$

Any subgroup H of S_n produces a partition π_H of $[n]$ through the orbits: two elements $j, k \in [n]$ lie in a block of π_H if and only if $j = s(k)$ for some $s \in H$. Any permutation $s \in S_n$ generates a subgroup of S_n, and hence there is a corresponding partition of $[n]$.

A *cycle* is a permutation that has at most one block of size greater than 1; we call this block the *support* of the cycle, which we take to be $\emptyset$ for the identity permutation ι. A cycle c is displayed as

$$c = (i_1\, i_2\, \ldots\, i_k),$$

where $c(i_1) = i_2, \ldots, c(i_{k-1}) = i_k, c(i_k) = i_1$. Two cycles are said to be disjoint if their supports are disjoint. Disjoint cycles commute. The *length* of a cycle is the size of the largest block minus 1; thus, the length of the cycle $(1\,2\,3\,5)$ is 3, and the length of a *transposition* $(a\,b)$ is 1. If $s \in S_n$, then a *cycle of* s is a cycle that coincides with s on some subset of $[n]$ and is the identity outside it. Then s is the product, in any order, of its distinct cycles. For example, the permutation

$$1 \mapsto 1,\, 2 \mapsto 5,\, 3 \mapsto 2,\, 4 \mapsto 6,\, 5 \mapsto 3,\, 6 \mapsto 4$$

is written as

$$\begin{pmatrix} 1 & 2 & 3 & 4 & 5 & 6 \\ 1 & 5 & 2 & 6 & 3 & 4 \end{pmatrix}$$

and has the cycle decomposition

$$(2\,5\,3)(4\,6),$$

not writing the identity cycle. The *length* $l(s)$ of a permutation s is the sum of the lengths of its cycles, and the *signature* of s is given by

$$\epsilon(s) = (-1)^{l(s)}. \tag{6.2}$$

Multiplying s by a transposition t either splits a cycle of s into two or joins two cycles into one:

$$\begin{aligned} (1\,j)(1\,2\,3\,\ldots\,j\,\ldots\,m) &= (1\,2\,3\,\ldots\,j-1)(j\,j+1\,\ldots\,m), \\ (1\,j)(1\,2\,3\,\ldots\,j-1)(j\,j+1\,\ldots\,m) &= (1\,2\,3\,\ldots\,j\,\ldots,m), \end{aligned} \tag{6.3}$$

with the sum of the cycle lengths either decreasing by 1 or increasing by 1,

$$l(ts) = l(s) \pm 1 \qquad \text{if } t \text{ is a transposition and } s \in S_n. \qquad (6.4)$$

Consequently, $\epsilon(ts) = -\epsilon(s)$ if t is a transposition. Since every cycle is a product of transpositions,

$$(1\,2\ldots k) = (1\,2)(2\,3)\ldots(k-1\,k),$$

so is every permutation, and so

$$\epsilon(s) = (-1)^k, \quad \text{if } s \text{ is a product of } k \text{ transpositions}.$$

The permutation s is said to be *even* if $\epsilon(s)$ is 1, and *odd* if $\epsilon(s) = -1$. We then have

$$\epsilon(rs) = \epsilon(r)\epsilon(s) \quad \text{for all } r, s \in S_n.$$

Thus, for any field $\mathbb{F}$, the homomorphism $\epsilon : S_n \to \{1, -1\} \subset \mathbb{F}^\times$ provides a one-dimensional, hence irreducible, representation of S_n on $\mathbb{F}$.

Returning to partitions, let $B_1, \ldots, B_m$ be the string of blocks of a partition $\pi \in \mathbb{P}_n$, listed in order of decreasing size:

$$|B_1| \geq |B_2| \geq \cdots \geq |B_m|.$$

Then

$$\lambda(\pi) = (|B_1|, \ldots, |B_m|) \qquad (6.5)$$

is called the *shape* of π. We denote by $\overline{\mathbb{P}}_n$ the set of all shapes of all the elements in $\mathbb{P}_n$.

A shape, in general, is simply a finite nondecreasing sequence of positive integers. Shapes are displayed visually as *Young diagrams* in terms of rows of empty boxes. For example, the diagram

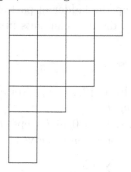

displays the shape $(4, 3, 3, 2, 1, 1)$.

Consider shapes λ and λ' in $\overline{\mathbb{P}}_n$. If $\lambda' \neq \lambda$, then there is a smallest j for which $\lambda'_j \neq \lambda_j$. If, for this j, $\lambda'_j > \lambda_j$, then we say that $\lambda' > \lambda$ in *lexicographic* order. This is an order relation on the partitions of n. The largest element is (n) and the smallest element is $(1, 1, \ldots, 1)$. Here is an ordering of $\overline{\mathbb{P}}_3$ displayed in decreasing lexicographic order:

$$ \tag{6.6} $$

There is also a natural partial order on $\mathbb{P}_n$, with $\pi_1 \leq \pi_2$ meaning that π_1 refines the blocks of π_2:

$$\pi_1 \leq \pi_2 \text{ if for any block } A \in \pi_1 \text{ there is a block } B \in \pi_2 \text{ with } A \subset B, \tag{6.7}$$

or, equivalently, each block of π_2 is the union of some of the blocks in π_1. Thus, $\pi_1 \leq \pi_2$ if π_1 is a "finer" partition than π_2. For example,

$$\{\{2,3\}, \{5\}, \{1\}, \{4\}, \{6\}\} \leq \{\{2,5,3\}, \{1\}, \{4,6\}\}$$

in $\mathbb{P}_6$. The "smallest" partition in this order is $\{\{1\}, \ldots, \{n\}\}$, and the "largest" is $\{[n]\}$:

$$\underline{0} = \{\{1\}, \ldots, \{n\}\} \qquad \text{and} \qquad \underline{1} = \{[n]\}. \tag{6.8}$$

For $\pi_1, \pi_2 \in \mathbb{P}_n$, define the *interval* $[\pi_1, \pi_2]$ to be

$$[\pi_1, \pi_2] = \{\pi \in \mathbb{P}_n : \pi_1 \leq \pi \leq \pi_2\}. \tag{6.9}$$

If we coalesce two blocks of a partition π to obtain a partition π_1, then we say that π_1 *covers* π. Clearly, π_1 covers π if and only if $\pi_1 \neq \pi$ and $[\pi, \pi_1] = \{\pi, \pi_1\}$. Climbing up the ladder of partial order one step at a time shows that for any $\pi_L \leq \pi_U$, distinct elements in $\mathbb{P}_n$, there is a sequence of partitions $\pi_1, \ldots, \pi_j$ with

$$\pi_L = \pi_1 \leq \cdots \leq \pi_j = \pi_U,$$

where π_i covers π_{i-1} for each $i \in \{2, \ldots, j\}$. Notice that *at each step up the number of blocks decreases by 1*.

If $\pi \in \mathbb{P}_n$ can be reached from $\underline{0}$ in l steps, each carrying it from one partition to a covering partition, then l is given by

$$l(\pi) = \sum_{B \in \pi} (|B| - 1) = n - |\pi|, \tag{6.10}$$

which is independent of the particular sequence of partitions used to go from $\underline{0}$ to π.

Proposition 6.1 *For any positive integer n and distinct partitions $\pi_1, \pi_2 \in \mathbb{P}_n$, if $\pi_1 \le \pi_2$, then $\lambda(\pi_1) < \lambda(\pi_2)$. In particular, if S is a nonempty subset of $\mathbb{P}_n$ and π is the element of largest shape in S, then π is a maximal element in S relative to the partial order $\le$.*

Proof. Let $B_1, \ldots, B_m$ be the blocks of a partition $\pi \in \mathbb{P}_n$, with $|B_1| \ge \cdots \ge |B_m|$, and let $\alpha_i = |B_i|$ for $i \in [m]$. Thus, $\lambda(\pi) = (\alpha_1, \ldots, \alpha_m)$. Let π' be the partition obtained from π by coalescing B_j and B_k for some $j > k$ in $[m]$. Then $\lambda(\pi') = (\alpha'_1, \ldots, \alpha'_{m-1})$, where

$$\alpha'_i = \begin{cases} \alpha_i & \text{if } \alpha_i > \alpha_j + \alpha_k, \\ \alpha_j + \alpha_k & \text{if } i \text{ is the smallest integer for which } \alpha_i \le \alpha_j + \alpha_k, \\ \alpha_{i+1} & \text{for all other } i. \end{cases}$$

$$(6.11)$$

From the second line above, if r is the smallest integer for which $\alpha_r \le \alpha_j + \alpha_k$, then

$$\alpha'_r = \alpha_j + \alpha_k \ge \alpha_r,$$

and, from the first line, $\alpha'_i = \alpha_i$ for $i < r$. This means $\lambda(\pi') > \lambda(\pi)$.

For any distinct $\pi_1, \pi_2 \in \mathbb{P}_n$ with $\pi_1 \le \pi_2$, there is a sequence of partitions $\tau_1, \ldots, \tau_N \in \mathbb{P}_n$ with τ_i obtained by coalescing two blocks of τ_{i-1}, for $i \in \{2, \ldots, N\}$, and $\tau_1 = \pi_1$ and $\tau_N = \pi_2$. Then $\lambda(\pi_1) = \lambda(\tau_1) < \cdots < \lambda(\tau_N) = \lambda(\pi_2)$. $\boxed{\text{QED}}$

6.2 Complements and Young Tableaux

The partial ordering $\le$ of partitions makes $\mathbb{P}_n$ a *lattice*: partitions π_1 and π_2 have a greatest lower bound as well as a least upper bound, which we denote

$$\pi_1 \wedge \pi_2 = \inf\{\pi_1, \pi_2\} \qquad \text{and} \qquad \pi_1 \vee \pi_2 = \sup\{\pi_1, \pi_2\}. \qquad (6.12)$$

More descriptively, $\pi_1 \wedge \pi_2$ consists of all the nonempty intersections $B \cap C$, with B a block of π_1 and C a block of π_2. Two elements $i, j \in [n]$ lie in the same block of $\pi_1 \vee \pi_2$ if and only if there is a sequence

$$i = i_0, i_1, \ldots, i_m = j,$$

where consecutive elements lie in a common block of either π_1 or π_2. In other words, two elements lie in the same block of $\pi_1 \vee \pi_2$ if one can travel from one element to the other by moving in steps, each of which stays inside either a block of π_1 or a block of π_2.

As in the lattice of left ideals of a semisimple ring, in the *partition lattice* $\mathbb{P}_n$ every element π has a complement π_c satisfying

$$\pi \wedge \pi_c = \underline{0} \qquad \text{and} \qquad \pi \vee \pi_c = \underline{1}, \tag{6.13}$$

and, as with ideals, the complement is not generally unique.

A Young tableau is a wonderfully compact device encoding a partition of $[n]$ along with a choice of the complement. It is a matrix of the form

$$\begin{array}{ccccccc}
a_{11} & \cdots & & \cdots & \cdots & a_{1\lambda_1} \\
a_{21} & \cdots & \cdots & a_{2\lambda_2} \\
\vdots & \vdots & \vdots \\
a_{m1} & \cdots & a_{m\lambda_m}.
\end{array} \tag{6.14}$$

We will take the entries a_{ij} as all distinct and drawn from $\{1, ..., n\}$. Thus, officially, a *Young tableau*, of size $n \in \{1, 2, 3, ...\}$ and *shape* $(\lambda_1, ..., \lambda_m) \in \overline{\mathbb{P}}_n$, is an injective mapping

$$T : \{(i, j) : i \in [m], j \in [\lambda_i]\} \to [n] : (i, j) \mapsto a_{ij}. \tag{6.15}$$

Note that, technically, a Young tableau is a bit more than a partition of $[n]$, as it comes with a specific ordering of the elements in each block of such a partition. The shape of a Young tableau is, however, the same as the shape of the corresponding partition.

The plural of "Young tableau" is "Young tableaux." In gratitude to Volker Börchers and Stefan Gieseke's LaTeX package youngtab, we will sometimes use the terms Youngtab and the plural Youngtabs.

It is convenient to display Youngtabs using boxes; for example,

1	2	4	5
3	6		
7			

.

Let $\mathbb{T}_n$ denote the set of all Youngtabs with n entries.

Each Youngtab specifies two partitions of $[n]$, one formed by the rows and the other by the columns:

$$\text{Rows}(T) = \{\text{rows of } T\},$$
$$\text{Cols}(T) = \{\text{columns of } T\}, \tag{6.16}$$

where, of course, each row and each column is viewed as a *set*. Here is a simple but essential observation about $\text{Rows}(T)$ and $\text{Cols}(T)$:

A block $R \in \text{Rows}(T)$ intersects a block $C \in \text{Cols}(T)$ in at most one element.

In fact, something stronger is true: if you pick any two entries in the Youngtab T, then you can travel from one to the other by moving successively horizontally along rows and vertically along columns (in the Youngtab, simply move from one entry back to the first entry in that row, then move up or down the first column till you reach the row containing the other entry, and then move horizontally along the row.) Thus,

$$\text{Rows}(T) \text{ and } \text{Cols}(T) \text{ are complements of each other in } \mathbb{P}_n. \tag{6.17}$$

A Young tableau thus provides an efficient package, keeping track of two complementary partitions of $[n]$. The complement provided by a Young tableau has special and useful features. Here is a summary of observations about complements in the lattice $\mathbb{P}_n$:

Theorem 6.1 *Let $\pi \in \mathbb{P}_n$ be a partition of $[n]$, and let*

$$\pi^{\perp} = \{\pi_1 \in \mathbb{P}_n : \pi \wedge \pi_1 = \underline{0}\}. \tag{6.18}$$

Any element of $\pi^{\perp}$ with largest shape in lexicographic order is also a maximal element of $\pi^{\perp}$ in the partial order on $\mathbb{P}_n$. Every maximal element π_c in $\pi^{\perp}$ is a complement of π, in the sense that it satisfies

$$\pi \wedge \pi_c = \underline{0},$$
$$\pi \vee \pi_c = \underline{1}. \tag{6.19}$$

If T is any Young tableau for which $\text{Rows}(T) = \pi$, then $\text{Cols}(T)$ is an element of largest shape in $\pi^{\perp}$, and similarly, with rows and columns interchanged.

A *Young complement* of π is a complement of largest shape. As explained in the preceding theorem, such a complement can be obtained from a Youngtab whose rows (or columns) form the partition π.

Proof. Consider a maximal element π_c of the set $\pi^\perp$ given in (6.18). Let i and j be any elements of $[n]$; we will show that i and j lie in the same block of $\pi \vee \pi_c$. This would mean that $\pi \vee \pi_c$ is $\underline{1}$. Let $i \in B_1$ and $j \in B_2$, where B_1 and B_2 are blocks of π_c; assume $B_1 \neq B_2$, because otherwise i and j both lie in the block of $\pi \vee \pi_c$ that contains B_1. Maximality of π_c in $\pi^\perp$ implies that two blocks of π_c cannot be coalesced while still retaining the first condition on π_c in (6.19); in particular, B_1 and B_2 each contains an element such that these two elements lie in the same block B of π. Thus, $B_1 \cup B_2 \cup B$ lies inside one block of $\pi \vee \pi_c$, and hence so do the elements i and j. This proves $\pi \vee \pi_c = \underline{1}$.

Let T be a Young tableau with $\mathrm{Rows}(T) = \pi$. We have already noted in (6.17) that $\pi_{yc} = \mathrm{Cols}(T)$ is a complementary partition to π in $\mathbb{P}_n$. Let R_1, ..., R_m be the blocks of π listed in decreasing order of size:

$$|R_1| \geq \cdots \geq |R_m|.$$

(Think of these as the rows of T from the top row to the bottom row.) Let C_1, ..., C_q be the blocks of π_{yc}, formed as follows: C_1 contains exactly one element from each R_j, and, for every $i \in \{2, \ldots, q\}$, C_i contains exactly one element from every nonempty $R_j - \cup_{k<i}C_k$.

Consider any $\pi_0 \in \pi^\perp$, and let B_1, ..., B_s be the blocks of π_0 listed in decreasing order. Our goal is to show that $\lambda(\pi_0) \leq \lambda(\pi_{yc})$. Each block of π_0 intersects each R_j in at most one element, and so the largest block B_1 contains at most m elements. Thus, $\lambda_1(\pi_0) \leq m = \lambda_1(\pi_{yc})$. If $\lambda_1(\pi_0) < \lambda_1(\pi_{yc})$, then $\pi_0 \leq \pi_{yc}$, and we are finished. Suppose then that $\lambda_1(\pi_0) = \lambda_1(\pi_{yc})$. Then B_1 intersects each R_j in exactly one element, and so $R_j - B_1$ is empty if and only if $R_j - C_1$ is empty, for any $j \in [m]$. Let i be the largest positive integer in $[s]$ for which (1) $\lambda_i(\pi_0) = \lambda_i(\pi_{yc})$ and (2) $|\{j \in [m] : R_j - \cup_{k\leq i}B_k = \emptyset\}| = |\{j \in [m] : R_j - \cup_{k\leq i}C_k = \emptyset\}|$. If $i = s$, then $\pi_0 \leq \pi_{yc}$ and again we would be finished. So suppose $i < s$. Now B_{i+1} contains at most one element from each nonempty $R_j - \cup_{k\leq i}B_k$ and C_{i+1} contains exactly one element from each nonempty $R_j - \cup_{k\leq i}C_k$. From condition 2 it follows that $|B_{i+1}| \leq |C_{i+1}|$, and the definition of i then implies that $|B_{i+1}| < |C_{i+1}|$. Thus, in all cases, $\lambda(\pi_0) \leq \lambda(\pi_{yc})$, proving that π_{yc} is an element of largest shape in $\pi^\perp$. $\boxed{\text{QED}}$

6.3 Symmetries of Partitions

The action of S_n on $[n]$ induces an action on the set $\mathbb{P}_n$ of all partitions of $[n]$: a permutation $s \in S_n$ carries the partition π to the partition $s(\pi)$, whose blocks are $s(B)$, with B running over the blocks of π. For example:

$$(13)(245) \cdot \{\{2, 5, 3\}, \{1\}, \{4, 6\}\} = \{\{4, 2, 1\}, \{3\}, \{5, 6\}\}.$$

Define the *fixing subgroup* Fix_π of a partition $\pi \in \mathbb{P}_n$ to consist of all permutations that carry each block of π into itself:

$$\text{Fix}_\pi = \{s \in S_n : s(B) = B \quad \text{for all } B \in \pi\}. \tag{6.20}$$

Theorem 6.2 *The mapping*

$$\text{Fix} : \mathbb{P}_n \to \{\text{subgroups of } S_n\} : \pi \mapsto \text{Fix}_\pi \tag{6.21}$$

is injective and order-preserving when the subgroups of S_n are ordered by inclusion. The mapping Fix *from $\mathbb{P}_n$ to its image inside the lattice of subgroups of S_n is an isomorphism:*

$$\text{Fix}_{\pi_1} \subset \text{Fix}_{\pi_2} \text{ if and only if } \pi_1 \leq \pi_2.$$

Furthermore, Fix *also preserves the lattice operations*

$$
\begin{aligned}
\text{Fix}_{\pi_1 \wedge \pi_2} &= \text{Fix}_{\pi_1} \cap \text{Fix}_{\pi_2}, \\
\text{Fix}_{\pi_1 \vee \pi_2} &= \text{the subgroup generated by } \text{Fix}_{\pi_1} \text{ and } \text{Fix}_{\pi_2}
\end{aligned}
\tag{6.22}
$$

for all $\pi_1, \pi_2 \in \mathbb{P}_n$.

There is an isomorphism of groups

$$\text{Fix}_\pi \to S_{\lambda_1(\pi)} \times \cdots \times S_{\lambda_m(\pi)}, \tag{6.23}$$

where $\left(\lambda_1(\pi), ..., \lambda_m(\pi)\right)$ is the shape of π. In particular, Fix_π is generated by the transpositions it contains.

Proof. A partition π is recovered from the fixing subgroup Fix_π as the set of orbits of Fix_π in $[n]$. Hence, $\pi \mapsto \text{Fix}_\pi$ is injective.

Suppose $\pi_1 \leq \pi_2$ in $\mathbb{P}_n$. Then any $B \in \pi_2$ is a union of blocks $B_1, ..., B_k \in \pi_1$, and so $s(B)$ is the union $s(B_1) \cup ... \cup s(B_k)$ for any $s \in S_n$; thus, $s(B) = B$ if $s \in \text{Fix}_{\pi_1}$. Hence, $\text{Fix}_{\pi_1} \subset \text{Fix}_{\pi_2}$.

Conversely, suppose $\text{Fix}_{\pi_1} \subset \text{Fix}_{\pi_2}$, and B is any block of π_2; then every $s \in \text{Fix}_{\pi_1}$ maps B into itself and so B is a union of blocks of π_1.

Let $s \in \text{Fix}_{\pi_1} \cap \text{Fix}_{\pi_2}$, and consider any block $B \in \pi_1 \wedge \pi_2$. Then $B = B_1 \cap B_2$ for some $B_1 \in \pi_1$ and $B_2 \in \pi_2$, and so $s(B) = s(B_1) \cap s(B_2) = B_1 \cap B_2 = B$. Hence, $\text{Fix}_{\pi_1} \cap \text{Fix}_{\pi_2} \subset \text{Fix}_{\pi_1 \wedge \pi_2}$. The reverse inclusion follows from the fact that Fix is order-preserving.

We turn next to (6.23). Let $B_1, ..., B_m$ be the blocks of a partition π, and let S_{B_j} be the group of permutations of the set B_j; then

$$\text{Fix}_\pi \to \prod_{j=1}^{m} S_{B_j} : s \mapsto (s|B_1, ..., s|B_m)$$

is clearly an isomorphism. Since each $S_{B_j} \simeq S_{|B_j|}$ is generated by its transpositions, so is Fix_π.

Now consider a transposition $(a\,b) \in \text{Fix}_{\pi_1 \vee \pi_2}$. If $\{a, b\}$ is in a block of π_1 or π_2, then s is in Fix_{π_1} or Fix_{π_2}. Suppose next that $a \in B_1 \in \pi_1$ and $b \in B_2 \in \pi_2$. Now two elements lie in the same block of $\pi_1 \vee \pi_2$ if and only if there is a sequence of elements starting from one and ending with the other:

$$a = i_1, i_2, ..., i_r = b,$$

with consecutive terms in the sequence always in the same block of either π_1 or of π_2. Consequently,

$$(i_k\, i_{k+1}) \in \text{Fix}_{\pi_1} \cup \text{Fix}_{\pi_2} \qquad \text{for all } k \in \{1, ..., r-1\}.$$

Let F be the subgroup of S_n generated by Fix_{π_1} and Fix_{π_2}. Observe that

$$(i_1\, i_2)(i_2\, i_3)(i_1\, i_2) = (i_1\, i_3) \in F,$$

and then

$$(i_1\, i_3)(i_3\, i_4)(i_1\, i_3) = (i_1\, i_4) \in F,$$

and thus, inductively,

$$(a\,b) = (i_1\, i_r) \in F.$$

Hence, every transposition in $\text{Fix}_{\pi_1 \vee \pi_2}$ is in F. Since $\text{Fix}_{\pi_1 \vee \pi_2}$ is generated by its transpositions, it follows that $\text{Fix}_{\pi_1 \vee \pi_2}$ is a subset of F. The reverse inclusion holds simply because Fix_{π_1} and Fix_{π_2} are both subsets of $\text{Fix}_{\pi_1 \vee \pi_2}$. This completes the proof of the second part of (6.22). $\boxed{\text{QED}}$

Recall from the proof of Theorem 6.1 how we can construct, for a partition $\pi \in \mathbb{P}_n$, a partition π_{yc} of largest shape satisfying $\pi \wedge \pi_{yc} = \underline{0}$. If π'_{yc} is another such partition, then the largest block C_1 of π_{yc} and the largest block C'_1 of π'_{yc} both contain exactly one element from each block of π; hence, there is a permutation $s_1 \in \text{Fix}_\pi$, which is a product of one transposition each for each block of π, that maps C'_1 to C_1. Next, removing C_1 and C'_1 from the picture, and arguing similarly for the next largest block C_2 of π_{yc} and the next largest block C'_2 of π'_{yc}, we have a permutation, again a product of transpositions preserving every block of π, that carries C'_2 to C_2. Proceeding in this way, we produce a permutation $s \in S_n$ which fixes each block of π and carries π'_{yc} to π_{yc}, with C_j going over to $s(C_j) = C'_j$. In summary, we have the following theorem:

Theorem 6.3 *Let $\pi \in \mathbb{P}_n$ and suppose $\pi_{yc}, \pi'_{yc} \in \mathbb{P}_n$ are Young complements of π:*

$$\pi \wedge \pi_{yc} = \underline{0} = \pi \wedge \pi'_{yc},$$
$$\lambda(\pi_{yc}) = \lambda(\pi'_{yc}) = \max\{\lambda(\pi_1) \ : \ \pi \wedge \pi_1 = \underline{0}\}. \qquad (6.24)$$

Let $C_1, ..., C_m$ be the distinct blocks of π_{yc}, ordered so that $|C_1| \geq \cdots \geq |C_m|$, and let $C'_1, ..., C'_m$ be the distinct blocks of π'_{yc}, also listed in decreasing order of size. Then there exists an $s \in \text{Fix}_\pi$ such that

$$s(C_j) = C'_j \quad \text{for all } j \in [m].$$

Conversely, if $s \in \text{Fix}_\pi$, then $s(\pi_{yc})$ is a Young complement of π.

Here is a useful consequence:

Theorem 6.4 *Suppose π_{yc} is a Young complement of $\pi \in \mathbb{P}_n$. Then, for any $s \in S_n$,*

$$\text{Fix}_\pi \cap s\text{Fix}_{\pi_{yc}}s^{-1} = \{\iota\} \quad \text{if } s \in \text{Fix}_\pi\text{Fix}_{\pi_{yc}},$$
$$\text{Fix}_\pi \cap s\text{Fix}_{\pi_{yc}}s^{-1} \neq \{\iota\} \quad \text{if } s \notin \text{Fix}_\pi\text{Fix}_{\pi_{yc}}, \qquad (6.25)$$

where ι is the identity permutation. The group $\text{Fix}_\pi \cap s\text{Fix}_{\pi_{yc}}s^{-1}$, as with all fixing subgroups, is generated by the transpositions it contains.

Thus, if T is any Young tableau with n entries, and $s \in S_n$, then

$$C_T \cap sR_Ts^{-1} = \{\iota\} \quad \text{if and only if } s \in C_TR_T, \qquad (6.26)$$

where R_T is the fixing subgroup for Rows(T) and C_T is the fixing subgroup for Cols(T). The group $C_T \cap sR_Ts^{-1}$, if nontrivial, contains a transposition.

Proof. Let $C_1, ..., C_q$ be the blocks of π_{yc} in decreasing order of size; then $s(C_1), ..., s(C_q)$ are the blocks of $s(\pi_{yc})$, also in decreasing order of size. From

$$\text{Fix}_{\pi \wedge s(\pi_{yc})} = \text{Fix}_{\pi} \cap s\text{Fix}_{\pi_{yc}}s^{-1}$$

we see that this subgroup is trivial if and only if $\pi \wedge s(\pi_{yc})$ is $\underline{0}$. Thus, this condition means $s(\pi_{yc})$, which has the same shape as π_{yc}, is also a Young complement of π. By Theorem 6.3 this holds if and only if there is an element $s_1 \in \text{Fix}_{\pi}$ such that $s_1 s(C_j) = C_j$ for each $j \in [q]$. The latter means $s_1 s$ is in the fixing subgroup of π_{yc}, and so the condition $\text{Fix}_{\pi} \cap s\text{Fix}_{\pi_{yc}}s^{-1} = \{\iota\}$ is equivalent to $s = s_1^{-1}s_2$ for some $s_1 \in \text{Fix}_{\pi}$ and $s_2 \in \text{Fix}_{\pi_{yc}}$. This establishes (6.25). The result (6.26) follows by specializing to $\pi = \text{Rows}(T)$ and $\pi_{yc} = \text{Cols}(T)$. $\boxed{\text{QED}}$

6.4   Conjugacy Classes to Young Tableaux

Any element in S_n can be expressed as a product of a unique set of disjoint cycles:

$$(a_{11}, \ldots, a_{1\lambda_1}) \ldots (a_{m1}, \ldots, a_{m\lambda_m}),$$

where the a_{ij} are distinct and run over $\{1, \ldots, n\}$. This permutation thus specifies a *partition*

$$(\lambda_1, \ldots, \lambda_m)$$

of n into positive integers $\lambda_1, \ldots, \lambda_m$:

$$\lambda_1 + \cdots + \lambda_m = n.$$

To make things definite, we require that

$$\lambda_1 \geq \lambda_2 \geq \cdots \geq \lambda_m.$$

The set of all such shapes $(\lambda_1, \ldots, \lambda_m)$ is naturally identifiable as the quotient

$$\overline{\mathbb{P}}_n \simeq \mathbb{P}_n/S_n. \tag{6.27}$$

This delineates the distinction between partitions of n and partitions of $[n]$.

Two permutations are conjugate if and only if they have the same cycle structure. Thus, the conjugacy classes of S_n correspond one-to-one to partitions of n.

The group S_n acts on the set of Youngtabs corresponding to each partition of n; if a Young tableau is viewed as a mapping T as in (6.15), the action is defined by composition with permutations:

$$S_n \times \mathbb{T}_n \to \mathbb{T}_n : (\sigma, T) \mapsto \sigma \circ T.$$

For example,

$$(134)(25)(67) \cdot \begin{array}{|c|c|c|c|} \hline 1 & 2 & 4 & 5 \\ \hline 3 & 6 \\ \cline{1-2} 7 \\ \cline{1-1} \end{array} \quad = \quad \begin{array}{|c|c|c|c|} \hline 3 & 5 & 1 & 2 \\ \hline 4 & 7 \\ \cline{1-2} 6 \\ \cline{1-1} \end{array}$$

For a tableau T, Young [81] introduced two subgroups of S_n:

$$R_T = \{\text{all } p \in S_n \text{ that preserve each row of } T\},$$
$$C_T = \{\text{all } q \in S_n \text{ that preserve each column of } T\}. \tag{6.28}$$

If we think in terms of the natural action of S_n on the set $\mathbb{P}_n$ of partitions of $[n]$, R_T is the fixing subgroup of the element $\mathrm{Rows}(T) \in \mathbb{P}_n$ and C_T is the fixing subgroup of $\mathrm{Cols}(T) \in \mathbb{P}_n$.

6.5   Young Tableaux to Young Symmetrizers

The *Young symmetrizer* for a Youngtab T is the element

$$y_T \overset{\text{def}}{=} c_T r_T = \sum_{q \in C_T, p \in R_T} (-1)^q qp \in \mathbb{Z}[S_n], \tag{6.29}$$

where

$$c_T = \sum_{q \in C_T} (-1)^q q,$$
$$r_T = \sum_{p \in R_T} p. \tag{6.30}$$

We have used, and will use, the notation

$$(-1)^q = \epsilon(q).$$

Observe that R_T acts with the trivial representation on the one-dimensional space $\mathbb{Q}r_T$, and C_T acts through the representation $\epsilon|C_T$ on the one-dimensional space $\mathbb{Q}c_T$. Indeed, c_T and r_T are, up to scalar multiples, idempotents in $\mathbb{Q}[S_n]$. Frobenius constructed y_T from c_T and r_T and showed that a certain scalar multiple of y_T is an indecomposable idempotent in $\mathbb{Q}[S_n]$.

Here is a formal statement of some of the basic observations about r_T, c_T, and y_T:

Proposition 6.2 *Let T be any Young tableau T with n entries. Then*

$$qy_T = (-1)^q y_T \quad \text{if } q \in C_T,$$
$$y_T p = y_T \quad \text{if } p \in R_T.$$
$$(6.31)$$

The row group R_T and column group C_T have a trivial intersection:

$$R_T \cap C_T = \{\iota\}, \tag{6.32}$$

where, as usual, ι denotes the identity permutation. Consequently, each element in the set

$$C_T R_T = \{qp : q \in C_T, p \in R_T\}$$

can be expressed in the form qp for a unique pair $(q, p) \in C_T \times R_T$. For any $s \in S_n$, the row and column symmetry groups behave as

$$R_{sT} = sR_T s^{-1} \quad \text{and} \quad C_{sT} = sC_T s^{-1}, \tag{6.33}$$

and the Young symmetrizer transforms to a conjugate:

$$y_{sT} = sy_T s^{-1}. \tag{6.34}$$

We leave the proof as Exercise 6.1.

6.6   Youngtabs to Irreducible Representations

We denote by $\iota \in S_n$ the identity permutation. Let R be any ring; then there is the "trace functional"

$$\text{Tr}_0 : R[S_n] \to R : x = \sum_{s \in S_n} x_s s \mapsto x_\iota.$$

Theorem 6.5 *Let T be a Young tableau for $n \in \{2, 3, ...\}$. Then, for the Young symmetrizer $y_T \in \mathbb{Z}[S_n]$, the trace $\mathrm{Tr}_0(y_T^2)$ is a positive integer γ_T, dividing $n!$. The element $e_T = \frac{1}{\gamma_T} y_T$ is an indecomposable idempotent in $\mathbb{Q}[S_n]$. The corresponding irreducible representation space $\mathbb{Q}[S_n]y_T$ has dimension d_T given by*

$$d_T = \frac{n!}{\gamma_T}. \tag{6.35}$$

There are elements $v_1, ..., v_{d_T} \in \mathbb{Z}[S_n]y_T$ that form a $\mathbb{Q}$-basis of $\mathbb{Q}[S_n]y_T$.

Proof. The indecomposability criterion in Proposition 4.7 will be our key tool.

To simplify the notation in the proof, we drop all subscripts indicating the fixed tableau T; thus, we write y instead of y_T.

Fix $t \in S_n$, and let

$$z = yty. \tag{6.36}$$

Our first objective is to prove that z is an integer multiple of y.

Observe that

$$qzp = (-1)^q z \quad \text{for all } p \in R_T \text{ and } q \in C_T, \tag{6.37}$$

because $qy = (-1)^q y$ and $yp = y$. Writing z as

$$z = \sum_{s \in S_n} z_s s,$$

where each z_s is an integer, we see that, for $q \in C_T$ and $p \in R_T$,

$$z_{qp} = \text{coeff. of } \iota \text{ in } q^{-1}zp^{-1} = (-1)^q z_\iota$$

Using this, we can express z as

$$z = z_\iota y + \sum_{s \notin C_T R_T} z_s s. \tag{6.38}$$

Next we show that the second term on the right is 0. For this we recall from Theorem 6.4 that if $s \notin C_T R_T$, then $C_T \cap sR_T s^{-1}$ is nontrivial, and hence contains some transposition τ; thus, we have the following:

If $s \notin C_T R_T$, then there are transpositions $\sigma \in R_T$ and $\tau \in C_T$ such that

$$\tau^{-1} s \sigma^{-1} = s. \tag{6.39}$$

Consequently,

$$(\tau z \sigma)_s = z_s.$$

But since $\tau \in C_T$ and $\sigma \in R_T$, we have

$$\tau z \sigma = (-1)^\tau z = -z,$$

from which, specializing to the coefficient of s, we have

$$(\tau z \sigma)_s = -z_s.$$

Hence,

$$z_s = 0 \quad \text{if } s \notin C_T R_T.$$

Looking back at (6.38), we conclude that

$$z = z_\iota y. \tag{6.40}$$

Recalling the definition of z in (6.36), we see that yty is an integer multiple of y for every $t \in S_n$. Consequently,

$$yxy \text{ is a } \mathbb{Q}\text{-multiple of } y \text{ for every } x \in \mathbb{Q}[S_n]. \tag{6.41}$$

Specializing to the case $t = \iota$, we have

$$yy = \gamma y, \tag{6.42}$$

where

$$\gamma = (y^2)_\iota. \tag{6.43}$$

In particular, the multiplier γ is an integer. We will show shortly that γ is a positive integer dividing $n!$. Then

$$e \stackrel{\text{def}}{=} \gamma^{-1} y \tag{6.44}$$

is well defined and is clearly an idempotent in $\mathbb{Q}[S_n]$. By (6.41), exe is a $\mathbb{Q}$-multiple of e for all $x \in \mathbb{Q}[S_n]$. Hence, by the indecomposablity criterion in Proposition 4.7, e is an indecomposable idempotent.

It remains to prove that γ is a positive integer dividing $n!$. The $\mathbb{Q}$-linear map

$$T_y : \mathbb{Q}[S_n] \to \mathbb{Q}[S_n] : a \mapsto ay \tag{6.45}$$

acts on the subspace $\mathbb{Q}[S_n]y$ by multiplication by the constant γ. Moreover, T_y maps any complementary subspace to $\mathbb{Q}[S_n]y$ (indeed, its entire domain) into $\mathbb{Q}[S_n]y$. Consequently,

$$\text{Tr}\,(T_y) = \gamma \dim_{\mathbb{Q}}(\mathbb{Q}[S_n]y). \tag{6.46}$$

On the other hand, in terms of the standard basis of $\mathbb{Q}[S_n]$ given by the elements of S_n, the trace of T_y is

$$\text{Tr}\,(T_y) = n! y_\iota = n!, \tag{6.47}$$

since from the definition of y it is clear that

$$y_\iota = 1.$$

Thus,

$$\gamma \dim_{\mathbb{Q}}(\mathbb{Q}[S_n]y) = n!. \tag{6.48}$$

Hence, γ is a positive integer dividing $n!$.

To finish, note that the elements ty, with t running over S_n, span $\mathbb{Z}[S_n]y$. Consequently, a subset of them form a $\mathbb{Q}$-basis of the vector space $\mathbb{Q}[S_n]y$. $\boxed{\text{QED}}$

We can upgrade to a general field. If $\mathbb{F}$ is any field, there is the natural ring homomorphism

$$\mathbb{Z} \to \mathbb{F} : m \mapsto m_{\mathbb{F}} \stackrel{\text{def}}{=} m1_{\mathbb{F}},$$

which is injective if $\mathbb{F}$ has characteristic 0, and which induces an injection of $\mathbb{Z}_p = \mathbb{Z}/p\mathbb{Z}$ onto the image $\mathbb{Z}_{\mathbb{F}}$ of $\mathbb{Z}$ in $\mathbb{F}$ if the characteristic of $\mathbb{F}$ is $p \neq 0$. To avoid too much notational distraction, we often sacrifice precision and denote $m1_{\mathbb{F}}$ as simply m instead of $m_{\mathbb{F}}$, bearing in mind that this might be the 0 element in $\mathbb{F}$. This induces a homomorphism of the corresponding group rings:

$$\mathbb{Z}[S_n] \to \mathbb{F}[S_n] : a \mapsto a_{\mathbb{F}}$$

for every $n \in \{1, 2, ...\}$. Again, we often simply write a instead of $a_{\mathbb{F}}$. For instance, the image of the Young symmetrizer $y_T \in \mathbb{Z}[S_n]$ in $\mathbb{F}[S_n]$ is denoted simply by y_T in the statement of the following result.

Theorem 6.6 *Let $n \in \{2, 3, \ldots\}$ and $\mathbb{F}$ be a field in which $n! \neq 0$. Let T be a Young tableau for n. Then $\gamma_T = \text{Tr}_0(y_T^2)$ is not zero in $\mathbb{F}$, and the element $e_T = \frac{1}{\gamma_T}y_T$, viewed as an element in $\mathbb{F}[S_n]$, is an indecomposable idempotent.*

The corresponding representation space $\mathbb{F}[S_n]y_T$ has dimension $d_{\mathbb{F},T}$ which satisfies

$$d_{\mathbb{F},T}1_\mathbb{F} = \frac{n!}{\gamma_T}1_\mathbb{F}. \tag{6.49}$$

If $\mathbb{F}$ has characteristic 0, then

$$d_{\mathbb{F},T} = d_{=}T\frac{n!}{\gamma_T} \tag{6.50}$$

does not depend on the field $\mathbb{F}$.

Proof. The argument is essentially a rerun of the proof of Theorem 6.5, mostly making sure we do not divide by 0 anywhere. In place of (6.41) we now have

$$y_T x y_T \text{ is an } \mathbb{F}\text{-multiple of } y_T \text{ for every } x \in \mathbb{F}[S_n]. \tag{6.51}$$

This again implies that $e_T = \gamma_T^{-1}y_T$ is an indecomposable idempotent, provided we make sure $\gamma_T = \mathrm{Tr}_0(y_T^2)$ is not 0 in $\mathbb{F}$. But γ_T is a divisor of $n!$, and hence is indeed not 0 in $\mathbb{F}$. Lastly, writing y for y_T and arguing as in (6.47), we work out the trace of

$$T_y : \mathbb{F}[S_n] \to \mathbb{F}[S_n] : a \mapsto ay \tag{6.52}$$

to be

$$\mathrm{Tr}\,(T_y) = n!y_\iota = n!, \tag{6.53}$$

by one count, and to be equal to $\gamma_T \dim_\mathbb{F}\big(\mathbb{F}[S_n]y\big)$ by another count; this shows that $\dim_\mathbb{F}\big(\mathbb{F}[S_n]y\big)$ equals $n!/\gamma_T$, both viewed as elements of $\mathbb{F}$. $\boxed{\text{QED}}$

6.7 Youngtab Apps

There is a whole jujitsu of Young tableau combinatorics which yield a powerful show of results. Here we go through just a few of these moves, extracting three "apps" that are often used. The standard, intricate and efficient, pathway to the results is from Weyl [79], who appears to credit von Neumann with this approach. We include alternative insights by way of proofs based on the viewpoint of partitions.

Proposition 6.3 *For Youngtabs T and T', each with n entries, if $\lambda(T') > \lambda(T)$ in lexicographic order, then:*

1. *There are two entries that both lie in one row of T' and in one column of T as well.*

2. *There exists a transposition σ lying in $R_{T'} \cap C_T$.*

In the language of partitions, if $\lambda(T') > \lambda(T)$, then $\mathrm{Cols}(T) \wedge \mathrm{Rows}(T') \neq \underline{0}$, and the nontrivial group

$$R_{T'} \cap C_T = \mathrm{Fix}_{\mathrm{Cols}(T) \wedge \mathrm{Rows}(T')} \tag{6.54}$$

is generated by the transpositions it contains.

Proof. Recall from Theorem 6.1 that the Young complement $\mathrm{Rows}(T)$ of $\mathrm{Cols}(T)$ is the partition of largest shape among all $\pi_1 \in \mathbb{P}_n$ for which $\mathrm{Cols}(T) \wedge \pi_1 = \underline{0}$. Now $\lambda(T') > \lambda(T)$ means that the shape of $\mathrm{Rows}(T')$ is larger than the shape of $\mathrm{Rows}(T)$, and so

$$\mathrm{Cols}(T) \wedge \mathrm{Rows}(T') \neq \underline{0}.$$

This just means that there is a column of T which intersects some row of T' in more than one element. Let i and j be two such elements. Then the transposition $(i\,j)$ lies in both $R_{T'}$ and C_T. Theorem 6.2 implies that the fixing subgroup (6.54) is generated by transpositions. $\boxed{\text{QED}}$

The more traditional argument follows:

Traditional Proof. Write λ' for $\lambda(T')$, and λ for $\lambda(T)$. Suppose λ' wins over λ completely in row 1: $\lambda_1' > \lambda_1$. Now $\lambda_1(S)$ is not just the number of entries in row 1 of a Young tableau S, it is also the number of columns of S. Therefore, there must exist two entries in the first row of T' that lie in the same column of T. Next suppose $\lambda_1' = \lambda_1$ and the elements of the first row of T' are distributed over different columns of T. Then we move all these elements "vertically" in T to the first row, obtaining a tableau T_1 whose first row is a permutation of the first row of T'. Having used only vertical moves, we have $T_1 = q_1 T$ for some $q_1 \in C_T$. We can replay the game now, focusing on row 2 downwards. Compare row 2 of T' with that of T_1. Again, if the rows are of equal length, then there is a vertical move in T_1 (which is therefore also a vertical move in T, because $C_{q_1 T} = C_T$) which produces a tableau $T_2 = q_2 q_1 T$, with $q_2 \in C_\mathrm{T}$, whose first row is the same as that of T_1, and whose second row is a permutation of the second row of T'. Proceeding this way, we reach the first j for which the jth row of T' has more elements than the jth row of T. Then each of the first $j - 1$ rows of T' is a permutation of the corresponding row of T_{j-1}; focusing on the Youngtabs made up of the remaining rows, recycling the argument we used for row 1, we see that there are two elements in the jth row of T' that lie in a single column in T_{j-1}. Since the columns of T_{j-1} are, as sets, identical to those of T, we have finished proving statement 1. Now, for statement 2, suppose a and b are distinct entries lying in one row of T' and in one column of T; then the transposition $(a\,b)$ lies in $R_{\mathrm{T}'} \cap C_\mathrm{T}$. $\boxed{\text{QED}}$

The next result says what happens with Youngtabs for a common partition.

Proposition 6.4 *Let T and T' be Young tableaux associated with a common partition λ. Let s be the element of S_n for which $T' = sT$. Then:*

1. *$s \notin C_\mathrm{T} R_\mathrm{T}$ if and only if there are two elements that are in one row of T' and also in one column of T.*

2. *$s \notin C_\mathrm{T} R_\mathrm{T}$ if and only if there is a transposition $\sigma \in R_\mathrm{T}$ and a transposition $\tau \in C_\mathrm{T}$, for which*
$$\tau s \sigma = s. \tag{6.55}$$

Conclusion 1, stated in terms of the row and column partitions, says that $\mathrm{Rows}(sT)$ and $\mathrm{Cols}(T)$ are Young complements of each other if and only if $s \in C_\mathrm{T} R_\mathrm{T}$.

Proof. The condition that there does not exist two elements that are in one row of $T' = sT$ and also in one column of T means that
$$\mathrm{Rows}(T') \wedge \mathrm{Cols}(T) = \underline{0},$$

which, since T' and T have the same shape, means that $\mathrm{Rows}(T')$ is a Young complement of $\mathrm{Cols}(T)$. From Theorem 6.3, $\mathrm{Rows}(T')$ is a Young complement for $\mathrm{Cols}(T)$ if and only if $s_1 \mathrm{Rows}(T') = \mathrm{Rows}(T)$ for some $s_1 \in \mathrm{Fix}_{\mathrm{Cols}(T)}$. Since $\mathrm{Rows}(T') = s\mathrm{Rows}(T)$, the condition is thus equivalent to the following:

There exists $s_1 \in \mathrm{Fix}_{\mathrm{Cols}(T)}$ for which $s_1 s \in \mathrm{Fix}_{\mathrm{Rows}(T)}$.

Thus, the condition that $\text{Cols}(T')$ is a Young complement to $\text{Rows}(T)$ is equivalent to $s \in \text{Fix}_{\text{Cols}(T)}\text{Fix}_{\text{Rows}(T)} = C_T R_T$.

For condition 2, recall that

$$
\begin{aligned}
\text{Fix}_{\text{Cols}(T) \wedge \text{Rows}(sT)} &= \text{Fix}_{\text{Cols}(T)} \cap \text{Fix}_{\text{Rows}(sT)} \\
&= \text{Fix}_{\text{Cols}(T)} \cap s\text{Fix}_{\text{Rows}(T)}s^{-1} \\
&= C_T \cap sR_Ts^{-1}
\end{aligned}
\tag{6.56}
$$

and the fixing subgroups are generated by the transpositions they contain. Therefore, $\text{Cols}(T)$ and $\text{Rows}(sT)$ are *not* Young complements if and only if there exists a transposition $\tau \in C_T$ such that $\sigma = s^{-1}\tau s$ is in R_T; being conjugate to a transposition, σ is also a transposition. $\boxed{\text{QED}}$

Here is a proof which bypasses the structure we have built for partitions:

Traditional Proof. Suppose $s = qp$, with $q \in C_T$ and $p \in R_T$. Consider two elements $s(i)$ and $s(j)$, with $i \neq j$, lying in the same row of T':

$$
T'_{ab} = s(i), \quad T'_{ac} = s(j).
$$

Thus, i and j lie in the same row of T:

$$
T_{ab} = i, \quad T_{ac} = j.
$$

The images $p(i)$ and $p(j)$ are also from the same row of T (hence different columns) and then $qp(i)$ and $qp(j)$ would be in different columns of T. Thus, the entries $s(i)$ and $s(j)$, lying in the same row of T', lie in different columns of T.

Conversely, suppose that if two elements lie in the same row of T', then they lie in different columns of T. We will show that the permutation $s \in S_n$ for which $T' = sT$ has to be in $C_T R_T$. Bear in mind that the sequence of row lengths for T' is the same as for T. The elements of row 1 of T' are distributed over distinct columns of T. Therefore, by moving these elements vertically, we can bring them all to the first row. This means that there is an element $q_1 \in C_T$ such that $T_1 = q_1 T$ and T' have the same *set* of elements for their first rows. Next, the elements of the second row of T' are distributed over distinct columns in T, and hence also in $T_1 = q_1 T$. Hence, there is a vertical move

$$
q_2 \in C_{q_1 T} = C_T,
$$

for which $T_2 = q_2 T_1$ and T' have the same set of first row elements and also the same set of second row elements.

Proceeding in this way, we obtain a $q \in C_T$ such that each row of T' is equal, as a *set*, to the corresponding row of qT:

$$\{T'_{ab} : 1 \leq b \leq \lambda_a\} = \{q(T_{ab}) : 1 \leq b \leq \lambda_a\} \qquad \text{for each } a.$$

But then we can permute horizontally: for each fixed a, permute the numbers T_{ab} so that the $q(T_{ab})$ match the T'_{ab}. Thus, there is a $p \in R_T$ such that

$$T' = qp(T).$$

Thus,

$$s = qp \in C_T R_T.$$

We turn to proving condition 2. Suppose $s \notin C_T R_T$. Then, by condition 1, there is a row a, and two entries $i = T_{ab}$ and $j = T_{ac}$, whose images $s(i)$ and $s(j)$ lie in a common column of T. Let $\sigma = (i\,j)$ and $\tau = \big(s(i)\,s(j)\big)$. Then $\sigma \in R_T$, $\tau \in C_T$, and

$$\tau s \sigma = s,$$

which is readily checked on i and j.

Conversely, suppose $\tau s \sigma = s$, where $\sigma = (i\,j) \in R_T$. Then i and j are in the same row of T, and so $s(i)$ and $s(j)$ are in the same row of T'. Now $s(i) = \tau(s(j))$ and $s(j) = \tau(s(i))$. Since $\tau \in C_T$ it follows that $s(i)$ and $s(j)$ are in the same column of T. $\boxed{\text{QED}}$

A Young tableau is *standard* if the entries in each row are in increasing order, from left to right, and the numbers in each column are also in increasing order, from top to bottom. For example,

1	2	7
3	4	
5	6	

Such a tableau must, of necessity, start with 1 in the top-left box, and each new row begins with the smallest number not already listed in any of the preceding rows. Numbers lying directly "south," directly "east," and southeast of a given entry are larger than this entry, and those to the north, west, and northwest are smaller.

In general, the boxes of a tableau are ordered in "book order": read the boxes from left to right along a row and then move down to the next row.

The Youngtabs, for a given partition, can be linearly ordered: if T and T' are standard, we declare that

$$T' > T$$

if the first entry T_{ab} of T that is different from the corresponding entry T'_{ab} of T' satisfies $T_{ab} < T'_{ab}$. The tableaux for a given partition can then be written in increasing/decreasing order. This is how they look for some partitions of 3:

$$\boxed{3\,2\,1} \; > \; \boxed{3\,1\,2} \; > \; \boxed{2\,3\,1} \; > \; \boxed{2\,1\,3} \; > \; \boxed{1\,3\,2} \; > \; \boxed{1\,2\,3}$$

For the partition $(2, 1)$ the Youngtabs descend as follows:

With this ordering we have the following result which states a condition for Young complementarity in terms of Youngtabs, not the partitions:

Proposition 6.5 *If T and T' are standard Young tableaux with a common partition, and $T' > T$, then there are two entries in some row of T that lie in one column of T'. Consequently, there exists a transposition σ lying in $R_T \cap C_{T'}$.*

Proof. Let $x = T_{ab}$ be the first entry of T that is less than the corresponding entry $y = T'_{ab}$. The entry x appears somewhere in the tableau T'. Because ab is the *first* location where T differs from T', and $T_{ab} = x$, we see that x cannot appear prior to the location T'_{ab}. But x being $< y = T'_{ab}$, it can also not appear directly south, east, or southeast of T'_{ab}. Thus, x must appear in T' in a row below the ath row and in a column $c < b$. Thus, the numbers T_{ac} (which equals T'_{ac}) and $T_{ab} = x$, appearing in the ath row of T, appear in the cth column of T'. $\boxed{\text{QED}}$

6.8 Orthogonality

We have seen that Youngtabs correspond to irreducible representations of S_n via indecomposable idempotents. Which Youngtabs correspond to inequivalent representations? Here is the first step to answering this question:

Theorem 6.7 *Suppose T and T' are Young tableaux with n entries, where $n \in \{2, 3, \ldots\}$; then*

$$y_{T'}y_T = 0 \qquad \text{if } \lambda(T') > \lambda(T) \text{ in lexicographic order.} \qquad (6.57)$$

Proof. Suppose $\lambda(T') > \lambda(T)$. Then by Proposition 6.3, there is a transposition $\sigma \in R_{T'} \cap C_T$. Then

$$y_{T'}y_T = y_{T'}\sigma\sigma y_T = (y_{T'})(-y_T) = -y_{T'}y_T.$$

Thus, $y_{T'}y_T$ is 0. $\boxed{\text{QED}}$

Here is the corresponding result for *standard* Youngtabs with common shape:

Theorem 6.8 *If T and T' are standard Young tableaux associated with a common partition of $n \in \{2, 3, \ldots\}$, then*

$$y_T y_{T'} = 0 \qquad \text{if } T' > T. \qquad (6.58)$$

Proof. By Proposition 6.5, there is a transposition $\sigma \in R_T \cap C_{T'}$. Then

$$y_T y_{T'} = y_T \sigma \sigma y_{T'} = (y_T)(-y_{T'}) = -y_T y_{T'},$$

and so $y_T y_{T'}$ is 0. $\boxed{\text{QED}}$

6.9 Deconstructing $\mathbb{F}[S_n]$

As a first consequence of orthogonality of the Young symmetrizers, we are able to distinguish between inequivalent irreducible representations of S_n:

Theorem 6.9 *Let T and T' be Young tableaux with n entries. Let $\mathbb{F}$ be a field in which $n! \neq 0$. Then the left ideals $\mathbb{F}[S_n]y_T$ and $\mathbb{F}[S_n]y_{T'}$ in $\mathbb{F}[S_n]$ are isomorphic as $\mathbb{F}[S_n]$-modules if and only if T and T' have the same shape.*

Proof. Suppose first that $\lambda(T) \neq \lambda(T')$. In Proposition 4.11 we showed that, for any finite group G and field $\mathbb{F}$ in which $|G|1_{\mathbb{F}} \neq 0$, idempotents y_1 and y_2 in $\mathbb{F}[G]$ generate nonisomorphic left ideals if $y_1\mathbb{F}[G]y_2 = 0$. Thus, it will suffice to verify that $y_{T'}sy_T$ is 0 for all $s \in S_n$. This is equivalent to checking that $y_{T'}sy_T s^{-1}$ is 0, which, by (6.34), is equivalent to $y_{T'}y_{sT}$ being 0. Since

T' and T have different shapes, we can assume that $\lambda(T') > \lambda(T)$. Then also $\lambda(T') > \lambda(sT)$, because sT and T have, of course, the same shape. Then the orthogonality result (6.57) implies that $y_{T'}y_{sT}$ is indeed 0.

Now suppose T and T' have the same shape. Then there is an $s \in S_n$ such that $T' = sT$. Recall that $y_{sT} = sy_T s^{-1}$. So there is the mapping

$$f : \mathbb{F}[S_n]y_T \to \mathbb{F}[S_n]y_{T'} : v \mapsto vs^{-1}.$$

This is clearly $\mathbb{F}[S_n]$-linear as well as a bijection, and hence is an isomorphism of $\mathbb{F}[S_n]$-modules. $\boxed{\text{QED}}$

Next, working with *standard* Youngtabs, we have the following consequence of orthogonality:

Theorem 6.10 *If $T_1, \ldots, T_m$ are all the standard Young tableaux associated with a common partition of n, then the sum $\sum_{j=1}^m \mathbb{F}[S_n]y_{T_j}$ is a direct sum if the characteristic of $\mathbb{F}$ does not divide $n!$.*

Proof. Order the T_j so that $T_1 < T_2 < \cdots < T_m$. Suppose $\sum_{j=1}^m \mathbb{F}[S_n]y_{T_j}$ is not a direct sum. Let r be the smallest element of $\{1, \ldots, n\}$ for which there exist $x_j \in \mathbb{F}[S_n]y_{T_j}$, for $j \in \{1, \ldots, r\}$, with $x_r \neq 0$, such that

$$\sum_{j=1}^r x_j = 0.$$

Multiplying on the right by y_{T_r} produces

$$\gamma_{T_r} x_r = 0,$$

because $y_{T_r}^2 = \gamma_{T_r} y_{T_r}$, and $y_{T_s} y_{T_r} = 0$ for $s < r$. Now γ_{T_r} is a divisor of $n!$, and so γ_{T_r} is not 0 in $\mathbb{F}$, and so

$$x_r = 0.$$

This contradiction proves that $\sum_{j=1}^m \mathbb{F}[S_n]y_{T_j}$ is a direct sum. $\boxed{\text{QED}}$

Finally, with all the experience and technology we have developed, we can take $\mathbb{F}[S_n]$ apart:

Theorem 6.11 *Let* $n \in \{2, 3, ...\}$, *and let* $\mathbb{F}$ *be a field in which* $n! 1_{\mathbb{F}} \neq 0$. *Denote by* $\mathbb{T}_n$ *the set of all Young tableaux with n entries, and demote by* $\overline{\mathbb{P}}_n$ *the set of all shapes of all partitions of n. Then for any* $p \in \overline{\mathbb{P}}_n$, *the sum*

$$A(p) = \sum_{T \in \mathbb{T}_n, \lambda(T) = p} \mathbb{F}[S_n] y_T \tag{6.59}$$

is a two-sided ideal in $\mathbb{F}[S_n]$ *that contains no other nonzero two-sided ideal. The mapping*

$$I : \prod_{p \in \overline{\mathbb{P}}_n} A(p) \to \mathbb{F}[S_n] : (a_p)_{p \in \overline{\mathbb{P}}_n} \mapsto \sum_{p \in \overline{\mathbb{P}}_n} a_p \tag{6.60}$$

is an isomorphism of rings.

Look back at the remark made immediately after the statement of Theorem 5.3. From this remark and (6.59) it follows that there is a subset Sh_p of $T \in \mathbb{T}_n$, all with fixed shape p, for which the simple modules $\mathbb{F}[S_n] y_T$ form a *direct sum* decomposition of $A(p)$:

$$A(p) = \bigoplus_{T \in \mathrm{Sh}_p} \mathbb{F}[S_n] y_T. \tag{6.61}$$

Proof. It is clear that $A(p)$ is a left ideal. To see that it is a right ideal we simply observe that if $\lambda(T) = p$, then for any $s \in S_n$

$$\mathbb{F}[S_n] y_T s = \mathbb{F}[S_n] s s^{-1} y_T s = \mathbb{F}[S_n] y_{s^{-1}T} \subset A(p),$$

where the last inclusion holds because $\lambda(s^{-1}T) = \lambda(T) = p$.

Now suppose p and p' are different partitions of n. Then for any tableaux T and T' with $\lambda(T) = p$ and $\lambda(T') = p'$, Theorem 6.9 says that $\mathbb{F}[S_n] y_T$ is not isomorphic to $\mathbb{F}[S_n] y_{T'}$, and so

$$\mathbb{F}[S_n] y_T \mathbb{F}[S_n] y_{T'} = 0,$$

because these two simple left ideals are not isomorphic (see Theorem 5.8, if you must). Consequently

$$A(p) A(p') = 0.$$

From this it follows that the mapping (6.60) preserves addition and multiplication.

For injectivity of I, let u_p be an idempotent generator of A_p for each $p \in \overline{\mathbb{P}}_n$. If

$$\sum_{p \in \overline{\mathbb{P}}_n} a_p = 0,$$

then multiplying on the right by u_p zeroes out all terms except the pth, which remains unchanged at a_p and hence is 0. Thus, I is injective.

On to surjectivity. It is time to recall (4.32); in the present context, it says that the number of nonisomorphic simple $\mathbb{F}[S_n]$-modules is at most the number of conjugacy classes in S_n, which is the same as $|P_n|$. So if L is any simple left ideal in $\mathbb{F}[S_n]$, then it must be isomorphic to any simple left ideal $\mathbb{F}[S_n]y_T$ lying inside $A(p)$, for exactly one $p \in \overline{\mathbb{P}}_n$, since such p are, of course, also $|\overline{\mathbb{P}}_n|$ in number. Then L is a right translate of this $\mathbb{F}[S_n]y_T$ and hence also lies inside $A(p)$. Therefore, the image of I is all of $\mathbb{F}[S_n]$.

Consequently, the image of I covers all of the group algebra $\mathbb{F}[S_n]$. $\boxed{\text{QED}}$

This is a major accomplishment. Yet there are unfinished tasks: what exactly is the value of the dimension of $\mathbb{F}[S_n]y_T$? And what is the character χ_T of the representation given by $\mathbb{F}[S_n]y_T$? We will revisit this place, enriched with more experience from a very different territory in Chap. 10, and gain an understanding of the character χ_T.

6.10 Integrality

Here is a dramatic consequence of our concrete picture of the representations of S_n through the modules $\mathbb{F}[S_n]y_T$:

Theorem 6.12 *Suppose $\rho : S_n \to \mathrm{End}_{\mathbb{F}}(E)$ is any representation of S_n on a finite-dimensional vector space E over a field $\mathbb{F}$ of characteristic 0, where $n \in \{2, 3, \ldots\}$. Then there is a basis in E relative to which, for any $s \in S_n$, the matrix $\rho(s)$ has all entries as integers. In particular, all characters of S_n are integers.*

Proof. First, by decomposing into simple pieces, we may assume that E is an irreducible representation. Then, thanks to Theorem 6.11, we may further take $E = \mathbb{F}[S_n]y_T$, for some Youngtab T, and ρ as the restriction ρ_T of the regular representation to this submodule of $\mathbb{F}[S_n]$.

The $\mathbb{Z}$-module $\mathbb{Z}[S_n]y_T$ is a submodule of the finitely generated free module $\mathbb{Z}[S_n]$, and hence is itself finitely generated and free (Theorem 12.4).

Fix a $\mathbb{Z}$-basis $v_1, ..., v_{d_T}$ of $\mathbb{Z}[S_n]y_T$. Multiplication on the left by a fixed $s \in S_n$ is a $\mathbb{Z}$-linear map of $\mathbb{Z}[S_n]y_T$ into itself and so has matrix $M_T(s)$, relative to the basis $\{v_i\}$, having all entries in $\mathbb{Z}$. Now $1 \otimes v_1, ..., 1 \otimes v_{d_T}$ is an $\mathbb{F}$-basis for the vector space $\mathbb{F}[S_n]y_T = \mathbb{F} \otimes_{\mathbb{Z}} \mathbb{Z}[S_n]y_T$ (see Theorem 12.11). Hence, the matrix for $\rho_T(s)$ is $M_T(s)$, which, as we noted, has entries that are all integers. $\boxed{\text{QED}}$

There is a more abstract reason, noted by Frobenius [29, Sect. 8], why characters of S_n have integer values: if $s \in S_n$ and k is prime to the order of s, then s^k is conjugate to s. See Weintraub [78, Theorem 7.1] for more details.

6.11 Rivals and Rebels

In contrast to our leisurely exploration, there are extremely efficient expositions of the theory of representations of S_n. Among these we mention the short and readable treatment of Diaconis [21, Chap. 7] and the characteristic-free development by James [50]. The long-established order of Young tableaux has been turned on its side by the sudden appearance of a method propounded by Okounkov and Vershik [64]; the book by Ceccherini-Silberstein et al. [10] is an extensive introduction to the Okounkov–Vershik theory, and a short self-contained exposition is available in the book by Hora and Obata [45, Chap. 9]. The study of Young tableaux is in itself an entire field which to the outsider has the feel of a secret society with a plethora of mysterious formulas, and rules and rituals with hyphenated parentage: the Murnaghan–Nakayama rule, the *jeu de taquin* of Schützenberger, the Littlewood–Richardson correspondence, and the Robinson–Schensted–Knuth algorithm. An initiation may be gained from the book by Fulton [37] (and an Internet search for "Schensted" is recommended). We have not covered the *hook length formula* which gives the dimension of irreducible representations of S_n; an unusual but simple proof of this formula is given by Glass and Ng [39].

6.12 Afterthoughts: Reflections

The symmetric group S_n is generated by transpositions, which are just the elements of order 2 in the group. There is a class of more geometric groups that are generated by elements of order 2. These are groups generated by reflections in finite-dimensional real vector spaces. In this section we will explore some aspects of such groups which resemble features we have studied for S_n.

Let E be a finite-dimensional real vector space, equipped with an inner product $\langle \cdot, \cdot \rangle$. A *hyperplane* in E is a codimension 1 subspace of E; equivalently, it is a subspace perpendicular to some nonzero vector v:

$$v^{\perp} = \{x \in E : \langle x, v \rangle = 0\}.$$

Reflection across this hyperplane is the linear map

$$R_{v^{\perp}} : E \to E$$

which fixes each point on $v^{\perp}$ and maps v to $-v$:

$$R_{v^{\perp}}(x) = x - 2\frac{\langle x, v \rangle}{\langle v, v \rangle}v \qquad \text{for all } x \in E.$$

A more elegant definition of reflection requires no inner product structure: a *reflection* across a codimension 1 subspace B in a general vector space V is a linear map $R : V \to V$ for which $R^2 = I$, the identity map on V, and $\ker(I - R) = B$.

By a *reflection group* in E let us mean a finite group of endomorphisms of E generated by a set of reflections across hyperplanes in E. Not all elements of such a group need be reflections. Let $\mathbb{H}_W$ be the set of all hyperplanes B such that the reflection R_B across B is in W. This is a finite set, of course. Let

$$\mathbb{P}_W = \{\pi : \pi \text{ is the intersection of a set of hyperplanes in } \mathbb{H}_W\}. \qquad (6.62)$$

This is a *hyperplane arrangement* (for the theory of hyperplane arrangements, see Orlik and Terao [63]). Observe that each $\pi \in \mathbb{P}_W$ is the intersection of all the hyperplanes of $\mathbb{H}_W$ that contain π as a subset:

$$\pi = \bigcap\{B \in \mathbb{H}_W : \pi \subset B\}. \qquad (6.63)$$

The set $\mathbb{P}_W$ is partially ordered by *reverse* inclusion:

$$\pi_1 \leq \pi_2 \text{ means } \pi_2 \subset \pi_1.$$

The least element $\underline{0}$ and the largest element $\underline{1}$ are

$$\underline{0} = E \quad \text{and} \quad \underline{1} = \cap_{B \in \mathbb{H}_W} B,$$

where E is viewed as the intersection of the empty family of hyperplanes in E (although, in general, $\cap \emptyset$ is fallacious territory in set theory!). Moreover, if $\pi_1, \pi_2 \in \mathbb{P}_W$, then

$$
\begin{aligned}
\pi_1 \vee \pi_2 &\overset{\text{def}}{=} \sup\{\pi_1, \pi_2\} = \pi_1 \cap \pi_2, \\
\pi_1 \wedge \pi_2 &\overset{\text{def}}{=} \inf\{\pi_1, \pi_2\} = \cap\{\pi \in \mathbb{P}_W : \pi \supset \pi_1, \pi_2\}.
\end{aligned}
\tag{6.64}
$$

Here, by definition, inf S is the largest element $\leq$ to all elements of S, and it exists, being just the intersection of the subspaces in S. For example, if B_1 and B_2 are distinct hyperplanes, then $B_1 \wedge B_2$ is E. Thus, $\mathbb{P}_W$ is a lattice, the *intersection lattice* for W.

Let us compare the intersection lattice $\mathbb{P}_W$ with the partition lattice $\mathbb{P}_n$ we have used for S_n. In the lattice $\mathbb{P}_n$, an atom is a partition that contains one two-element set and all others are one-element sets. The analog in the lattice $\mathbb{P}_W$ is the hyperplanes of $\mathbb{H}_W$. Relation (6.63) means that each element $\pi \in \mathbb{P}_W$ is the supremum of the atoms that are below it:

$$
\pi = \sup\{B \in \mathbb{H}_W : B \leq \pi\}.
\tag{6.65}
$$

The analog for $\mathbb{P}_n$ also holds: any partition $\pi \in \mathbb{P}_n$ is the supremum of the atoms that lie below it.

For a subspace $\pi \in \mathbb{P}_W$, let π_c be the intersection of the hyperplanes in $\mathbb{H}_W$ which do not contain π:

$$
\pi_c = \bigcap\{B \in \mathbb{H}_W : \pi \not\subset B\}.
\tag{6.66}
$$

Using (6.63), we then have

$$
\pi \vee \pi_c = \bigcap_{B \in \mathbb{H}_W} B = \underline{1}.
\tag{6.67}
$$

Moreover, since there is no hyperplane which contains both π and π_c, the infimum of $\{\pi, \pi_c\}$ is E:

$$
\pi \wedge \pi_c = E = \underline{0}.
\tag{6.68}
$$

For this lattice complementation we also have

$$
\begin{aligned}
\pi_1 &\leq \pi_2 \Rightarrow (\pi_2)_c \leq (\pi_1)_c, \\
(\pi_c)_c &= \pi.
\end{aligned}
\tag{6.69}
$$

Now consider symmetries of $\mathbb{P}_W$: for each $\pi \in \mathbb{P}_W$ we have the subgroup of all $s \in S$ which fix each point in π:

$$\mathrm{Fix}_\pi = \{s \in W : s|\pi = \mathrm{id}_\pi\}. \tag{6.70}$$

The mapping

$$\mathrm{Fix} : \mathbb{P}_W \to \{\text{subgroups of } W\}$$

is clearly order-preserving:

$$\text{if } \pi_1 \leq \pi_2 \text{ then } \mathrm{Fix}_{\pi_1} \subset \mathrm{Fix}_{\pi_2}. \tag{6.71}$$

Remarkably, Fix_π is generated by the order 2 elements it contains, these being the reflections across the hyperplanes containing π (see Humphreys [46, Sect. 1.5]). Consequently, π may be recovered from Fix_π as the intersection of the fixed point sets of all reflections $r \in \mathrm{Fix}_\pi$:

$$\pi = \cap_{r \in \mathrm{Fix}_\pi, r^2 = I} \ker(I - r). \tag{6.72}$$

We can now summarize our observations into the following analog of Theorem 6.2:

Theorem 6.13 *The mapping*

$$\mathrm{Fix} : \mathbb{P}_W \to \{\text{subgroups of } W\} : \pi \mapsto \mathrm{Fix}_\pi$$

is injective and order-preserving when the subgroups of W are ordered by inclusion. The mapping Fix from $\mathbb{P}_W$ to its image inside the lattice of subgroups of W is an order-preserving isomorphism:

$$\mathrm{Fix}_{\pi_1} \subset \mathrm{Fix}_{\pi_2} \text{ if and only if } \pi_1 \leq \pi_2.$$

Furthermore, Fix also preserves the lattice operations

$$\begin{aligned} \mathrm{Fix}_{\pi_1 \wedge \pi_2} &= \mathrm{Fix}_{\pi_1} \cap \mathrm{Fix}_{\pi_2}, \\ \mathrm{Fix}_{\pi_1 \vee \pi_2} &= \text{the subgroup generated by } \mathrm{Fix}_{\pi_1} \text{ and } \mathrm{Fix}_{\pi_2} \end{aligned} \tag{6.73}$$

for all $\pi_1, \pi_2 \in \mathbb{P}_W$.

The group Fix_π is generated by the reflections it contains.

As in the case of S_n, we also have

$$\mathrm{Fix}_{s(\pi)} = s\mathrm{Fix}_\pi s^{-1} \tag{6.74}$$

for all $\pi \in \mathbb{P}_W$ and $s \in W$.

We step off this train of thought at this point, having seen that the method of using partitions, and beyond that the Young tableaux, have reflections beyond the realm of the symmetric groups.

Exercises

6.1. Prove Proposition 6.2.

6.2. Work out the Young symmetrizers for all the Youngtabs for S_3. Decompose $\mathbb{F}[S_3]$ into a direct sum of simple left ideals. Work out the irreducible representations given by these ideals.

6.3. Let G be a finite group and $\mathbb{F}$ be the field of fractions of a principal ideal domain R. If $\rho : G \to \mathrm{End}_\mathbb{F}(V)$ is a representation of G on a finite-dimensional vector space V over $\mathbb{F}$, show that there is a basis of V such that for every $g \in G$, the matrix of $\rho(g)$ relative to this basis has entries all in R. (You can use Theorem 12.5.)

6.4. For H being any subgroup of S_n, let Orb_H be the set of all orbits of H in $[n]$; in detail, $\mathrm{Orb}_H = \{\{h(j) : h \in H\} : j \in [n]\}$. Then $\mathrm{Orb} :$ {subgroups of S_n} $\to \mathbb{P}_n$ is an order-preserving map, where subgroups are ordered by inclusion, and the set $\mathbb{P}_n$ of all partitions of $[n]$ is ordered so that $\pi_1 \leq \pi_2$ if each block in π_1 is contained inside some block of π_2. For any partition $\pi \in \mathbb{P}_n$, let Fix_π be the subgroup of S_n consisting of all $s \in S_n$ for which $s(B) = B$ for all blocks $B \in \pi$. Show that for $\pi \in \mathbb{P}_n$ and H being any subgroup of S_n (a) if $\mathrm{Fix}_\pi \subset H$, then $\pi \leq \mathrm{Orb}_H$ and (b) if $H \subset \mathrm{Fix}_\pi$, then $\mathrm{Orb}_H \leq \pi$.

6.5. For any positive integer n, and any $k \in [n] = \{1, \ldots, n\}$, the *Jucys–Murphy* element X_k in $R[S_n]$ is defined to be

$$X_k = (1\,k) + \cdots + (k-1\,k), \tag{6.75}$$

with $X_1 = 0$, and R is any commutative ring. Show that for $k > 1$, the element X_k commutes with every element of $R[S_{k-1}]$, where we view S_{k-1} as a subset of S_n in the natural way. Show that $X_1, \ldots, X_n$ generate a commutative subalgebra of $R[S_n]$. For the standard Young tableau

$$T = \begin{array}{|c|c|c|} \hline 1 & 2 & 5 \\ \hline 3 & 4 \\ \cline{1-2} \end{array}$$

work out $X_4 y_T$. The Jucys–Murphy elements play an important role in the Okounkov–Vershik theory [64].

Chapter 7

Characters

The *character* of a representation ρ of a group G on a finite-dimensional vector space E, over a field $\mathbb{F}$, is the function χ_ρ on G given by

$$\chi_\rho : G \to \mathbb{F} : g \mapsto \mathrm{Tr}\big(\rho(g)\big). \tag{7.1}$$

This logically sensible definition turns history on its head, because Frobenius [29] first constructed characters directly from groups, and certain associated determinants, and not from their representations. We will explore the group determinant of Dedekind and Frobenius in Sect. 7.7, but only after developing the properties of characters as defined by (7.1). In this chapter we will explore the multifaceted nature of group characters through a lush landscape of results, virtually all of which were discovered by Frobenius in his pioneering works [29, 30, 31, 32, 33, 34, 35, 36]

Let us recall some of the terminology and basic facts. If E is the space on which the representation ρ acts, then χ_ρ is also denoted χ_E, although this is a slight abuse of notation. A *character* of G is the character of some finite-dimensional representation of G, and a *complex* character is the character of a complex representation. An *irreducible* or *simple* character is the character of an irreducible representation. A character is always a *central function*:

$$\chi_\rho(ghg^{-1}) = \chi_\rho(h) \qquad \text{for all } g, h \in G. \tag{7.2}$$

A different face of conjugation invariance is expressed by the fact that

$$\chi_{\rho_1} = \chi_{\rho_2}$$

A.N. Sengupta, *Representing Finite Groups: A Semisimple Introduction*,
DOI 10.1007/978-1-4614-1231-1_7, © Springer Science+Business Media, LLC 2012

whenever ρ_1 and ρ_2 are equivalent representations. We proved this in Proposition 1.2. The character χ_ρ extends naturally to a linear function

$$\chi_\rho : \mathbb{F}[G] \to \mathbb{F} : \sum_g x_g g \mapsto \sum_g x_g \chi_\rho(g)$$

which is central in the sense that

$$\chi_\rho(ab) = \chi_\rho(ba) \qquad \text{for all } a, b \in \mathbb{F}[G]. \tag{7.3}$$

There is generally no need to distinguish between χ viewed as a function on $\mathbb{F}[G]$ and as a function on G. The following properties of characters follow directly from the definitions of direct sums and tensor products of representations:

$$\chi_{E \oplus F} = \chi_E + \chi_F, \tag{7.4}$$
$$\chi_{E \otimes F} = \chi_E \chi_F \tag{7.5}$$

for any finite-dimensional representations E and F. Thus, if E decomposes as

$$E = \bigoplus_{i=1}^m n_i E_i,$$

where E_i are representations and $n_i E_i$ is the direct sum of n_i copies of E_i, then

$$\chi_E = \sum_{i=1}^s n_i \chi_{E_i}. \tag{7.6}$$

7.1   The Regular Character

We will work with a finite group G and a field $\mathbb{F}$.

The *regular representation* ρ_{reg} of a finite group G is its representation through left multiplications on the group algebra $\mathbb{F}[G]$: with $g \in G$ is associated $\rho_{\text{reg}}(g) : \mathbb{F}[G] \to \mathbb{F}[G] : x \mapsto gx$. We denote the character of this representation by χ_{reg}:

$$\chi_{\text{reg}} \stackrel{\text{def}}{=} \text{character of the regular representation.} \tag{7.7}$$

As usual, we may view this as a function on $\mathbb{F}[G]$:

$$\chi_{\text{reg}}(x) = \text{trace of the linear map } \mathbb{F}[G] \to \mathbb{F}[G] : y \mapsto xy \tag{7.8}$$

for all $x \in \mathbb{F}[G]$.

Let us work out χ_{reg} on any element

$$b = \sum_{h \in G} b_h h \in \mathbb{F}[G].$$

For any $g \in G$ we have

$$bg = \sum_{h \in G} b_h hg = b_e g + \sum_{w \in G, w \neq g} b_{wg^{-1}} w,$$

and so, in terms of the basis of $\mathbb{F}[G]$ given by the elements of G, left multiplication by b has a matrix with b_e running down the main diagonal. Hence,

$$\chi_{\text{reg}}(b) = |G| b_e. \tag{7.9}$$

We can rewrite (7.9) as

$$\frac{1}{|G|} \text{Tr} \left(\rho_{\text{reg}}(b) \right) = b_e \qquad \text{if } |G| \neq 0 \text{ in } \mathbb{F}. \tag{7.10}$$

The map

$$\text{Tr}_e : \mathbb{F}[G] \to \mathbb{F} : b \mapsto b_e,$$

is itself also called a *trace*, and is a central function on $\mathbb{F}[G]$. Unlike χ_{reg}, the trace Tr_e is both meaningful and useful even if $|G| 1_{\mathbb{F}}$ is 0 in $\mathbb{F}$.

In Chap. 4 we saw that there is a maximal string of nonzero central idempotent elements $u_1, \ldots, u_s$ in $\mathbb{F}[G]$ such that the map

$$I : \prod_{i=1}^{s} \mathbb{F}[G] u_i \to \mathbb{F}[G] : (a_1, \ldots, a_s) \mapsto a_1 + \cdots + a_s \tag{7.11}$$

is an isomorphism of algebras, where $\mathbb{F}[G] u_i$ is a two-sided ideal in $\mathbb{F}[G]$ and is an algebra in itself, having u_i as a multiplicative identity. The statement that I in (7.11) preserves multiplication encodes the observation that

$$\mathbb{F}[G] u_i \mathbb{F}[G] u_j = 0 \qquad \text{if } i \neq j.$$

If $|G| 1_{\mathbb{F}} \neq 0$, then, on picking a simple left ideal L_i of $\mathbb{F}[G]$ lying inside $\mathbb{F}[G] u_i$ for each i, every irreducible representation of G, viewed as an $\mathbb{F}[G]$-module, is isomorphic to some L_i, and

$$\mathbb{F}[G] u_i = \underbrace{L_i \oplus \cdots \oplus L_i}_{d_i \text{ copies}}$$

for some positive integer d_i every $i \in [s]$. Let χ_i be the character of the restriction of the regular representation to the subspace L_i:

$$\chi_i(g) = \mathrm{Tr}\left(\rho_{\mathrm{reg}}(g)|L_i\right). \tag{7.12}$$

If $|G|1_{\mathbb{F}} \neq 0$, then every finite-dimensional representation of G is isomorphic to a finite direct sum of copies of the L_i, and so in this case every character χ of G is a linear combination of the form

$$\chi = \sum_{i=1}^{s} n_i \chi_i, \tag{7.13}$$

where n_i is the number of copies of L_i in a direct sum decomposition of the representation for χ into irreducible components.

Thus, if $|G|1_{\mathbb{F}} \neq 0$ in $\mathbb{F}$, then

$$\chi_{\mathrm{reg}} = \sum_{i=1}^{s} d_i \chi_i, \tag{7.14}$$

where d_i is the number of copies of L_i in a direct sum decomposition of $\mathbb{F}[G]$ into simple left ideals. We know that

$$d_i = \dim_{\mathbb{D}_i} L_i,$$

where $\mathbb{D}_i$ is the division ring

$$\mathbb{D}_i = \mathrm{End}_{\mathbb{F}[G]u_i} L_i.$$

If $\mathbb{F}$ is also algebraically closed, then $\mathbb{D}_i = \mathbb{F}$ and d_i equals $\dim_{\mathbb{F}} L_i$ for all $i \in [s]$. For more on these facts, including proofs, see Theorem 4.4.

Recalling (7.8), and noting that

$$a_j \mathbb{F}[G]u_i = 0 \quad \text{if } a_j \in \mathbb{F}[G]u_j \text{ and } j \neq i,$$

we have

$$\chi_i(a_j) = 0 \qquad \text{if } a_j \in \mathbb{F}[G]u_j \text{ and } j \neq i. \tag{7.15}$$

Thus,

$$\chi_i \big| \mathbb{F}[G]u_j = 0 \quad \text{if } j \neq i \tag{7.16}$$

Equivalently,

$$\chi_i(u_j) = 0 \quad \text{if } j \neq i, \tag{7.17}$$

where, as usual, u_j is the generating idempotent for $\mathbb{F}[G]u_j$. On the other hand,

$$\chi_i(u_i) = \dim_{\mathbb{F}} L_i \tag{7.18}$$

because the central element u_i acts as the identity on $L_i \subset \mathbb{F}[G]u_i$. In fact, we have

$$\chi_{\text{reg}}(yu_i) = d_i\chi_i(y) \quad \text{for all } y \in G. \tag{7.19}$$

Lemma 7.1 *If L is an irreducible representation of a finite group G over an algebraically closed field $\mathbb{F}$ whose characteristic does not divide $|G|$, then $\dim_{\mathbb{F}} L$ is also not divisible by the characteristic of $\mathbb{F}$.*

There will be a remarkably sharpened version of this result later in Theorem 7.12.

Proof. Let $P : L \to L$ be a linear projection map with one-dimensional range, so that the trace of P is 1. Then by Schur's lemma, the $\mathbb{F}[G]$-linear map $P_1 = \sum_{g \in G} gPg^{-1} : L \to L$ is a scalar multiple cI of the identity, and so, taking the trace, we have $|G| \cdot 1_{\mathbb{F}}$ (which, by assumption, is not 0) equals $c \dim_{\mathbb{F}} L$. Hence, $\dim_{\mathbb{F}} L$ is not 0 in $\mathbb{F}$. $\boxed{\text{QED}}$

   One aspect of the importance and utility of characters is codified in the following fundamental observation:

Theorem 7.1 *Suppose G is a finite group and $\mathbb{F}$ is a field; assume that either (1) $\mathbb{F}$ has characteristic 0 or (2) $|G|1_{\mathbb{F}} \neq 0$ and $\mathbb{F}$ is algebraically closed. Then the irreducible characters of G over the field $\mathbb{F}$ are linearly independent.*

Proof. Let $\chi_1, \ldots, \chi_s$ be the distinct irreducible characters of G for representations on vector spaces over the field $\mathbb{F}$. Recall the central idempotents $u_1, \ldots, u_s \in \mathbb{F}[G]$, such that χ_i is the character of the restriction of the regular representation of G to any simple submodule of $\mathbb{F}[G]u_i$ for each $i \in [s]$. Suppose now that

$$\sum_{i=1}^{s} c_i\chi_i - 0, \tag{7.20}$$

where $c_1, \ldots, c_s \in \mathbb{F}$. Then, applying (7.20) to u_j and using (7.17) and (7.18), we find that

$$c_j \dim_{\mathbb{F}} L_j = 0.$$

Thus, since hypotheses 1 and 2 both imply that each $\dim_{\mathbb{F}} L_j$ is not 0 in $\mathbb{F}$, it follows that each c_j is 0. $\boxed{\text{QED}}$

Linear independence encodes the following important fact about characters:

Theorem 7.2 *Suppose G is a finite group and $\mathbb{F}$ is an algebraically closed field of characteristic 0. Two finite-dimensional representations of G, over $\mathbb{F}$, have the same character if and only if they are equivalent.*

Proof. Let $L_1, \ldots, L_s$ be a maximal collection of inequivalent irreducible representations of G. If E is a representation of G, then E is equivalent to a direct sum

$$E \simeq \bigoplus_{i=1}^{s} n_i L_i, \qquad (7.21)$$

where $n_i L_i$ is a direct sum of n_i copies of L_i. Then

$$\chi_E = \sum_{i=1}^{s} n_i \chi_i.$$

The coefficients n_i are uniquely determined by χ_E, and hence so is the decomposition (7.21) up to isomorphism. $\boxed{\text{QED}}$

7.2   Character Orthogonality

The character, being a trace, has interesting and useful features inherited from the nature of the trace functional. We will explore some of these properties in this section. A note of warning: we will use the bra-ket formalism introduced at the end of Sect. 1.6. When working with a vector space V, and its dual V', we will often denote a typical element of V by $|v\rangle$ and a typical element of V' by $\langle f|$, with the evaluation of $\langle f|$ on $|v\rangle$ denoted by

$$\langle f|v\rangle.$$

Assume that G is a finite group and $\mathbb{F}$ is a field. Let

$$T : E \to F$$

be an $\mathbb{F}$-linear map between simple $\mathbb{F}[G]$-modules. Then the G-symmetrized version

$$T_1 = \sum_{g \in G} gTg^{-1}$$

satisfies

$$hT_1 = T_1 h \quad \text{for all } h \in G.$$

Then, by Schur's lemma, T_1 it is either 0 or an isomorphism. A general linear map $T : E \to F$, viewed as a matrix relative to bases in E and F, is a linear combination of matrices that all entries as 0 except for one, which is 1; we specialize T to such a matrix. We choose now a special form for the map T. Pick a basis $|e_1\rangle, \ldots, |e_m\rangle$ of the vector space E, a basis $|f_1\rangle, \ldots, |f_n\rangle$ of F, and, for any particular choice of $j \in [n]$ and $k \in [m]$, let T be given by

$$T = |f_j\rangle\langle e_k| : |v\rangle \mapsto \langle e_k|v\rangle|f_j\rangle = v_k|f_j\rangle,$$

where v_k is the kth component of $|v\rangle$ written out in the basis $|e_1\rangle, \ldots, |e_m\rangle$. (If bra-kets bother you, write e_i for $|e_i\rangle$ and e_i' for the dual basis element $\langle e_i|$, and similarly for $|f_i\rangle$ and $\langle f_i|$.)

Symmetrizing T, we obtain

$$T_1 = \sum_{g \in G} \rho_F(g)|f_j\rangle\langle e_k|\rho_E(g)^{-1}. \tag{7.22}$$

If ρ_E and ρ_F are inequivalent representations of G, then T_1 is 0, and so

$$\langle f_j|T_1|e_k\rangle = 0,$$

which simply says that the jth component of $T_1|e_j\rangle$ is 0. Substituting in the expression (7.22) for T_1 gives

$$\sum_{g \in G} \rho_F(g)_{jj}\rho_E(g^{-1})_{kk} = 0. \tag{7.23}$$

Summing over j as well as k produces

$$\sum_{g \in G} \chi_F(g)\chi_E(g^{-1}) = 0. \tag{7.24}$$

Writing this as a sum over conjugacy classes produces the identity (7.26) below. This is one of several orthogonality relations discovered by Frobenius. Here is an official summary:

Theorem 7.3 *If ρ_1 and ρ_2 are inequivalent irreducible representations of a finite group on vector spaces over any field $\mathbb{F}$, then*

$$\sum_{g \in G} \chi_{\rho_1}(g)\chi_{\rho_2}(g^{-1}) = 0. \qquad (7.25)$$

Equivalently,

$$\sum_{C \in \mathcal{C}} |C|\chi_{\rho_1}(C)\chi_{\rho_2}(C^{-1}) = 0, \qquad (7.26)$$

where $\mathcal{C}$ is the set of all conjugacy classes in G and $f(C)$ denotes the constant value of a central function f on a conjugacy class C.

Why the term "orthogonality"? The answer is seen by noticing that, when working with complex representations, relation (7.25) can be viewed as saying that the vectors

$$(\chi_E(g))_{g \in G} \in \mathbb{C}^G$$

are orthogonal to each other for inequivalent choices of the irreducible representation E.

Next we use Schur's lemma in the case where the representations are the same. Consider an $\mathbb{F}$-linear map

$$T : E \to E,$$

where E is a simple $\mathbb{F}[G]$-module. Forming the symmetrized version just as above, we have, again by Schur's lemma,

$$\sum_{g \in G} gTg^{-1} = cI, \qquad (7.27)$$

for some scalar $c \in \mathbb{F}$, provided, of course, we assume now that $\mathbb{F}$ is algebraically closed (or at least that $\mathbb{F}$ is a splitting field for G). The value of c is obtained by taking the trace of both sides in (7.27):

$$|G|\mathrm{Tr}\,(T) = c \dim_{\mathbb{F}} E. \qquad (7.28)$$

Picking a T whose trace is 1 shows that $\dim_{\mathbb{F}} E \neq 0$ in the field $\mathbb{F}$, provided $|G|1_{\mathbb{F}} \neq 0$. Thus, if $\mathbb{F}$ is algebraically closed and $|G|1_{\mathbb{F}} \neq 0$, then

$$\sum_{g \in G} gTg^{-1} = \frac{|G|\mathrm{Tr}\,(T)}{\dim_{\mathbb{F}} E}I. \qquad (7.29)$$

Using a basis $|e_1\rangle, \ldots, |e_m\rangle$ of E, we take T to be

$$T_{jk} = |e_j\rangle\langle e_k|,$$

and this gives

$$\sum_{g \in G} \rho_E(g) |e_j\rangle\langle e_k| \rho_E(g)^{-1} = c_{jk} I, \qquad (7.30)$$

where

$$c_{jk} \dim_{\mathbb{F}} E = |G| \operatorname{Tr}(T_{jk}) = \delta_{jk} |G|. \qquad (7.31)$$

Bracketing (7.30) between $\langle e_j | \ldots | e_k \rangle$, we have

$$\sum_{g \in G} \langle e_j | \rho_E(g) | e_j \rangle \langle e_k | \rho_E(g)^{-1} | e_k \rangle = c_{jk} \delta_{jk}.$$

Summing over j and k produces

$$\sum_{g \in G} \chi_E(g) \chi_E(g^{-1}) = |G|.$$

Here is a clean summary of our conclusions:

Theorem 7.4 *If ρ is an irreducible representation of a finite group on a vector space over an algebraically closed field $\mathbb{F}$ in which $|G| 1_{\mathbb{F}} \neq 0$, then*

$$\sum_{g \in G} \chi_\rho(g) \chi_\rho(g^{-1}) = |G|. \qquad (7.32)$$

Equivalently,

$$\sum_{C \in \mathcal{C}} |C| \chi_\rho(C) \chi_\rho(C^{-1}) = |G|, \qquad (7.33)$$

where $\mathcal{C}$ is the set of all conjugacy classes in G and $f(C)$ denotes the constant value of a central function f on a conjugacy class C.

As is often the case, the condition that $\mathbb{F}$ is algebraically closed can be replaced by the requirement that $\mathbb{F}$ be a splitting field for G.

The two results we have proven here so far can be combined into one: if ρ_1 and ρ_2 are irreducible representations , then

$$\frac{1}{|G|} \sum_{g \in G} \chi_{\rho_1}(g) \chi_{\rho_2}(g^{-1}) = \begin{cases} 1 & \text{if } \rho_1 \text{ is equivalent to } \rho_2, \\ 0 & \text{if } \rho_1 \text{ is not equivalent to } \rho_2, \end{cases} \qquad (7.34)$$

provided that the underlying field $\mathbb{F}$ is algebraically closed and $|G| 1_{\mathbb{F}} \neq 0$. Here is another perspective on this:

Theorem 7.5 *Suppose ρ_1 and ρ_2 are representations of a finite group G on finite-dimensional vector spaces E_1 and E_2, respectively, over a field $\mathbb{F}$ in which $|G|1_{\mathbb{F}} \neq 0$. Then*

$$\frac{1}{|G|} \sum_{g \in G} \chi_{\rho_1}(g)\chi_{\rho_2}(g^{-1}) = \dim_{\mathbb{F}} \operatorname{Hom}_{\mathbb{F}[G]}(E_1, E_2), \qquad (7.35)$$

where $\operatorname{Hom}_{\mathbb{F}[G]}(E_1, E_2)$ is the vector space of all $\mathbb{F}[G]$-linear maps $E_1 \to E_2$.

Before heading into the proof, observe that if ρ_1 and ρ_2 are inequivalent irreducible representations, then, by Schur's lemma, $\operatorname{Hom}_{\mathbb{F}[G]}(E_1, E_2)$ is 0, whereas if ρ_1 and ρ_2 are equivalent irreducible representations, then, again by Schur's lemma, $\operatorname{Hom}_{\mathbb{F}[G]}(E_1, E_2)$ is one-dimensional if $\mathbb{F}$ is algebraically closed. The version we now have works even if ρ_1 and ρ_2 are not irreducible and shows that, in fact, the averaged character product on the left in (7.35) takes into account the multiplicities of irreducible constituents of E_1 and E_2.

Proof. The key point is that the G-symmetrization $T \to T_0$ in (7.36) is a projection map onto $\operatorname{Hom}_{\mathbb{F}[G]}(E_1, E_2)$ and the trace of this projection gives the dimension of $\operatorname{Hom}_{\mathbb{F}[G]}(E_1, E_2)$. In more detail, consider the map

$$\Pi_0 : \operatorname{Hom}_{\mathbb{F}}(E_1, E_2) \to \operatorname{Hom}_{\mathbb{F}}(E_1, E_2) : T \mapsto T_0 = \frac{1}{|G|} \sum_{g \in G} \rho_{E_2}(g)^{-1} T \rho_{E_1}(g).$$

$$(7.36)$$

Clearly, T_0 lies in the subspace $\operatorname{Hom}_{\mathbb{F}[G]}(E_1, E_2)$ inside $\operatorname{Hom}_{\mathbb{F}}(E_1, E_2)$. Moreover, if T is already in this subspace, then $T_0 = T$. Thus, $\Pi_0^2 = \Pi_0$ and is a projection map with image $\operatorname{Hom}_{\mathbb{F}[G]}(E_1, E_2)$. Every element $T \in \operatorname{Hom}_{\mathbb{F}}(E_1, E_2)$ splits uniquely as a sum:

$$T = \underbrace{\Pi_0(T)}_{\in \operatorname{Im}(\Pi_0)} + \underbrace{(1 - \Pi_0)(T)}_{\in \ker(\Pi_0)}.$$

Thus,

$$\operatorname{Hom}_{\mathbb{F}}(E_1, E_2) = \operatorname{Hom}_{\mathbb{F}[G]}(E_1, E_2) \oplus \ker \Pi_0.$$

Form a basis of $\operatorname{Hom}_{\mathbb{F}}(E_1, E_2)$ by pooling together a basis of $\operatorname{Hom}_{\mathbb{F}[G]}(E_1, E_2)$ with a basis of $\ker \Pi_0$; relative to this basis, the matrix of P_0 is diagonal, with an entry of 1 for each basis vector of $\operatorname{Hom}_{\mathbb{F}[G]}(E_1, E_2)$ and 0 for all other entries. Hence,

$$\operatorname{Tr}(\Pi_0) = \dim_{\mathbb{F}} \operatorname{Hom}_{\mathbb{F}[G]}(E_1, E_2). \qquad (7.37)$$

Now let us calculate the trace on the left more concretely. If E_1 or E_2 is $\{0\}$, then the result is trivial, so we assume that neither space is 0. Choose a basis $|e_1\rangle, \ldots, |e_m\rangle$ in E_1 and a basis $|f_1\rangle, \ldots, |f_n\rangle$ in E_2. The elements

$$T_{jk} = |f_j\rangle\langle e_k| : E_1 \to E_2 : |v\rangle \mapsto \langle e_k|v\rangle\langle f_j|,$$

where $\langle e_k|v\rangle$ is the kth component of $|v\rangle$ in the basis $\{|e_i\rangle\}$, form a basis of $\mathrm{Hom}_{\mathbb{F}}(E_1, E_2)$. The image of T_{jk} under the projection Π_0 is

$$
\begin{aligned}
\Pi_0(T_{jk}) &= \frac{1}{|G|} \sum_{g \in G} \rho_{E_2}(g)^{-1} |f_j\rangle\langle e_k| \rho_{E_1}(g) \\
&= \sum_{1 \le i \le m, 1 \le l \le n} \frac{1}{|G|} \sum_{g \in G} \langle f_l|\rho_{E_2}(g)^{-1}|f_j\rangle\langle e_k|\rho_{E_1}(g)|e_i\rangle \, |f_l\rangle\langle e_i|.
\end{aligned}
\tag{7.38}
$$

Thus, the T_{jk} component of $\Pi_0(T_{jk})$ is

$$\frac{1}{|G|} \sum_{g \in G} \langle f_j|\rho_{E_2}(g)^{-1}|f_j\rangle\langle e_k|\rho_{E_1}(g)|e_k\rangle$$

and so the trace of Π_0 is found by summing over j and k:

$$\mathrm{Tr}\,(\Pi_0) = \frac{1}{|G|} \sum_{g \in G} \chi_{\rho_2}(g^{-1})\chi_{\rho_1}(g). \tag{7.39}$$

Combining this with (7.37) brings us to our goal: (7.35). $\boxed{\text{QED}}$

The roles of characters and conjugacy classes can be interchanged to reveal another orthogonality identity:

Theorem 7.6 *Let $\mathcal{R}$ be a maximal set of inequivalent irreducible representations of a finite group G over an algebraically closed field $\mathbb{F}$ in which $|G|1_{\mathbb{F}} \ne 0$. Then*

$$\sum_{\rho \in \mathcal{R}} \chi_\rho(C')\chi_\rho(C^{-1}) = \frac{|G|}{|C|}\delta_{C,C'} \tag{7.40}$$

for any conjugacy classes C and C' in G.

Proof. Let $\chi_1, \ldots, \chi_s$ be all the distinct irreducible characters of G, over $\mathbb{F}$, and let $C_1, \ldots, C_s$ be all the distinct conjugacy classes in G. From Theorems 7.4 and 7.3 we have

$$\sum_{j=1}^s \frac{|C_j|}{|G|} \chi_i(C_j)\chi_k(C_j^{-1}) = \delta_{ik}. \tag{7.41}$$

Let us read this as a matrix equation: let A and B be $s \times s$ matrices specified by

$$A_{ij} = \frac{|C_j|}{|G|}\chi_i(C_j) \quad \text{and} \quad B_{jk} = \chi_k(C_j^{-1})$$

for all $i, j, k \in [s]$. Then relation (7.41) means AB is the identity matrix I, and hence BA is also I. Thus,

$$\sum_{j=1}^{s} B_{ij}A_{jk} = \delta_{ik},$$

which means that

$$\sum_{j=1}^{s} \chi_j(C_i^{-1})\frac{|C_k|}{|G|}\chi_j(C_k) = \delta_{ik}$$

for all $i, k \in [s]$. If we write C' for C_i and C for C_k, a small amount of rearrangement brings us to our destination: (7.40). $\boxed{\text{QED}}$

The argument given above is a slight reformulation of the proof given by Frobenius himself. You can explore a longer but more insightful alternative route in Exercise 7.2.

Here is a nice consequence, which can be seen by other means as well:

Theorem 7.7 *Let G be a finite group, and $\mathbb{F}$ be an algebraically closed field in which $|G|1_{\mathbb{F}} \neq 0$. If $g_1, g_2 \in G$ are such that $\chi(g_1) = \chi(g_2)$ for every irreducible character χ of G over $\mathbb{F}$, then g_1 and g_2 belong to the same conjugacy class.*

Proof. Let C be the conjugacy class of g_1 and let C' be that of g_2. Let $\chi_1, \ldots, \chi_s$ be all the distinct irreducible characters of G over $\mathbb{F}$. By hypothesis, $\chi_i(C') = \chi_i(C)$ for all $i \in [s]$. Using (7.40), we have

$$\frac{|G|}{|C|} = \sum_{i=1}^{s} \chi_i(C)\chi_i(C^{-1}) = \sum_{i=1}^{s} \chi_i(C')\chi_i(C^{-1}) = \frac{|G|}{|C|}\delta_{C,C'},$$

which implies that $\delta_{C,C'} = 1$; thus, $C = C'$. $\boxed{\text{QED}}$

Before we look at yet another consequence of Schur's lemma for characters, it will be convenient to introduce a certain product of functions on G called *convolution*. Let G be a finite group and $\mathbb{F}$ be any field. Recall that an element $\sum_{g \in G} x_g g$ of the group algebra $\mathbb{F}[G]$ is just a different expression for the function $G \to \mathbb{F} : g \mapsto x_g$. It is, however, also useful to relate functions $G \to \mathbb{F}$ to elements of $\mathbb{F}[G]$ in a less obvious way. Assume $|G|1_\mathbb{F} \neq 0$ and associate with a function $f : G \to \mathbb{F}$ the element

$$\underline{f} = \frac{1}{|G|} \sum_{g \in G} f(g)g^{-1}. \tag{7.42}$$

The association

$$\mathbb{F}^G \to \mathbb{F}[G] : f \mapsto \underline{f}$$

is clearly an isomorphism of $\mathbb{F}$ vector spaces. Let us see what in $\mathbb{F}^G$ corresponds to the product structure on $\mathbb{F}[G]$. If $f_1, f_2 : G \to \mathbb{F}$, then a simple calculation produces

$$\underline{f_1}\,\underline{f_2} = \underline{f_1*f_2}, \tag{7.43}$$

where $f_1 * f_2$ is the *convolution* of the functions f_1 and f_2, specified by

$$f_1*f_2(h) = \frac{1}{|G|} \sum_{g \in G} f_1(g)f_2(hg^{-1}) \tag{7.44}$$

for all $h \in G$. Of course, all this makes sense only when $|G|1_\mathbb{F} \neq 0$. (If $|G|$ were divisible by the character of the field $\mathbb{F}$, then one could still define a convolution by dropping the dividing factor $|G|$. One other caveat: we put a twist in (7.42) with the g^{-1} on the right, which resulted in what may be a somewhat uncomfortable twist in definition (7.44) of the convolution.)

Here is a stronger form of the character orthogonality relations, expressed in terms of the convolution of characters:

Theorem 7.8 *Let E and F be irreducible representations of a finite group G over an algebraically closed field in which $|G|1_\mathbb{F} \neq 0$. Then*

$$\chi_E * \chi_F = \begin{cases} \frac{1}{\dim_\mathbb{F} E}\chi_E & \text{if } E \text{ and } F \text{ are equivalent,} \\ 0 & \text{if } E \text{ and } F \text{ are not equivalent.} \end{cases} \tag{7.45}$$

Explicitly,

$$\frac{1}{|G|} \sum_{h \in G} \chi_E(gh^{-1})\chi_F(h) = \begin{cases} \frac{1}{\dim_\mathbb{F} E}\chi_E(g) & \text{if } E \text{ and } F \text{ are equivalent,} \\ 0 & \text{if } E \text{ and } F \text{ are not equivalent.} \end{cases} \tag{7.46}$$

More generally, if $\chi_1, \ldots, \chi_k$ are characters of irreducible representations of G, over the field $\mathbb{F}$, then

$$\sum_{\{(a_1,\ldots,a_k) \in G^k : a_1,\ldots,a_k = c\}} \chi_1(a_1), \ldots, \chi_k(a_k) = \begin{cases} \left(\frac{|G|}{d_1}\right)^{k-1} \chi_1(c) & \text{if all } \chi_j \text{ are} \\ & \text{equal to } \chi_1, \\ 0 & \text{otherwise} \end{cases}$$

$$(7.47)$$

for any $c \in G$, with $d_1 = \chi_1(e)$ being the dimension of the representation space of the character χ_1.

As in the first character orthogonality result, Proposition 7.3, the second case in (7.45) holds without any conditions on the field $\mathbb{F}$.

Proof. Suppose first E and F are inequivalent representations. In this case the argument is a rerun, with a simple modification, of the proof of the first character orthogonality relation, Proposition 7.3. Fix bases $|e_1\rangle, \ldots, |e_m\rangle$ in E and $|f_1\rangle, \ldots, |f_n\rangle$ in F, and let

$$T_{jk} = |f_j\rangle\langle e_k|.$$

Then

$$\sum_{g \in G} \rho_F(g^{-1}) T_{jk} \rho_E(h) \rho_E(g)$$

is an $\mathbb{F}[G]$-linear map $E \to F$ and hence, by Schur's lemma, is 0; bracketing between $\langle f_j|$ and $|e_k\rangle$ gives

$$\sum_{g \in G} \langle f_j | \rho_F(g^{-1}) | f_j \rangle \langle e_k | \rho_E(h) \rho_E(g) | e_k \rangle = 0.$$

Summing over j and k produces

$$\sum_{g \in G} \chi_F(g^{-1}) \chi_E(hg) = 0,$$

which is the second case in (7.45). Now suppose E and F are equivalent, and so we simply set $F = E$. Recall from (7.29) the identity

$$\sum_{g \in G} \rho_E(g^{-1}) T \rho_E(g) = \frac{|G| \operatorname{Tr}(T)}{\dim_{\mathbb{F}} E} I, \tag{7.48}$$

which is valid for all $T \in \text{End}_{\mathbb{F}}(E)$. Apply this to $|e_j\rangle\langle e_k|\rho_E(h)$ for T to obtain

$$\sum_{g \in G} \rho_E(g^{-1})|e_j\rangle\langle e_k|\rho_E(hg) = \frac{|G|\langle e_k|\rho_E(h)|e_j\rangle}{\dim_{\mathbb{F}} E}I.$$

Bracketing this between $\langle e_j|$ and $|e_k\rangle$ gives

$$\sum_{g \in G} \rho_E(g^{-1})_{jj}\rho_E(hg)_{kk} = \frac{|G|}{\dim_{\mathbb{F}} E}\rho_E(h)_{kj}\delta_{jk}.$$

Summing over j and k produces

$$\sum_{g \in G} \chi_E(g^{-1})\chi_E(hg) = \frac{|G|}{\dim_{\mathbb{F}} E}\chi_E(h).$$

Iterating this, we obtain the general formula (7.47). $\boxed{\text{QED}}$

7.3 Character Expansions

From Theorem 7.1 we know that the irreducible characters of a finite group G are linearly independent if the underlying field $\mathbb{F}$ is algebraically closed and $|G|1_{\mathbb{F}} \neq 0$. The following makes this result more meaningful:

Theorem 7.9 *Let G be a finite group and $\mathbb{F}$ be a field; assume that $|G|1_{\mathbb{F}} \neq 0$ and $\mathbb{F}$ is algebraically closed. Then the distinct irreducible characters form a basis of the vector space of all central functions on G with values in $\mathbb{F}$.*

As usual, this would work with algebraic closedness replaced by the requirement that $\mathbb{F}$ is a splitting field for G. This result also implies Theorem 7.7, which we proved earlier directly from the orthogonality relations.

Proof. Viewing a function on G as an element of $\mathbb{F}[G]$, we see that the subspace of central functions corresponds precisely to the center Z of $\mathbb{F}[G]$. As we saw in Theorem 4.9 and the discussion preceding it, under the given hypotheses, $\dim_{\mathbb{F}} Z$ is exactly the number of distinct irreducible characters of G. Since these characters are linearly independent, we conclude that they form a basis of the vector space of central functions $G \to \mathbb{F}$. $\boxed{\text{QED}}$

When the underlying field $\mathbb{F}$ is a subfield of the complex field $\mathbb{C}$, we denote by $L^2(G)$ the vector space of all functions $G \to \mathbb{F}$, equipped with the Hermitian inner product specified by

$$\langle f_1, f_2 \rangle_{L^2} = \frac{1}{|G|} \sum_{g \in G} \overline{f_1(g)} f_2(g) \tag{7.49}$$

for $f_1, f_2 : G \to \mathbb{F} \subset \mathbb{C}$. (For a general field we can consider the bilinear form given by $\sum_{g \in G} f_1(g^{-1}) f_2(g)$.)

From character orthogonality (7.34) we know that the irreducible complex characters are orthonormal,

$$\langle \chi_j, \chi_k \rangle_{L^2} = \delta_{jk},$$

whereas from Theorem 7.9 we know that they form a basis of the space of central functions. Thus, we have the following theorem:

Theorem 7.10 *For a finite group G, the irreducible complex characters form an orthonormal basis of the vector space of all central functions $G \to \mathbb{C}$ with respect to the inner product $\langle \cdot, \cdot \rangle_{L^2}$ in (7.49).*

Let us note the following result which can be a quick way of checking irreducibility:

Proposition 7.1 *A complex character χ is irreducible if and only if $\|\chi\|_{L^2} = 1$.*

Proof. Suppose χ decomposes as

$$\chi = \sum_{i=1}^{s} n_i \chi_i,$$

where $\chi_1, \ldots, \chi_s$ are the irreducible complex characters. Then

$$\|\chi\|_{L^2}^2 = \sum_{i=1}^{s} n_i^2,$$

and so the norm of χ is 1 if and only if all n_i are 0, except for one, which equals 1. $\boxed{\text{QED}}$

Here is an immediate application:

Proposition 7.2 *Let $\rho_1, \ldots, \rho_s$ be a maximal collection of inequivalent irreducible complex representations of a finite group, and denote by E_i the representation space of ρ_i. Then, for any positive integer n and for each $i = (i_1, \ldots, i_n) \in \{1, \ldots, s\}^n$, the representation $\rho_i = \rho_{i_1} \otimes \cdots \otimes \rho_{i_n}$ of G^n on $E_i = E_{i_1} \otimes \cdots \otimes E_{i_n}$ given by*

$$\rho_i(g_1, \ldots, g_n) = \rho_{i_1}(g_1) \ldots \rho_{i_n}(g_n)$$

is irreducible, and the representations ρ_i, with i running over $[s]^n$, form a maximal collection of inequivalent complex representations of G^n.

Proof. Write χ_j for χ_{E_j} for any $j \in [s]$. Then for any $i = (i_1, \ldots, i_n) \in [s]^n$,

$$\chi_i = \chi_{i_1} \otimes \cdots \otimes \chi_{i_n} : G^n \to \mathbb{C} : (g_1, \ldots, g_n) \mapsto \chi_{i_1}(g_1), \ldots, \chi_{i_n}(g_n)$$

is the character of the tensor product representation of G^n on $E_{i_1} \otimes \cdots \otimes E_{i_n}$. The functions χ_i are orthonormal in $L^2(G^n)$, and s^n in number. Now s^n is the number of conjugacy classes in G^n. Hence, $E_{i_1} \otimes \cdots \otimes E_{i_n}$ runs over all the irreducible representations of G^n as $(i_1, \ldots, i_n)$ runs over $[s]^n$. $\boxed{\text{QED}}$

The appearance of the Hermitian inner product $\langle \cdot, \cdot \rangle_{L^2}$ may be a bit unsettling: where did it come from? Is it somehow "natural"? The key feature that makes this pairing of functions on G so useful is its invariance:

Proposition 7.3 *For any finite group G, identify $L^2(G)$ with the group algebra $\mathbb{C}[G]$ by the linear isomorphism*

$$I : L^2(G) \to \mathbb{C}[G] : f \mapsto I(f) = \underline{f},$$

where

$$\underline{f} = \frac{1}{|G|} \sum_{h \in G} f(h^{-1}) h.$$

Then the regular representation ρ_{reg} of G corresponds to the representation $R_{\text{reg}} = I^{-1} \rho_{\text{reg}} I$ on $L^2(G)$ given explicitly by

$$(R_{\text{reg}}(g)f)(h) = f(hg) \tag{7.50}$$

for all $g, h \in G$, and $f \in L^2(G)$. Moreover, R_{reg} is a unitary representation of G on $L^2(G)$:

$$\langle R_{\text{reg}}(g)f_1, R_{\text{reg}}(g)f_2 \rangle_{L^2} = \langle f_1, f_2 \rangle_{L^2} \tag{7.51}$$

for all $g \in G$ and all $f_1, f_2 \in L^2(G)$.

The proof is straightforward verification, which we leave as an exercise.

There is still one curiosity that has not been satisfied: does the G-invariance of the inner product pin it down uniquely up to multiples? Briefly, the answer is "nearly"; explore this in Exercise 7.10 (and look back at Exercise 1.18 for some related ideas.)

7.4   Comparing Z-Bases

We will work with a finite group G and an algebraically closed field $\mathbb{F}$ in which $|G|1_{\mathbb{F}} \neq 0$.

We have seen two natural bases for the center Z of $\mathbb{F}[G]$. One consists of all the conjugacy class sums

$$z_C = \sum_{g \in C} g, \tag{7.52}$$

with C running over $\mathcal{C}$, the set of all conjugacy classes in G (take a quick look back at Theorem 3.3). The other consists of $u_1, \ldots, u_s$, which form the maximal set of nonzero orthogonal central idempotents in $\mathbb{F}[G]$ adding up to 1 (for this, see Proposition 4.6). Our goal in this section is to express these two bases in terms of each other by using the simple characters of G.

Pick a simple left ideal L_i in the two-sided ideal $\mathbb{F}[G]u_i$, for each $i \in [s]$, and let χ_i be the character of ρ_i, the restriction of the regular representation to the submodule $L_i \subset \mathbb{F}[G]$. Then $\chi_1, \ldots, \chi_s$ are all the distinct irreducible characters of G. Multiplication by u_i acts as the identity on the block $\mathbb{F}[G]u_i$ and is 0 on all other blocks $\mathbb{F}[G]u_j$ for $j \neq i$. Moreover,

$$\mathbb{F}[G]u_i \simeq L_i^{d_i},$$

where

$$d_i = \dim_{\mathbb{F}} L_i.$$

From this we see that $\chi_{\text{reg}}(gu_j)$ is the trace of a block-diagonal matrix, with one $d_j \times d_j$ block given by $\rho_j(g)$ and all other blocks being 0; hence,

$$\chi_{\text{reg}}(gu_j) = \chi_j(g)d_j, \tag{7.53}$$

for all $g \in G$ and $j \in [s]$, with χ_{reg} being the character of the regular representation, given explicitly by

$$\chi_{\text{reg}}(g) = \begin{cases} |G| & \text{if } g = e, \\ 0 & \text{if } g \neq e. \end{cases} \tag{7.54}$$

We are ready to prove the basis conversion result:

Theorem 7.11 *Let $\chi_1, \ldots, \chi_s$ be all the distinct irreducible characters of a finite group G over an algebraically closed field $\mathbb{F}$ in which $|G|1_{\mathbb{F}} \neq 0$, and let $d_j = \chi_j(e)$ be the dimension of the representation space for χ_j. Then the elements*

$$u_i = \sum_{g \in G} \frac{d_i}{|G|} \chi_i(g^{-1}) g = \sum_{C \in \mathcal{C}} \frac{d_i}{|G|} \chi_i(C^{-1}) z_C, \qquad (7.55)$$

for $i \in [s]$, form the maximal set of nonzero orthogonal central idempotents adding up to 1 in $\mathbb{F}[G]$, where $\mathcal{C}$ is the set of all conjugacy classes in G and $\chi_i(C^{-1})$ denotes the value of χ_i on any element in the conjugacy class $C^{-1} = \{c^{-1} : c \in C\}$. In the other direction,

$$z_C = \sum_{j=1}^{s} \frac{|C|}{d_j} \chi_j(C) u_j \qquad (7.56)$$

for every $C \in \mathcal{C}$.

Proof. Writing u_i as

$$u_i = \sum_{g \in G} u_i(g) g$$

and applying χ_{reg} to $g^{-1} u_i$, we have

$$u_i(g)|G| = \chi_{\text{reg}}(g^{-1} u_i) = \chi_i(g^{-1}) d_i. \qquad (7.57)$$

Thus,

$$u_i = \sum_{g \in G} \frac{d_i}{|G|} \chi_i(g^{-1}) g, \qquad (7.58)$$

and the sum can be condensed into a sum over conjugacy classes since $\frac{d_i}{|G|} \chi_i(g^{-1})$ is constant when g runs over a conjugacy class.

To prove (7.56), note first that since $u_1, \ldots, u_s$ is a basis of Z, we can write

$$z_C = \sum_{j=1}^{s} \lambda_j u_j, \qquad (7.59)$$

for some $\lambda_1, \ldots, \lambda_s \in \mathbb{F}$. To find the value of λ_j, apply the character χ_j to z_C:

$$\chi_j(z_C) = \sum_{g \in C} \chi_j(g) = |C| \chi_j(C). \qquad (7.60)$$

Because $\chi_j(u_i) = \delta_{ij} d_j$, we see from (7.59) that $\chi_j(z_C)$ is also $\lambda_j d_j$. Hence, we have (7.56). $\boxed{\text{QED}}$

More insight into (7.56) will be revealed in (7.80).

We will put the basis change formulas to use in the next two sections to explore two very different paths.

7.5   Character Arithmetic

In this section we venture out very briefly in a direction quite different from that which we have been exploring in this chapter. Our first objective is to prove the following remarkable result:

Theorem 7.12 *The dimension of any irreducible representation of a finite group G is a divisor of $|G|$ if the underlying field $\mathbb{F}$ for the representation is algebraically closed and has characteristic 0.*

We will work with a finite group G, of order $n = |G|$, and a field $\mathbb{F}$ which is algebraically closed and has characteristic 0. Being a field of characteristic 0, $\mathbb{F}$ contains a copy of $\mathbb{Z}$ and hence also a copy of the rationals $\mathbb{Q}$. Being algebraically closed, such a field also contains n distinct nth roots of unity. Moreover, these roots form a multiplicative group which has generators called *primitive nth roots of unity* (for $\overline{\mathbb{Q}}$ these are $e^{2\pi ki/n}$, with $k \in [n]$ coprime to n).

A key fact to be used is the arithmetic feature of characters we noted in Theorem 1.5: the value of any character of G is a sum of nth roots of unity. We will first reformulate this slightly using some new terminology.

A polynomial $p(X)$ is said to be *monic* if it is of the form $\sum_{k=0}^{m} p_k X^k$ with $p_m = 1$ and $m \geq 1$. An element $\alpha \in \mathbb{F}$ is an *algebraic integer* if $p(\alpha) = 0$ for some monic polynomial $p(X) \in \mathbb{Z}[X]$. Here are two useful basic facts:

1. The sum or product of two algebraic integers is an algebraic integer, and so the set of all algebraic integers is a ring.

2. If $x \in \mathbb{Q}$ is an algebraic integer, then $x \in \mathbb{Z}$.

Proofs are given in Sect. 12.7.

With this language and technology at hand, here is a restatement of Theorem 1.5:

Theorem 7.13 *Suppose G is a group containing n elements and $\mathbb{F}$ is a field of characteristic 0 containing n distinct nth roots of unity. Then for any representation ρ of G on a finite-dimensional vector space over $\mathbb{F}$ and for any $g \in G$, the value $\chi_\rho(g)$ is a linear combination of $1, \eta, \ldots, \eta^{n-1}$ with integer coefficients, where η is a primitive nth root of unity; thus, $\chi_\rho(g) \in \mathbb{Z}[\eta]$ viewed as a subring of $\mathbb{F}$. In particular, $\chi_\rho(g)$ is an algebraic integer.*

We can turn now to proving Theorem 7.12.

Proof of Theorem 7.12. Let $u_1, \ldots, u_s$ be the maximal set of nonzero orthogonal central idempotents adding up to 1 in $\mathbb{F}[G]$; we will work with any particular u_i. From formula (7.55) we have

$$\frac{n}{d_i} u_i = \sum_{g \in G} \chi_i(g^{-1}) g. \tag{7.61}$$

On the right we have an element of $\mathbb{F}[G]$ in which all coefficients are in the ring $\mathbb{Z}[\eta]$. The interesting observation here is that multiplication by n/d_i carries $u_i h$ into a linear combination of the elements $u_i g$ with coefficients in $\mathbb{Z}[\eta]$:

$$\frac{n}{d_i} u_i h = u_i \frac{n}{d_i} u_i h = \sum_{g \in G} \chi_i(g^{-1}) u_i g h.$$

Thus, on the $\mathbb{Z}$-module F consisting of all linear combinations of the elements $u_i g$ with coefficients in $\mathbb{Z}[\eta]$, multiplication by n/d_i acts as a $\mathbb{Z}$-linear map $F \to F$. Then (do Exercise 7.3 and find that) there is a monic polynomial $p(X)$ such that $p(n/d_i) = 0$. Thus, n/d_i is an algebraic integer. But then, by fact 2 above, it must be an integer in $\mathbb{Z}$, which means that d_i divides n.

QED

Here is some more on characters with an arithmetic flavor:

Theorem 7.14 *Suppose G is a finite group and $\mathbb{F}$ is an algebraically closed field in which $|G| 1_{\mathbb{F}} \neq 0$. Let χ be the character of an irreducible representation of G on a vector space of dimension d over the field $\mathbb{F}$. Then*

$$\frac{|C|}{d} \chi(C)$$

is an algebraic integer, for any conjugacy class C in G.

Proof. Let $u_1, \ldots, u_s$ be the maximal set of nonzero orthogonal central idempotents adding up to 1 in $\mathbb{F}[G]$, and let $C_1, \ldots, C_s$ be all the distinct conjugacy classes in G. Let

$$z_i = z_{C_i} = \sum_{g \in C_i} g.$$

Recall from (7.56) that

$$z_i = \sum_{j=1}^{s} \frac{|C_i|}{d_j} \chi_j(C_i) u_j,$$

from which we have

$$z_i u_k = \frac{|C_i|}{d_k} \chi_k(C_i) u_k.$$

Then

$$\frac{|C_i|}{d_k} \chi_k(C_i) z_j u_k = z_j z_i u_k$$

$$= \sum_{k=1}^{s} \kappa_{i,m\,j} z_m u_k, \tag{7.62}$$

where the structure constants $\kappa_{i,m\,j}$ are integers specified by

$$z_i z_j = \sum_{m=1}^{s} \kappa_{i,m\,j} z_m, \tag{7.63}$$

and given more specifically by

$$\kappa_{i,m\,j} = |\{(a,b) \in C_i \times C_j : ab = h\}| \quad \text{for any fixed } h \in C_m. \tag{7.64}$$

(We encountered these in (3.7) and will work with them again shortly.) The equality of the first term and the last term in (7.62) implies that, for each fixed $i, k \in [s]$, multiplication by $\frac{|C_i|}{d_k} \chi_k(C_i)$ is a $\mathbb{Z}$-linear map of the $\mathbb{Z}$-module spanned by the elements $z_m u_k$ with m running over $[s]$:

$$\sum_{m=1}^{s} \mathbb{Z} z_m u_k \to \sum_{m=1}^{s} \mathbb{Z} z_m u_k : x \mapsto \frac{|C_i|}{d_k} \chi_k(C_i) x. \tag{7.65}$$

Then, just as in the proof of Theorem 7.12, Exercise 7.3 implies that $\frac{|C_i|}{d_k} \chi_k(C_i)$ is an algebraic integer. $\boxed{\text{QED}}$

We will return to a simpler proof in the next section which will give an explicit monic polynomial (7.76), with integer coefficients, of which the quantities $\frac{|C|}{d} \chi(C)$ are solutions.

7.6 Computing Characters

In his classic work, Burnside [9, Sect. 223] describes an impressive method of working out all irreducible complex characters of a finite group directly from the multiplication table for the group, without ever having to work out any irreducible representations! This is an amazing achievement, viewed from the logical pathway we have followed. However, from the viewpoint of the historical pathway, this is only natural, because Frobenius [29, (8)] effectively *defined* characters by this method using just the group multiplication table.

We will work with a finite group G and an algebraically closed field $\mathbb{F}$ in which $|G|1_{\mathbb{F}} \neq 0$.

Under our hypotheses on $\mathbb{F}$, the number of conjugacy classes in G is s, the number of distinct irreducible representations of G. Let $C_1, \ldots, C_s$ be the distinct elements of $\mathcal{C}$. Let $\rho_1, \ldots, \rho_s$ be a maximal collection of inequivalent irreducible representations of G, and let χ_j be the character of ρ_j and d_j be the dimension of ρ_j. Let z_i be the sum of the elements in the conjugacy class C_i:

$$z_i = \sum_{g \in C_i} g \quad \text{for } i \in \{1, \ldots, s\}.$$

Recall the basis change formula (7.66):

$$z_j = \sum_{i=1}^{s} \frac{|C_j|}{d_i} \chi_i(C_j) u_i \tag{7.66}$$

for every $j \in \{1, \ldots, s\}$. For any $z \in Z$, the center of $\mathbb{F}[G]$, let $M(z)$ be the linear map

$$M(z) : Z \to Z : w \mapsto zw. \tag{7.67}$$

This is just the restriction of the regular representation to Z. The idea is to extract information by looking at the matrix of $M(z)$ first for the basis $z_1, \ldots, z_s$, and then for the basis $u_1, \ldots, u_s$.

Now take a quick look back at Proposition 3.1: the structure constants $\kappa_{j,ik} \in \mathbb{F}$ are specified by the requirement that

$$z_k z_j = \sum_{l=1}^{s} \kappa_{k,ij} z_i \quad \text{for all } j, k \in [s]. \tag{7.68}$$

Another way to view the structure constants $\kappa_{j,ik}$ is given by

$$\kappa_{k,ij} = |\{(a,b) \in C_k \times C_j : ab = c\}| \qquad (7.69)$$

for any fixed choice of c in C_i. Clearly, at least in principle, the structure constants can be worked out from the multiplication table for the group G. Then, *relative to the basis* $z_1, \ldots, z_s$, *the matrix* $M(k)$ *of* $M(z_k)$ *has the* (i,j)*th entry given by* $\kappa_{k,ij}$:

$$M(k) = \begin{bmatrix} \kappa_{k,11} & \kappa_{k,12} & \cdots & \kappa_{k,1s} \\ \kappa_{k,21} & \kappa_{k,22} & \cdots & \kappa_{k,2s} \\ \vdots & \vdots & \cdots & \vdots \\ \kappa_{k,s1} & \kappa_{k,s2} & \cdots & \kappa_{k,ss} \end{bmatrix}. \qquad (7.70)$$

Now consider the action of $M(z_k)$ on u_j,

$$M(z_k)u_j = z_k u_j = \frac{|C_k|}{d_j}\chi_j(C_k)u_j, \qquad (7.71)$$

by using (7.66). Thus, the elements $u_1, \ldots, u_s$ are *eigenvectors* for $M(z_k)$, with u_j having eigenvalue $\frac{|C_k|}{d_j}\chi_j(C_k)$.

Recalling formula (7.55),

$$u_j = \sum_{k=1}^{s} \frac{d_j}{|G|}\chi_j(C_k^{-1})z_k,$$

we can display u_j as a column vector, with respect to the basis $z_1, \ldots, z_s$, as

$$\vec{u_j} = \begin{bmatrix} \frac{d_j}{|G|}\chi_j(C_1^{-1}) \\ \vdots \\ \frac{d_j}{|G|}\chi_j(C_s^{-1}) \end{bmatrix}. \qquad (7.72)$$

Then, in matrix form,

$$M(k)\vec{u_j} = \frac{|C_k|}{d_j}\chi_j(C_k)\vec{u_j}. \qquad (7.73)$$

Thus, for each fixed $j \in [s]$, the vector $\vec{u_j}$ is a simultaneous eigenvector of the s matrices $M(1), \ldots, M(s)$.

A program that computes eigenvectors and eigenvalues can then be used to work out the values $\frac{|C_j|}{d_i}\chi_i(C_j)$. Next recall the character orthogonality relation (7.25), which we can write as

$$\sum_{k=1}^{s} |C_k|\chi_i(C_k)\chi_i(C_k^{-1}) = |G|, \tag{7.74}$$

and then as

$$\sum_{k=1}^{s} \frac{1}{|C_k|} \frac{|C_k|}{d_i}\chi_i(C_k)\frac{|C_k^{-1}|}{d_i}\chi_i(C_k^{-1}) = \frac{|G|}{d_i^2}. \tag{7.75}$$

Thus, once we have computed the eigenvalue $\frac{|C|}{d_i}\chi_i(C)$ for each conjugacy class C and each $i \in [s]$, we can determine $|G|/d_i^2$ and hence the values $d_1, \ldots, d_s$. Finally, we can compute the values $\chi_i(C)$ of the characters χ_i on all the conjugacy classes C as

$$\chi_i(C) = \frac{1}{|C|}d_i\frac{|C|}{d_i}\chi_i(C).$$

An unpleasant feature of this otherwise wonderful procedure is that for $\mathbb{F} = \mathbb{C}$ the eigenvalues, which are complex numbers, are determined only approximately by a typical matrix algebra program that computes eigenvalues. Dixon [23] showed how character values can be computed exactly once they are known to close enough approximation (this was explored in Exercise 1.21). Dixon also provided a method of computing the characters exactly by using reduction mod p, for large enough prime p. These ideas have been coded in programs such as GAP that compute group characters.

There is one pleasant theoretical consequence of the exploration of the matrices M_k; this is Frobenius's simple proof of Theorem 7.14:

Simple proof of Theorem 7.14. As usual, let G be a finite group, $\mathbb{F}$ be an algebraically closed field in which $|G|1_{\mathbb{F}} \neq 0$, $C_1, \ldots, C_s$ be all the distinct conjugacy classes in G, $\chi_1, \ldots, \chi_s$ be the distinct irreducible characters of G, over the field $\mathbb{F}$, and d_j be the dimension of the representation for the character χ_j. Then, as we have seen already, the matrices $M(k)$, with *integer entries* as given in (7.70), have the eigenvalues $\frac{|C_k|}{d_j}\chi_j(C_k)$. Thus, these eigenvalues are solutions for $\lambda \in \mathbb{F}$ of the characteristic equation

$$\det(\lambda I - M(k)) = 0, \tag{7.76}$$

which is clearly a monic polynomial. Because all entries of the matrix $M(k)$ are integers, all coefficients in the polynomial in λ on the left side of (7.76) are also integers. Hence, each $\frac{|C_k|}{d_j}\chi_j(C_k)$ is an algebraic integer. $\boxed{\text{QED}}$

Here is a simple example, going back to Burnside [9, Section 222] and Frobenius and Schur [36], of the interplay between the properties of a group and of its characters.

Theorem 7.15 *If G is a finite group for which every complex character is real-valued, then $|G|$ is even.*

Proof. Suppose $|G|$ is odd. Then, since the order of every element of G is a divisor of $|G|$, there is no element of order 2 in G, and so $g \neq g^{-1}$ for all $g \neq e$. If χ is a nontrivial irreducible complex character of G, then

$$\sum_{g \in G} \chi(g) = 0,$$

by orthogonality with the trivial character. Since χ is, by hypothesis, real-valued, we have

$$\chi(g) = \chi(g^{-1}) \text{ for all } g \in G,$$

and then

$$0 = \sum_g \chi(g) = \chi(e) + \sum_{g \in S}(\chi(g) + \chi(g^{-1})) = d + 2\sum_{g \in S}\chi(g),$$

where d is the dimension of the representation for χ, and S is a set containing half the elements of $G - \{e\}$. But then $d/2$ is both a rational and an algebraic integer and hence (see Proposition 12.3) it is actually an integer in $\mathbb{Z}$. Thus, d is even. $\boxed{\text{QED}}$

For a restatement, with an elementary proof, do Exercise 7.11.

7.7 Return of the Group Determinant

Let G be a finite group with n elements and $\mathbb{F}$ be a field. Dedekind's group determinant, described in his letters [19] to Frobenius, is the determinant of the $|G| \times |G|$ matrix

$$[X_{ab^{-1}}]_{a,b \in G},$$

where X_g is a variable associated with each $g \in G$. Let F_G be the matrix formed in the case where the variables are chosen so that $X_a = X_b$ when a and b are in the same conjugacy class. The matrix F_G was introduced by Frobenius [35, (11)]. For more history, aside from the original works of Frobenius [29, 30, 31, 32, 33, 34, 35, 36] and Dedekind [19], see the books by Hawkins [42, Chap. 10] and Curtis [14] and the article by Lam [53]; Hawkins [43] also presents an enjoyable and enlightening analysis of letters from Frobenius to Dedekind.

Let $\mathbb{F}$ be a field and R be the regular representation of G; thus, for $g \in G$,

$$R(g) : \mathbb{F}[G] \to \mathbb{F}[G] : y \mapsto gy.$$

Then the (a, b)th entry of the matrix of $R(g)$, relative to the basis of $\mathbb{F}[G]$ given by the elements of G, is

$$R(g)_{ab} = \begin{cases} 1 & \text{if } gb = a, \\ 0 & \text{if } gb \neq a, \end{cases}$$

which means $R(g)_{ab} = 1$ if $g = ab^{-1}$, and $R(g)_{ab} = 0$ otherwise. Then the matrix for $\sum_{g \in G} R(g) X_g$ has (a, b)th entry $X_{ab^{-1}}$. Thus,

$$F_G = \sum_{g \in G} R(g) X_g. \tag{7.77}$$

Since X_g has a common value, call it X_C, for all g in a conjugacy class C, we can rewrite F_G as

$$F_G = \sum_{C \in \mathcal{C}} R(z_C) X_C, \tag{7.78}$$

where $\mathcal{C}$ is the set of conjugacy classes in G, and z_C is the conjugacy class sum

$$z_C = \sum_{g \in C} g. \tag{7.79}$$

Now suppose the field $\mathbb{F}$ is such that $|G| 1_\mathbb{F} \neq 0$. Then there are simple left ideals $L_1, \ldots, L_s$ in $\mathbb{F}[G]$ such that every simple left ideal in $\mathbb{F}[G]$ is isomorphic, as a left $\mathbb{F}[G]$-module, to L_i for exactly one $i \in [s]$, and the $\mathbb{F}$-algebra $\mathbb{F}[G]$ is isomorphic to the product of subalgebras $A_1, \ldots, A_s$, where A_i is the sum of all left ideals isomorphic to L_i. Assume, moreover, that $\mathbb{F}$ is a *splitting field* for G in that $\text{End}_{\mathbb{F}[G]}(L_i)$ consists of just the constant maps

$x \mapsto cx$ for $c \in \mathbb{F}$. For instance, $\mathbb{F}$ could be algebraically closed. Then A_i is the direct sum of d_i simple left ideals, where $d_i = \dim_{\mathbb{F}} L_i$. For any element z in the center Z of $\mathbb{F}[G]$, the endomorphism $R(z)$ acts as multiplication by a scalar $c_z \in \mathbb{F}$ on each L_i. Denoting by χ_i the character of the regular representation restricted to L_i, we have

$$\chi_i(z) \overset{\text{def}}{=} \mathrm{Tr}\,(R(z)|L_i) = \mathrm{Tr}\,(c_z I_{L_i}) = c_z d_i,$$

where I_{L_i} is the identity mapping on L_i. Hence,

$$c_z = \frac{1}{d_i}\chi_i(z).$$

Taking z_C for z shows that

$$R(z_C)|L_i = \frac{|C|}{d_i}\chi_i(C)I_i, \tag{7.80}$$

where $\chi_i(C)$ is the value of the character χ_i on any element in C (and not to be confused with $\chi_i(z_C)$ itself). Consequently,

$$F_G|L_i = \sum_{C \in \mathcal{C}} \frac{|C|}{d_i}\chi_i(C)X_C I_i. \tag{7.81}$$

Thus, F_G can be displayed as a giant block-diagonal matrix, with each $i \in [s]$ contributing d_i blocks, each such block being the scalar matrix in (7.81). Taking the determinant, we have

$$\det F_G = \prod_{i=1}^{s} \left(\sum_{C \in \mathcal{C}} \frac{|C|}{d_i}\chi_i(C)X_C \right)^{d_i^2}. \tag{7.82}$$

The entire universe of representation theory grew as a flower from Frobenius's meditation on Dedekind's determinant. Formula (7.82) (Frobenius [29, (22)] and [30]) shows how all the characters of G are encoded in the determinant.

For S_3, with conjugacy classes labeled by variables Y_1 (identity element), Y_2 (transpositions), and Y_3 (three-cycles), (7.82) reads

$$
\begin{vmatrix}
Y_1 & Y_3 & Y_3 & Y_2 & Y_2 & Y_2 \\
Y_3 & Y_1 & Y_3 & Y_2 & Y_2 & Y_2 \\
Y_3 & Y_3 & Y_1 & Y_2 & Y_2 & Y_2 \\
Y_2 & Y_2 & Y_2 & Y_1 & Y_3 & Y_3 \\
Y_2 & Y_2 & Y_2 & Y_3 & Y_1 & Y_3 \\
Y_2 & Y_2 & Y_2 & Y_3 & Y_3 & Y_1
\end{vmatrix}
\tag{7.83}
$$
$$
= (Y_1 - Y_3)^4 (Y_1 + 3Y_2 + 2Y_3)(Y_1 - 3Y_2 + 2Y_3),
$$

which you can verify directly at your leisure/pleasure.

7.8 Orthogonality of Matrix Elements

In this section we will again use the bra-ket formalism from the end of Sect. 1.6. By a *matrix element* for a group G we mean a function on G of the form

$$
G \to \mathbb{F} : g \mapsto \langle e' | \rho(g) | e \rangle,
$$

where ρ is a representation of G on a vector space E over a field $\mathbb{F}$, and $|e\rangle \in E$ and $\langle e' |$ is a vector in the dual space E'. (Note that "matrix element" does not mean the entry in some matrix.)

In this section we will explore some straightforward extensions of the orthogonality relations from characters to matrix elements.

Theorem 7.16 *If ρ_E and ρ_F are inequivalent irreducible representations of a finite group G on vector spaces E and F, respectively, then the matrix elements of ρ and ρ' are orthogonal in the sense that*

$$
\sum_{g \in G} \langle f' | \rho_F(g) | f \rangle \langle e' | \rho_E(g^{-1}) | e \rangle = 0
\tag{7.84}
$$

for all $\langle f' | \in F^*$, $\langle e' | \in E^*$ and all $|e\rangle \in E$, $|f\rangle \in F$.

Proof. The linear map

$$
T_1 = \sum_{g \in G} \rho_F(g) | f \rangle \langle e' | \rho_E(g^{-1}) : E \to F
$$

is $\mathbb{F}[G]$-linear and hence is 0 by Schur's lemma. $\boxed{\text{QED}}$

Now assume that $\mathbb{F}$ is algebraically closed and has characteristic 0. Let E be a fixed irreducible representation of G. Then Schur's lemma implies that for any $T \in \mathrm{End}_{\mathbb{F}}(E)$ the symmetrized operator T_0 on the left in (7.85) is a multiple of the identity. The value of this multiplier is easily obtained by comparing traces,

$$\frac{1}{|G|} \sum_{g \in G} g T g^{-1} = T_0 = \frac{1}{\dim_{\mathbb{F}} E} \mathrm{Tr}(T) I, \qquad (7.85)$$

noting that both sides have trace equal to $\mathrm{Tr}(T)$.

Working with a basis $\{e_i\}_{i \in I}$ of E, with dual basis $\{\langle e^j |\}_{j \in I}$ satisfying

$$\langle e^j | e_i \rangle = \delta_i^j,$$

we then have

$$\langle e^j | T_0 | e_i \rangle = \frac{1}{\dim_{\mathbb{F}} E} \mathrm{Tr}(T) \delta_i^j \quad \text{for all } i, j \in I. \qquad (7.86)$$

Taking for T the particular operator

$$T = \rho_E(h) |e_k\rangle \langle e^l|$$

shows that

$$\frac{1}{|G|} \sum_{g \in G} \langle e^j | \rho_E(gh) | e_k \rangle \langle e^l | \rho_E(g^{-1}) | e_i \rangle = \frac{1}{\dim_{\mathbb{F}} E} \rho_E(h)_k^l \delta_i^j \quad \text{for all } i, j \in I. \qquad (7.87)$$

A look back at (7.44) provides an interpretation of this in terms of convolution.

We can summarize our observations by specializing to $\mathbb{F} = \mathbb{C}$:

Theorem 7.17 *Let $E_1, \ldots, E_s$ be a collection of irreducible representations of a finite group G, over an algebraically closed field $\mathbb{F}$ in which $|G|1_{\mathbb{F}} \neq 0$, such that every irreducible complex representation of G is equivalent to E_i for exactly one $i \in [s]$. For each $r \in [s]$, choose a basis $\{|e(r)_i\rangle : i \in [d_r]\}$, where $d_r = \dim_{\mathbb{F}} E_r$, and let $\{\langle e(r)^i| : i \in [d_r]\}$ be the corresponding dual basis in E_r'. Let $\rho_{r,ij}$ be the matrix element:*

$$\rho_{r,ij} : G \to \mathbb{C} : g \mapsto \langle e(r)^i | \rho_{E_r}(g) | e(r)_j \rangle.$$

Then the scaled matrix elements

$$d_r^{1/2} \rho_{r,ij}, \tag{7.88}$$

with $i, j \in [d_r]$, and r running over $[s]$, form an orthonormal basis of $L^2(G)$. Moreover, the convolution of matrix elements of an irreducible representation E is a multiple of a matrix element for the same representation, the multiplier being 0 or $1/\dim_{\mathbb{C}} E$.

Proof. From the orthogonality relation (7.84) and the identity (7.85), it follows that the functions in (7.88) are orthonormal in $L^2(G)$. The total number of these functions is

$$\sum_{r=1}^{s} d_r^2.$$

But this is precisely the number of elements in G, which is also the same as $\dim L^2(G)$. Thus, the functions (7.88) form a basis of $L^2(G)$. The convolution result follows from (7.87) on replacing g by gh^{-1}. $\boxed{\text{QED}}$

7.9   Solving Equations in Groups

We close our exploration of characters with an application with which Frobenius [29] began his development of the notion of characters. This is the task of counting solutions of equations in a group.

Theorem 7.18 *Let $C_1, \ldots, C_m$ be distinct conjugacy classes in a finite group G. Then*

$$|\{(c_1, \ldots, c_m) \in C_1 \times \cdots \times C_m \,|\, c_1 \ldots c_m = e\}|$$

$$= \frac{|C_1| \ldots |C_m|}{|G|} \sum_{i=1}^{s} \frac{1}{d_i^{m-2}} \chi_i(C_1) \ldots \chi_i(C_m), \tag{7.89}$$

where $\chi_1, \ldots, \chi_s$ are all the distinct irreducible characters of G, over an algebraically closed field $\mathbb{F}$ in which $|G|1_{\mathbb{F}} \neq 0$, d_i is the dimension of the representation for the character χ_i, and $\chi_i(C)$ is the constant value of χ_i on C. Moreover,

$$|\{(c_1, \ldots, c_m) \in C_1 \times \cdots \times C_m \,|\, c_1 \ldots c_m = c\}|$$

$$= \frac{|C_1| \ldots |C_m|}{|G|} \sum_{i=1}^{s} \frac{1}{d_i^{m-1}} \chi_i(C_1), \ldots, \chi_i(C_m) \chi_i(c^{-1})$$

$$\tag{7.90}$$

for any $c \in G$. The left sides of (7.89) and (7.90), integers as they stand, are viewed as elements of $\mathbb{F}$, by multiplication with $1_\mathbb{F}$.

As always, the algebraic closedness for $\mathbb{F}$ may be weakened to the requirement that it is a splitting field for G.

Proof. Let $z_i = \sum_{g \in C_i} g$ be the element in the center Z of $\mathbb{F}[G]$ corresponding to the conjugacy class C_i. Recall the trace functional Tr_e on $\mathbb{F}[G]$ given by $\mathrm{Tr}_e(x) = x_e$, the coefficient of e in $x = \sum_g x_g g \in \mathbb{F}[G]$. Clearly,

$$\mathrm{Tr}_e(z_1 \ldots z_m) = |\{(c_1, \ldots, c_m) \in C_1 \times \cdots \times C_m \,|\, c_1 \ldots c_m = e\}|, \qquad (7.91)$$

where the right side is taken as an element in $\mathbb{F}$. This is the key observation; the rest of the argument is a matter of working out the trace on the left from the trace of the regular representation, decomposed into simple submodules. Using the regular representation R, given by

$$R(x) : \mathbb{F}[G] \to \mathbb{F}[G] : y \mapsto xy \qquad \text{for all } x \in \mathbb{F}[G],$$

we have

$$\mathrm{Tr}\, R(x) = |G| \mathrm{Tr}_e(x) \qquad \text{for all } x \in \mathbb{F}[G].$$

So

$$\mathrm{Tr}_e(z_1 \ldots z_m) = \frac{1}{|G|} \mathrm{Tr}\, R(z_1 \ldots z_m). \qquad (7.92)$$

Now recall relation (7.80),

$$R(z_j)|L_i = \frac{|C_j|}{d_i} \chi_i(C_j) I_i, \qquad (7.93)$$

where I_i is the identity map on L_i, and $L_1, \ldots, L_s$ are simple left ideals in $\mathbb{F}[G]$ such that every simple left ideal in $\mathbb{F}[G]$ is isomorphic to L_i for exactly one $i \in [s]$. As we know from the structure of $\mathbb{F}[G]$, this algebra is the direct sum

$$\mathbb{F}[G] = \bigoplus_{i=1}^{s} (L_{i1} \oplus \ldots \oplus L_{id_i}),$$

where each L_{ik} is isomorphic, as a left $\mathbb{F}[G]$-module, to L_i. On each of the d_i subspaces L_{ik}, each of dimension d_i, the endomorphism $R(z_j)$ acts by multiplication by the scalar $\frac{|C_j|}{d_i} \chi_i(C_j)$. Consequently,

$$\mathrm{Tr}\, R(z_1 \ldots z_m) = \sum_{i=1}^{s} d_i \left(\prod_{j=1}^{m} \frac{|C_j| \chi_i(C_j)}{d_i} \right) d_i. \qquad (7.94)$$

Combining this with the relationship between Tr_e and Tr given in (7.92), along with the counting formula (7.91), yields the number of $(c_1, \ldots, c_m) \in C_1 \times \cdots \times C_m$ with $c_1 \ldots c_m = e$.

Now for any $c \in G$, let

$$P(c) = \{(c_1, \ldots, c_m) \in C_1 \times \ldots \times C_m : c_1 \ldots c_m = c\}.$$

Then for any $h \in G$, the map

$$(g_1, \ldots, g_m) \mapsto (hg_1 h^{-1}, \ldots, hg_m h^{-1})$$

gives a bijection between $P(c)$ and $P(hch^{-1})$. Moreover, the union of the sets $P(c')$ with c' running over the conjugacy class $C(c)$ of c is in bijection with the set

$$\{(c_1, \ldots, c_m, d) \in C_1 \times \cdots \times C_m \times C(c^{-1}) : c_1 \ldots c_m d = e\}.$$

Comparing the cardinalities, we have

$$|C(c)|\,|P(c)| = \frac{|C_1| \ldots |C_m||C_{c^{-1}}|}{|G|} \sum_{i=1}^{s} \frac{1}{d_i^{m-1}} \chi_i(C_1) \ldots \chi_i(C_m)\chi_i(c^{-1})$$

Since $|C(c)|$ equals $|C(c^{-1})|$, this establishes the formula (7.90) for $|P(c)|$. $\boxed{\text{QED}}$

Frobenius [29] also determined the number of solutions to commutator equations in terms of characters:

Theorem 7.19 *Let G be a finite group and χ be the character of an irreducible representation of G on a vector space, of dimension d, over an algebraically closed field $\mathbb{F}$ in which $|G|1_{\mathbb{F}} \neq 0$. Then*

$$\sum_{b \in G} \chi(ab^{-1}hb) = \frac{|G|}{d}\chi(a)\chi(h) \tag{7.95}$$

for all $a, h \in G$, and

$$\sum_{a,b \in G} \chi(aba^{-1}b^{-1}c) = \left(\frac{|G|}{d}\right)^2 \chi(c) \tag{7.96}$$

for all $c \in G$. Moreover,

$$|\{(a,b) \in G^2 : aba^{-1}b^{-1} = c\}| = \sum_{i=1}^{s} \frac{|G|}{d_i}\chi_i(c) \qquad (7.97)$$

for all $c \in G$, where $\chi_1, \ldots, \chi_s$ are all the distinct irreducible characters of G over the field $\mathbb{F}$, and the left side of (7.97) is taken as an element of $\mathbb{F}$ by multiplication with $1_{\mathbb{F}}$.

Proof. For any $a \in G$, let

$$z_a = \sum_{c \in C_a} c,$$

where C_a is the conjugacy class of a. Compare this with the sum

$$\sum_{g \in G} gag^{-1}.$$

Each term in this sum is repeated $|\text{Stab}_a|$ times, where Stab_a is the set $\{g \in G : gag^{-1} = a\}$, and

$$|\text{Stab}_a| = \frac{|G|}{|C_a|}.$$

Hence,

$$z_a = \frac{|C_a|}{|G|} \sum_{g \in G} gag^{-1}. \qquad (7.98)$$

Let R_χ denote an irreducible representation whose character is χ. Then, for any central element z in $\mathbb{F}[G]$, the endomorphism $R(z)$ is multiplication by the constant $\chi(z)/d$; moreover, if z_C is the sum $\sum_{g \in C} g$ for a conjugacy class C, then $\chi(z_C) = |C|\chi(C)$, where χ is the constant value of χ on C. Then

$$\chi(z_a z_h) = \text{Tr}\, R_\chi(z_a)R_\chi(z_h)$$
$$= \text{Tr}\, \left(\frac{|C_a|}{d}\chi(a)\frac{|C_h|}{d}\chi(h)I\right) \qquad (7.99)$$
$$= \frac{|C_a||C_h|}{d^2}\chi(a)\chi(h)d.$$

Now observe that

$$
\begin{aligned}
\chi(z_a z_h) &= \chi\left(\frac{|C_a|}{|G|}\frac{|C_b|}{|G|}\sum_{g,b\in G} gag^{-1}bhb^{-1}\right) \\
&= \frac{|C_a|}{|G|}\frac{|C_b|}{|G|}\chi\left(\sum_{g\in G}\sum_{b\in G} gabhb^{-1}g^{-1}\right) \qquad \text{(on replacing b by gb)} \\
&= \frac{|C_a|}{|G|}\frac{|C_b|}{|G|}|G|\sum_{b\in G}\chi(abhb^{-1}).
\end{aligned}
$$

$$(7.100)$$

Combining this with (7.99), we have

$$
\sum_{b\in G}\chi(abhb^{-1}) = \frac{|G|}{d}\chi(a)\chi(h). \tag{7.101}
$$

Taking ca for a and $h = a^{-1}$, and adding up over a as well, we have

$$
\sum_{a,b\in G}\chi(aba^{-1}b^{-1}c) = \frac{|G|}{d}\sum_a \chi(ca)\chi(a^{-1}) = \left(\frac{|G|}{d}\right)^2 \chi(c)
$$

upon using the character convolution formula in Theorem 7.8. Next, for the count,

$$
\begin{aligned}
|\{(a,b) : aba^{-1}b^{-1} = c\}| &= \sum_{a,b\in G} \mathrm{Tr}_e(aba^{-1}b^{-1}c^{-1}) \\
&= \frac{1}{|G|}\sum_{a,b}\sum_{i=1}^{s} d_i\chi_i(aba^{-1}b^{-1}c^{-1}) \\
&= \frac{1}{|G|}\sum_{i=1}^{s} d_i\frac{|G|^2}{d_i^2}\chi_i(c^{-1}) \\
&= \sum_{i=1}^{s}\frac{|G|}{d_i}\chi_i(c^{-1}).
\end{aligned}
$$

$$(7.102)$$

To finish, note that the replacement $(a,b)\mapsto(b,a)$ changes c to c^{-1}. $\boxed{\text{QED}}$

The previous results on commutator equations and product equations led to a count of solutions of equations that have topological significance, as we will discuss in Sect. 7.11.

Theorem 7.20 *Let G be a finite group and $\chi_1, \ldots, \chi_s$ be all the distinct irreducible characters of G over an algebraically closed field $\mathbb{F}$ in which $|G|1_{\mathbb{F}} \neq 0$. For positive integers n and k, and any conjugacy classes $C_1, \ldots, C_k$ in G, let*

$$M(C_1, \ldots, C_k)$$
$$= \{(\alpha, c_1, \ldots, c_k) \in G^{2n} \times C_1 \times \cdots \times C_k : K_n(\alpha)c_1 \ldots c_k = e\}, \tag{7.103}$$

where

$$K_n(a_1, b_1, \ldots, a_n, b_n) = a_1 b_1 a_1^{-1} b_1^{-1} \ldots a_n b_n a_n^{-1} b_n^{-1}.$$

Then

$$|M(C_1, \ldots, C_k)| = |G| \sum_{i=1}^{s} (|G|/d_i)^{2n-2} \left(\frac{|C_1|\chi_i(C_1)}{d_i} \cdots \frac{|C_k|\chi_i(C_k)}{d_i} \right), \tag{7.104}$$

where the left side is taken as an element of $\mathbb{F}$ by multiplication with $1_{\mathbb{F}}$.

The group G acts by conjugation on $M(C_1, \ldots, C_k)$, and so it seems natural to factor out one term $|G|$ on the right in (7.104); the terms in the sum are algebraic integers. A special case of interest is when $k = 1$ and $C_1 = \{e\}$; then

$$|K_n^{-1}(e)| = |G| \sum_{i=1}^{s} \left(\frac{|G|}{d_i} \right)^{2n-2}. \tag{7.105}$$

Proof. The key observation is that we can disintegrate $M(C_1, \ldots, C_k)$ by means of the projection maps

$$p_j : (a_1, b_1, \ldots, a_n, b_n, c_1, \ldots, c_k) \mapsto (a_j, b_j) \mapsto a_j b_j a_j^{-1} b_j^{-1}.$$

Take any point $h = (h_1, \ldots, h_n) \in G^n$ and consider the preimage in G^{2n} of h under the map

$$p : G^{2n} \to G^n : (a_1, b_1, \ldots, a_n, b_n) \mapsto \big(K_1(a_1, b_1), \ldots, K_1(a_n, b_n)\big).$$

Then $M(C_1, \ldots, C_k)$ is the union of the "fibers" $p^{-1}(h) \times \{(c_1, \ldots, c_k)\}$, with $(c_1, \ldots, c_k)$ running over all solutions in $C_1 \times \ldots \times C_k$ of

$$c_1 \ldots c_k = h_1 \ldots h_n. \tag{7.106}$$

The idea behind the calculation below is best understood by visualizing the set $M(C_1, \ldots, C_k)$ as a union of "fibers" over the points $(h, c_1, \ldots c_k)$ and then by viewing each fiber as essentially a product of sets of the form $K_1^{-1}(c_j)$.

From (7.97) we have

$$|p^{-1}(h_1,\ldots,h_n)| = \prod_{j=1}^{n}\left(\sum_{i=1}^{s}\frac{|G|}{d_i}\chi_i(h_j)\right) = \sum_{i_1,\ldots,i_n\in[s]}\frac{|G|^n}{d_{i_1}\ldots d_{i_n}}\chi_{i_1}(h_1)\ldots\chi_{i_n}(h_n)$$

and then on using the general character convolution formula (7.47), we have

$$\sum_{h_1\ldots h_n=c}|p^{-1}(h_1,\ldots.h_n)| = \sum_{i=1}^{s}\frac{|G|^n}{d_i^n}\frac{|G|^{n-1}}{d_i^{n-1}}\chi_i(c). \tag{7.107}$$

Now we need to sum this up over all solutions of $c_1\ldots c_k = c$ with $(c_1,\ldots,c_k)$ running over $C_1\times\cdots\times C_k$, as noted in (7.106). Using the count formula (7.90), we obtain

$$\frac{|C_1|\ldots|C_k|}{|G|}\sum_{j=1}^{s}\frac{\chi_j(C_1)\ldots\chi_j(C_k)}{d_j^{k-1}}\chi_j(c^{-1})\sum_{i=1}^{s}\frac{|G|^n}{d_i^n}\frac{|G|^{n-1}}{d_i^{n-1}}\chi_i(c). \tag{7.108}$$

Lastly, this needs to be summed over $c\in G$. Using the convolution formula

$$\sum_{c}\chi_j(c^{-1})\chi_i(c) = |G|\delta_{ij},$$

we finally arrive at

$$|M(C_1,\ldots C_k)| = |G|^{2n-1}|C_1|\ldots|C_k|\sum_{i=1}^{s}\frac{\chi_i(C_1)\ldots\chi_i(C_k)}{d_i^{2n+k-2}}. \tag{7.109}$$

$\boxed{\text{QED}}$

Next, we have what is perhaps an even more remarkable count, courtesy of Frobenius and Schur [36, Sect. 4]:

Theorem 7.21 *Let G be a finite group and χ_ρ be the character of an irreducible representation of G on a vector space of dimension d_ρ, over an algebraically closed field $\mathbb{F}$ in which $|G|1_{\mathbb{F}}\neq 0$. Let c_ρ be the Frobenius–Schur indicator of ρ, having value 0 if ρ is not isomorphic to the dual ρ', having value 1 if there is a nonzero G-invariant symmetric bilinear form on V, and having value -1 if there is a nonzero G-invariant skew-symmetric bilinear form on V. Then*

$$\frac{1}{|G|}\sum_{g\in G}\rho(g^2) = \frac{c_\rho}{d_\rho}I, \tag{7.110}$$

where I is the identity map on V, and so

$$\frac{1}{|G|} \sum_{g \in G} \chi_\rho(g^2 b) = \frac{c_\rho}{d_\rho} \chi(b) \tag{7.111}$$

for all $b \in G$. Moreover, if $\rho_1, \ldots, \rho_s$ are a maximal set of inequivalent irreducible representations of G over the field $\mathbb{F}$, then

$$|\{(g_1, \ldots, g_n) \in G^n : g_1^2 \ldots g_n^2 = e\}| = |G| \sum_{i=1}^{s} \left(c_i \frac{|G|}{d_i} \right)^{n-2}, \tag{7.112}$$

where $c_i = c_{\rho_i}$ and $d_i = d_{\rho_i}$, and the equality in (7.112) is with both sides taken as elements of $\mathbb{F}$.

We discussed the Frobenius–Schur indicator c_ρ in Theorem 1.4. Now we have a formula for it:

$$c_\rho = \frac{1}{|G|} \sum_{g \in G|} \chi_\rho(g^2), \tag{7.113}$$

where, recall, $c_\rho \in \{0, 1, -1\}$. For the division by d_ρ in (7.110), and elsewhere, recall from Lemma 7.1 that $d_\rho \neq 0$ in $\mathbb{F}$.

Proof. Fix a basis $u_1, \ldots, u_d$ of V. For any particular $a, b \in [d]$, let B be the bilinear form on V for which $B(u_i, u_j)$ is 0 except for $(i, j) = (a, b)$, in which case $B(u_a, u_b) = 1$. Now let S be the corresponding G-invariant bilinear form specified by

$$S(v, w) = \sum_{g \in G} B\big(\rho(g)v, \rho(g)w\big).$$

By Theorem 1.4,

$$S(v, w) = c_\rho S(w, v)$$

for all $v, w \in V$. Taking $v = u_i$ and $w = u_j$, we obtain

$$\sum_{g \in G} \rho(g)_{ai} \rho(g)_{bj} = c_\rho \sum_{g \in G} \rho(g)_{aj} \rho(g)_{bi}. \tag{7.114}$$

This holds for all $i, j, a, b \in [d]$. Taking $i = b$ and summing over i brings us to

$$\sum_{g \in G} [\rho(g)^2]_{aj} = c_\rho \sum_{g \in G} \chi_\rho(g) \rho(g)_{aj},$$

which means

$$\sum_{g \in G} \rho(g^2) = c_\rho \sum_{g \in G} \chi_\rho(g) \rho(g). \tag{7.115}$$

Taking the trace of this produces

$$\sum_{g \in G} \chi(g^2) = c_\rho \sum_{g \in G} \chi_\rho(g)^2. \tag{7.116}$$

If ρ is isomorphic to ρ', then

$$\chi(g) = \chi_\rho(g) = \chi_{\rho'}(g) = \chi_\rho(g^{-1}) = \chi(g^{-1})$$

for all $g \in G$, and so the sum $\sum_g \chi(g)^2$ is the same as $\sum_g \chi(g^{-1})\chi(g)$, which, in turn, is just $|G|$. Then (7.116) implies

$$c_\rho = \frac{1}{|G|} \sum_{g \in G} \chi(g^2). \tag{7.117}$$

If ρ is not isomorphic to its dual ρ', then, by definition, $c_\rho = 0$, and so from (7.116) we see that (7.117) still holds.

Since $\sum_{g \in G} g^2$ is in the center of $\mathbb{F}[G]$, and ρ is irreducible, Schur's lemma implies that $\sum_{g \in G} \rho(g^2)$ is a scalar multiple kI of the identity I, and the scalar k is obtained by comparing traces:

$$\sum_{g \in G} \rho(g^2) = kI, \tag{7.118}$$

where

$$k = \frac{1}{d} \mathrm{Tr} \sum_{g \in G} \rho(g^2) = \frac{1}{d} \sum_{g \in G} \chi(g^2) = \frac{|G| c_\rho}{d},$$

where we used formula (7.117) for the Frobenius–Schur indicator c_ρ. Recall from Theorem 7.12 that d is a divisor of $|G|$, and, in particular, is not 0 in $\mathbb{F}$. This proves (7.110):

$$\frac{1}{|G|} \sum_{g \in G} \rho(g^2) = \frac{c_\rho}{d} I. \tag{7.119}$$

Multiplying by $\rho(h)$ and taking the trace produces (7.111):

$$\frac{1}{|G|} \sum_{g \in G} \chi_\rho(g^2 b) = \frac{c_\rho}{d} \chi(h) \tag{7.120}$$

for all $h \in G$.

Now we can count, using the now familiar "delta function"

$$\mathrm{Tr}_e = \frac{1}{|G|}\chi_{\mathrm{reg}} = \frac{1}{|G}\sum_{i=1}^{s} d_i\chi_i,$$

where χ_{reg} is the character of the regular representation of G on $\mathbb{F}[G]$. Working in $\mathbb{F}$, we have

$$
\begin{aligned}
|\{(g_1,\ldots,g_n) \in G^n : g_1^2 \ldots g_n^2 = e\}| &= \sum_{g_1,\ldots,g_n \in G} \mathrm{Tr}_e(g_1^2 \ldots g_n^2) \\
&= \frac{1}{|G|}\sum_{i=1}^{s}\sum_{g_1,\ldots,g_n} d_i\chi_i(g_1^2 \ldots g_n^2) \\
&= \frac{1}{|G|}|G|^n \sum_{i=1}^{s} d_i \left(\frac{c_i}{d_i}\right)^n d_i \\
&= |G|^{n-1}\sum_{i=1}^{s} \frac{c_i^n}{d_i^{n-2}},
\end{aligned}
\tag{7.121}
$$

which implies (7.112). $\boxed{\mathrm{QED}}$

7.10 Character References

Among many sights and sounds we have bypassed in our exploration of character theory are (1) Burnside's $p^a q^b$ theorem [9, Corollary 29, Chap. XVI], a celebrated application of character theory to the structure of groups; (2) zero sets of characters; and (3) Galois-theoretic results for characters. Burnside's enormous work [9], especially Chap. XVI, contains a vast array of results, from the curious to the deep, in character theory. The book by Isaacs [48] is an excellent reference for a large body of results in character theory, covering Burnside's $p^a q^b$ theorem, zero sets of characters, Galois-theoretic results for characters, and much more. The book by Hill [44] explains several pleasant applications of character theory to the structure of groups. An encyclopedic account of character theory has been presented by Berkovic and Zhmud' [3].

7.11 Afterthoughts: Connections

The fundamental group $\pi_1(\Sigma, o)$ of a topological space Σ, with a chosen base point o, is the set of homotopy classes of loops based at o, taken as a group under composition/concatenation of paths. If Σ is an orientable surface of genus n with k disks cut out as holes on the surface, then $\pi_1(\Sigma, o)$ is generated by elements $A_1, B_1, \ldots, A_n, B_n, S_1, \ldots, S_k$, subject to the following relation:

$$A_1 B_1 A_1^{-1} B_1^{-1} \ldots A_n B_n A_n^{-1} B_n^{-1} S_1 \ldots S_k = I, \qquad (7.122)$$

where I is the identity element. Here the loops S_i go around the boundaries of the deleted disks. If G is any group, then a homomorphism

$$\phi : \pi_1(\Sigma, o) \to G$$

is completely specified by the values of ϕ on A_i, B_i, and S_j:

$$\big(\phi(A_1), \phi(B_1), \ldots, \phi(A_n), \phi_n(B_n), \phi(S_1), \ldots, \phi_n(S_n)\big),$$

which is a point in the set $M(C_1, \ldots, C_k)$ (defined in (7.103)) if the boundary "holonomies" $\phi(S_j)$ are restricted to lie in the conjugacy classes C_j. Thus, $M(C_1, \ldots, C_k)$ has a topological meaning. The group G acts on $M(C_1, \ldots, C_k)$ by conjugation and the quotient space $M(C_1, \ldots, C_k)/G$ appears in many different incarnations, including as the moduli space of flat connections on a surface and as the phase space of a three-dimensional gauge field theory called Chern–Simons theory. In these contexts G is a compact Lie group. The space $M(C_1, \ldots, C_k)/G$ is not generally a smooth manifold but is made up of strata, which are smooth spaces. The physical context of a phase space provides a natural measure of volume on $M(C_1, \ldots, C_k)/G$. The volume of this space was computed by Witten [80] (see also [72]). The volume formula is, remarkably or not, very similar to the Frobenius formula (7.104) for $|M(C_1, \ldots, C_k)|$. Witten also computed a natural volume measure for the case where the surface is not orientable, and this produces the analog of the Frobenius—Schur count formula (7.112). For other related results and exploration, see the article by Mulase and Penkava [61]. Zagier's appendix to the beautiful book by Lando and Zvonkin [54] also contains many interesting results in this connection.

Exercises

7.1. Let $u = \sum_{h \in G} u_h h$ be an idempotent in $A = \mathbb{F}[G]$, and let χ_u be the character of the regular representation of G restricted to Au:

$$\chi_u(x) = \text{trace of } Au \to Au : y \mapsto xy.$$

(a) Show that, for any $x \in G$,

$$\chi_u(x) = \text{trace of } A \to A : y \mapsto xyu.$$

(b) Check that for $x, g \in G$,

$$xgu = \sum_{h \in G} u(g^{-1}x^{-1}h)h.$$

(c) Conclude that

$$\chi_u(x) = \sum_{g \in G} u(g^{-1}x^{-1}g) \qquad \text{for all } x \in G. \tag{7.123}$$

Equivalently,

$$\sum_{x \in G} \chi_u(x^{-1})x = \sum_{g \in G} gug^{-1}. \tag{7.124}$$

(d) Show that the dimension of the representation on Au is

$$d_u = |G|u(1_G),$$

where 1_G is the unit element in G.

7.2. (This exercise follows an argument in the Appendix in [54] by Zagier.) Let G be a finite group and $\mathbb{F}$ be a field in which $|G|1_\mathbb{F} \neq 0$. For $(g, h) \in G \times G$, let $T_{(g,h)} : \mathbb{F}[G] \to \mathbb{F}[G]$ be specified by

$$T_{(g,h)}(a) = gah^{-1} \qquad \text{for } a \in \mathbb{F}[G] \text{ and } g, h \in G. \tag{7.125}$$

Compute the trace of $T_{(g,h)}$ using the basis of $\mathbb{F}[G]$ given by the elements of G to show that

$$\text{Tr}\, T_{(g,h)} = \begin{cases} 0 & \text{if } g \text{ and } h \text{ are not in the same conjugacy class,} \\ \frac{|G|}{|C|} & \text{if } g \text{ and } h \text{ both belong to the same conjugacy class } C. \end{cases} \tag{7.126}$$

Next recall that $\mathbb{F}[G]$ is the direct sum of maximal two-sided ideals $\mathbb{F}[G]u_j$, with j running over an index set $\mathcal{R}$; then

$$\operatorname{Tr} T_{(g,h)} = \sum_{j \in \mathcal{R}} \operatorname{Tr} \left(T_{(g,h)} | \mathbb{F}[G]u_j \right). \qquad (7.127)$$

Now assume that $\mathbb{F}$ is also algebraically closed; then we know that, picking a simple left ideal $L_j \subset \mathbb{F}[G]u_j$, there is an isomorphism

$$\rho_j : \mathbb{F}[G]u_j \to \operatorname{End}_{\mathbb{F}}(L_j),$$

where $\rho_j(x)y = xy$ for all $x \in \mathbb{F}[G]u_j$ and $y \in L_j$, and so

$$\operatorname{Tr} \left(T_{(g,h)} | \mathbb{F}[G]u_j \right) = \operatorname{Tr} \left(\rho_j \circ T_{(g,h)} | \mathbb{F}[G]u_r \circ (\rho_j)^{-1} \right).$$

Now use the identification

$$\operatorname{End}_{\mathbb{F}}(L_j) \simeq L_j \otimes L'_j,$$

where L'_j is the vector-space dual to L_j, to show that

$$\begin{aligned} \operatorname{Tr} \left(T_{(g,h)} | \mathbb{F}[G]u_j \right) &= \operatorname{Tr} \left(\rho_j(g) \right) \operatorname{Tr} \left(\rho_j(h^{-1}) \right) \\ &= \chi_j(g)\chi_j(h^{-1}). \end{aligned} \qquad (7.128)$$

Combine this with (7.127) and (7.126) to obtain the orthogonality relation (7.40).

7.3. Let M be a finitely generated $\mathbb{Z}$-module and $A : M \to M$ be a $\mathbb{Z}$-linear map. Show that there is a monic polynomial $p(X)$ such that $p(A) = 0$.

7.4. Let $\chi_1, \ldots, \chi_s$ be all the distinct irreducible characters of a finite group G over an algebraically closed field of characteristic 0, and let $\{C_1, \ldots, C_s\}$ be the conjugacy classes in G. Then show that

$$\chi_i(C_l^{-1}) = \frac{1}{|G|} \sum_{1 \le j,k \le s} \chi_i(C_j^{-1})\chi_i(C_k^{-1})\kappa_{jk,l} \qquad (7.129)$$

for all $i \in \{1, \ldots, s\}$, where $\kappa_{jk,l}$ are the structure constants of G.

7.5. Prove the character orthogonality relations from the orthogonality of matrix elements.

7.6. Let G be a finite group, $\mathbb{F}$ be a field in which $|G|1_{\mathbb{F}} \neq 0$, and ρ be a finite-dimensional representation of G on a vector space E over $\mathbb{F}$. Show that the dimension of the subspace E^{fix} of elements of E fixed under $\rho(g)$ for all $g \in G$ is

$$\dim_{\mathbb{F}} E^{\text{fix}} = \frac{1}{|G|} \sum_{g \in G} \chi_{\rho}(g). \qquad (7.130)$$

7.7. The *character table* of a finite group G that has s conjugacy classes is the $s \times s$ matrix $[\chi_i(C_j)]_{1 \leq i,j \leq s}$, where $C_1, \ldots, C_s$ are the conjugacy classes in G and $\chi_1, \ldots, \chi_s$ are the distinct irreducible complex characters of G. Show that the determinant of this matrix is nonzero.

7.8. Verify Dedekind's factorization of the group determinant for S_3:

$$\begin{vmatrix} X_1 & X_2 & X_3 & X_4 & X_5 & X_6 \\ X_3 & X_1 & X_2 & X_5 & X_6 & X_4 \\ X_2 & X_3 & X_1 & X_6 & X_4 & X_5 \\ X_4 & X_5 & X_6 & X_1 & X_2 & X_3 \\ X_5 & X_6 & X_4 & X_3 & X_1 & X_2 \\ X_6 & X_4 & X_5 & X_2 & X_3 & X_1 \end{vmatrix} \qquad (7.131)$$
$$= (u+v)(u-v)(u_1 u_2 - v_1 v_2),$$

where

$$u = X_1 + X_2 + X_3, \quad u_1 = X_1 + \omega X_2 + \omega^2 X_3, u_2 = X_1 + \omega^2 X_2 + \omega X_3,$$
$$v = X_4 + X_5 + X_6, \quad u_1 = X_4 + \omega X_5 + \omega^2 X_6, v_2 = X_4 + \omega^2 X_5 + \omega X_6,$$

where ω is a primitive cube root of unity.

7.9. Let G be a finite group and $\chi_1, \ldots, \chi_s$ be all the distinct irreducible characters of G over an algebraically closed field $\mathbb{F}$ in which $|G|1_{\mathbb{F}} \neq 0$. Prove the following identity of Frobenius [29, Sect. 5, (6)]:

$$\sum_{\{(t_1, \ldots, t_m) \in G^m : t_1 \ldots t_m = e\}} \chi(a_1 t_1, \ldots, a_m t_m) = \left(\frac{|G|}{d}\right)^{m-1} \chi(a_1) \ldots \chi(a_m) \qquad (7.132)$$

for all $a_1, \ldots, a_m \in G$. Use this to prove the counting formula:

$$\left|\{(t_1, \ldots, t_m) \in G^m \; : \; t_1, \ldots, t_m = c, \quad a_1 t_1 \ldots a_m t_m = c\}\right|$$

$$= \sum_{i=1}^{s} \left(\frac{|G|}{d_i}\right)^{m-2} \chi_i(a_1) \ldots \chi_i(a_m)$$

(7.133)

for all $a_1, \ldots, a_m \in G$.

7.10. Suppose a group G is represented irreducibly on a finite-dimensional vector space V over an algebraically closed field $\mathbb{F}$. Let $B : V \times V \to \mathbb{F}$ be a nonzero bilinear function which is G-invariant in the sense that $B(gv, gw) = B(v, w)$ for all vectors $v, w \in V$ and $g \in G$. Show that

(a) B is nondegenerate. (Hint: View B as a linear map $V \to V'$ and use Schur's lemma.)

(b) If B_1 is also a G-invariant bilinear form on V, then $B_1 = cB$ for some $c \in \mathbb{F}$.

(c) If G is a finite group, and $\mathbb{F} = \mathbb{C}$, then either B or $-B$ is positive-definite. (A bilinear form D on V is positive-definite if $D(v, v) > 0$ for all nonzero $v \in V$.)

7.11. Let $\rho_1, \ldots, \rho_s$ be a maximal set of inequivalent irreducible representations of a finite group G over an an algebraically closed field $\mathbb{F}$ in which $|G|1_\mathbb{F} \neq 0$. Let $\mathcal{C}$ be the set of all conjugacy classes in G. Let ρ' denote the representation dual to ρ, so that for the characters we have $\chi_{\rho'}(g) = \chi_\rho(g^{-1})$ for all $g \in G$. By computing both sides of the identity

$$\sum_{i=1}^{s} \sum_{C \in \mathcal{C}} \frac{|C|}{|G|} \chi_{\rho_i}(C) \chi_{\rho_i'}(C^{-1}) = \sum_{c \in \mathcal{C}} \sum_{i=1}^{s} \frac{|C|}{|G|} \chi_{\rho_i}(C) \chi_{\rho_i}\left((C^{-1})^{-1}\right)$$

show that the number of irreducible representations that are isomorphic to their duals is equal to the number of conjugacy classes C for which $C^{-1} = C$:

$$\left|\{i \in [s] : \rho_i \simeq \rho_i'\}\right| = \left|\{C \in \mathcal{C} : C = C^{-1}\}\right|.$$

(7.134)

(For a different, combinatorial proof of this, see the book by Hill [44].) Now suppose $n = |G|$ is odd. If $C = C^{-1}$ is a conjugacy class containing an element a, then $gag^{-1} = a^{-1}$ for some $g \in G$, and $g^n a g^{-n} = a^{-1}$, since n is odd, and so $a = a^{-1}$, which can only hold if $a = e$. Thus, when $|G|$ is odd, there is exactly one conjugacy class that is equal to its own inverse, and hence there is exactly one irreducible representation, over $\mathbb{F}$, that is equivalent to its dual.

7.12. Let G be a finite group, $\mathbb{F}$ be a field, and T be the representation of G on $\mathbb{F}[G]$ given by

$$T(g)x = gxg^{-1} \qquad \text{for all } x \in \mathbb{F}[G] \text{ and } g \in G.$$

Compute the character χ_T of T. Next, for the character χ of a representation of G over $\mathbb{F}$, find a meaning for the sum $\sum_{C \in \mathcal{C}} \chi(C)$, where $\mathcal{C}$ is the set of all conjugacy classes in G.

Chapter 8

Induced Representations

A representation of a group G produces, by restriction, a representation of any subgroup H. Remarkably, there is a procedure that runs in the opposite direction, producing a representation of G from a representation of H. This method, introduced by Frobenius [33], is called *induction*, and is a powerful technique for constructing and analyzing the structure of representations.

8.1 Constructions

Consider a finite group G, a subgroup H, and a representation ρ of H on a finite-dimensional vector space E over a field $\mathbb{F}$. Among all functions on G with values in E we single out those which transform in a nice way in relation to H; specifically, let E_1 be the set of all maps $\psi : G \to E$ for which

$$\psi(ah) = \rho(h^{-1})\psi(a) \qquad \text{for all } a \in G \text{ and } h \in H. \tag{8.1}$$

If ψ satisfies (8.1), we say that it is *equivariant* with respect to ρ and the action of H on G by right multiplication: $G \times H \to G : (g, h) \mapsto gh$.

It is clear that E_1 is a subspace of the finite-dimensional vector space $\mathrm{Map}(G, E)$ of all maps $G \to E$. Now the space $\mathrm{Map}(G, E)$ carries a natural representation of G,

$$G \times \mathrm{Map}(G, E) \to \mathrm{Map}(G, E) : (a, \psi) \mapsto L_a\psi,$$

where

$$L_a\psi(b) = \psi(a^{-1}b) \quad \text{for all } a, b \in G, \tag{8.2}$$

A.N. Sengupta, *Representing Finite Groups: A Semisimple Introduction*,
DOI 10.1007/978-1-4614-1231-1_8, © Springer Science+Business Media, LLC 2012

and this representation preserves the subspace E_1. This representation of G on E_1 is the *induced representation* of ρ on G. We will denote it by $i_H^G \rho$.

Creating good notation for the induced representation is a challenge, and it is best to be flexible. If E is the original representation space of H, then sometimes it is more convenient to denote the induced representation by E^G (which is why we denote the set of all functions $G \to E$ by $\mathrm{Map}(G, E)$).

A function $\psi : G \to E$ is, by set-theoretic definition, a set of ordered pairs $(a, v) \in G \to E$, with a unique v paired with any given a. The condition (8.1) on ψ requires that if $(a, v) \in \psi$ then $\bigl(ah, \rho(h^{-1})v\bigr)$ is also in ψ. In physics there is a useful notion of "a quantity which transforms" according to a specified rule; here we can think of ψ as such a quantity which, when "realized" by means of a is "measured" as the vector v, but when the "frame of reference" a is changed to ah the measured vector is $\rho(h^{-1})v$.

It will often be convenient to work with a set of elements $g_1, \ldots, g_m \in G$, where $g_1 H, \ldots, g_m H$ are all the distinct left cosets of H in G. Such a set $\{g_1, \ldots, g_m\}$ is called a *complete set of left coset representatives* of H in G.

It is useful to note that an element $\psi \in E_1$ is completely determined by listing its values at elements $g_1, \ldots, g_m \in G$, which form a complete set of left coset representatives. Moreover, we can arbitrarily assign the values of ψ at the points $g_1, \ldots, g_m$. In other words, the mapping

$$E_1 \to E^m : \psi \mapsto \bigl(\psi(g_1), \ldots, \psi(g_m)\bigr) \tag{8.3}$$

is an isomorphism of vector spaces (Exercise 8.1).

The isomorphism (8.3) makes it clear that *the dimension of the induced representation* is given by

$$\dim i_H^G \rho = |G/H|(\dim \rho). \tag{8.4}$$

Think of a function $\psi : G \to E$ as a formal sum

$$\psi = \sum_{g \in G} \psi(g) g.$$

More officially, we can identify the vector space $\mathrm{Map}(G, E)$ with the tensor product $E \otimes \mathbb{F}[G]$:

$$\mathrm{Map}(G, E) \to E \otimes \mathbb{F}[G] : \psi \mapsto \sum_{g \in G} \psi(g) \otimes g.$$

The subspace E_1 corresponds to those elements $\sum_g v_g \otimes g$ that satisfy

$$\sum_g v_g \otimes g = \sum_g \rho(h^{-1})v_g \otimes gh \qquad \text{for all } h \in H. \qquad (8.5)$$

The representation i_H^G is then specified quite simply:

$$i_H^G(g)(v_a \otimes a) = v_a \otimes ga. \qquad (8.6)$$

The induced representation is meaningful even if the field $\mathbb{F}$ is replaced by a commutative ring R. Let E be an $R[H]$-module. View $R[G]$ as a right $R[H]$-module. Let

$$E^G = R[G] \otimes_{R[H]} E \qquad (8.7)$$

be the tensor product $R[G] \otimes E$ quotiented by the submodule spanned by elements of the form $(xb) \otimes v - x \otimes (bv)$ with $x, b \in R[H]$, $v \in V$. Now view this *balanced tensor product* as a left $R[G]$-module by specifying the action of $R[G]$ through

$$a(x \otimes v) = (ax) \otimes v \qquad \text{for all } x, a \in R[G], \ v \in E. \qquad (8.8)$$

For more, consult the discussion following the definition in (12.54). Notice the mapping

$$j : E \to E^G : v \mapsto e \otimes v, \qquad (8.9)$$

where e, the identity in G, is viewed as $1e \in R[G]$. Then by the balanced tensor product property, we have

$$j(hv) = h \otimes v = h(e \otimes v) = hj(v) \qquad (8.10)$$

for all $h \in H$ and $v \in E$, and so j is $R[H]$-linear (with E^G viewed, by restriction, as a left $R[H]$-module for the moment).

Pick, as before, $g_1, \ldots, g_m \in G$ forming a complete set of left coset representatives of H in G. Then you can check quickly that $\{g_1, \ldots, g_m\}$ is a basis for $R[G]$, viewed as a right $R[H]$-module (Exercise 8.3). A consequence (details being outsourced to Theorem 12.10) is that

$$E^G = g_1 R[G] \otimes_{R[H]} E \oplus \cdots \oplus g_m R[G] \otimes_{R[H]} E. \qquad (8.11)$$

In fact, every element of E^G can then be expressed as $\sum_i g_i \otimes v_i$ with $v_i \in E$ uniquely determined. This shows the equivalence with the approach used in (8.5), with E_1 being isomorphic to E^G as $R[G]$-modules.

We now have several distinct definitions of E^G, all of which are identifiable with each other. This is an expression of the essential *universality* of the induction process that we will explore later in Sect. 8.4.

8.2   The Induced Character

We will work with G, H, and E as in the preceding section: H is a subgroup of the finite group G, and E is an $\mathbb{F}[H]$-module. As before,

$$E^G = \mathbb{F}[G] \otimes_{\mathbb{F}[H]} E$$

is an $\mathbb{F}[G]$-module, and there is the $\mathbb{F}[H]$-linear map

$$j : E \to E^G : v \mapsto 1e \otimes v.$$

Set

$$E_0 = j(E),$$

which is a sub-$\mathbb{F}[H]$-module of E^G. Pick $g_1, \ldots, g_m \in G$ forming a complete set of left coset representatives of H in G. Then

$$E^G = g_1 E_0 \oplus \ldots \oplus g_m E_0,$$

where $g_i E_0$ is $i_H^G \rho(g_i) E_0$. The map

$$L_g : E^G \to E^G : v \mapsto i_H^G \rho(g) v$$

carries the subspace $g_i E_0$ bijectively onto $g g_i E_0$. Thus, $g g_i E_0$ equals $g_i E_0$ if and only if $g_i^{-1} g g_i$ is in H. Consequently, the map L_g has zero trace if g is not conjugate to any element in H. If g is conjugate to an element h of H, then

$$\operatorname{Tr}(L_g) = n_g \operatorname{Tr}(L_h|E_0) = n_g \chi_\rho(h), \qquad (8.12)$$

where n_g is the number of i for which $g_i^{-1} g g_i$ is in H.

We can summarize these observations in the following theorem:

Theorem 8.1 *Let H be a subgroup of a finite group G and $i_H^G \rho$ be the induced representation of G from a representation ρ of H on a finite-dimensional vector space E over a field $\mathbb{F}$. Let $g_1, \ldots, g_m \in G$ be such that $g_1 H, \ldots, g_m H$ are all the distinct left cosets of H in G. Then the character of $i_H^G \rho$ is given by*

$$(i_H^G \chi_\rho)(g) = \sum_{j=1}^{m} \chi_\rho^0(g_j^{-1} g g_j) \qquad \text{for all } g \in G, \qquad (8.13)$$

where χ_ρ^0 is equal to the character χ_ρ of ρ on $H \subset G$ and is 0 outside H. If $|H|$ is not divisible by the characteristic of the field $\mathbb{F}$, then

$$(i_H^G \chi_\rho)(g) = \frac{1}{|H|} \sum_{a \in G} \chi_\rho^0(a^{-1} g a) \qquad \text{for all } g \in G. \qquad (8.14)$$

The division by $|H|$ in (8.14) is needed because each g_i for which $g_i^{-1}gg_i$ is in H is counted $|g_iH|$ $(=|H|)$ times in the sum on the right of (8.14):

$$\chi_\rho^0((g_ih)^{-1}g(g_ih)) = \chi_\rho^0(g_i^{-1}gg_i).$$

In the special case when H is a *normal* subgroup of G, the element $g_j^{-1}gg_j$ lies in H if and only if g is in H. Hence, we have the following:

Proposition 8.1 *For a normal subgroup H of a finite group G, and a finite-dimensional representation ρ of G, the character of the induced representation $i_H^G\rho$ is 0 outside the normal subgroup H.*

8.3   Induction Workout

As usual, we will work with a subgroup H of a finite group G, and a representation ρ of H on a finite-dimensional vector space E over a field $\mathbb{F}$. Fix $g_1, \ldots, g_m \in G$ forming a complete set of left coset representatives of H in G. For this section we will use the induced representation space E_1, which, recall, is the space of all maps $\psi : G \to E$ for which

$$\psi(ah) = \rho(h^{-1})\psi(a) \quad \text{for all } a \in G \text{ and } h \in H.$$

Then the induction process produces the representation ρ_1 of G on E_1 given by

$$\rho_1(a)\psi : b \mapsto \psi(a^{-1}b),$$

and E_1 is isomorphic to E^m via the map

$$E_1 \to E^m : \psi \mapsto (\psi(g_1), \ldots, \psi(g_m)).$$

Let us work out the representation ρ_1 as it appears in E^m; we will denote the representation on E_1 again by ρ_1. For any $g \in G$ we have

$$(\rho_1(g)\psi(g_1), \ldots, \rho_1(g)\psi(g_m)) = (\psi(g^{-1}g_1), \ldots, \psi(g^{-1}g_m)). \tag{8.15}$$

Now for each i the element $g^{-1}g_i$ falls into a unique coset g_jH; that is, there is a unique j for which $g_j^{-1}g^{-1}g_i = h \in H$. Note that

$$h^{-1} = g_i^{-1}gg_j.$$

Then, for such i and j, we have

$$\psi(g^{-1}g_i) = \psi(g_j h) = \rho(h^{-1})\psi(g_j).$$

Thus, the action of $\rho_1(g)$ is

$$\rho_1(g) : \begin{bmatrix} \psi_1 \\ \vdots \\ \psi_m \end{bmatrix} \mapsto \begin{bmatrix} \sum_j \rho^0(g_1^{-1}gg_j)\psi_j \\ \vdots \\ \sum_j \rho^0(g_m^{-1}gg_j)\psi_j \end{bmatrix},$$

where ρ^0 is ρ on H and is 0 outside H. Note that in each of the sums $\sum_j$, all except possibly one term is 0. The matrix of $\rho_1(g)$ is

$$\rho_1(g) = \begin{bmatrix} \rho^0(g_1^{-1}gg_1) & \rho^0(g_1^{-1}gg_2) & \cdots & \rho^0(g_1^{-1}gg_m) \\ \vdots & \vdots & \cdots & \vdots \\ \rho^0(g_m^{-1}gg_1) & \rho^0(g_m^{-1}gg_2) & \cdots & \rho^0(g_m^{-1}gg_m) \end{bmatrix}. \tag{8.16}$$

Note again in this big matrix that each row and each column has exactly one nonzero entry. Moreover, if H is a normal subgroup and $h \in H$, then the matrix in (8.16) for $\rho_1(h)$ is a block-diagonal matrix, with each diagonal block being ρ evaluated on one of the G-conjugates of h lying inside H.

Let us see how this works out for S_3 (which is the same as the dihedral group D_3). The elements of S_3 are as follows:

$$\iota, \quad c = (123), \quad c^2 = (132), \quad r = (12), \quad rc = (23), \quad rc^2 = (13),$$

where ι is the identity element. Thus, r and c generate S_3 subject to the relations

$$r^2 = c^3 = \iota, \quad rcr^{-1} = c^2.$$

The subgroup $C = \{\iota, c, c^2\}$ is normal. The group S_3 decomposes into cosets

$$S_3 = C \cup rC.$$

Consider the one-dimensional representation ρ of C on $\mathbb{Q}[\omega]$, where ω is a primitive cube root of 1, specified by

$$\rho(c) = \omega.$$

Let ρ_1 be the induced representation; by (8.4) its dimension is

$$\dim \rho_1 = |S_3/C|(\dim \rho) = 2.$$

We can write out the matrices for $\rho_1(c)$ and $\rho_1(r)$:

$$\rho_1(c) = \begin{bmatrix} \rho^0(\iota^{-1}c\iota) & \rho^0(\iota^{-1}cr) \\ \rho^0(r^{-1}c\iota) & \rho^0(r^{-1}cr) \end{bmatrix} = \begin{bmatrix} \omega & 0 \\ 0 & \omega^2 \end{bmatrix},$$

$$\rho_1(r) = \begin{bmatrix} \rho^0(\iota^{-1}r\iota) & \rho^0(r^{-1}r\iota) \\ \rho^0(r^{-1}r\iota) & \rho^0(r^{-1}rr) \end{bmatrix} = \begin{bmatrix} 0 & 1 \\ 1 & 0 \end{bmatrix},$$
(8.17)

Looking back at (2.7), we recognize this as an irreducible representation of D_3 given geometrically as follows: $\rho_1(c)$ arises from conjugation of a rotation by 120° and r arises by reflection across a line. Note that restricting ρ_1 to C does not simply return ρ; in fact, $\rho_1|C$ decomposes as a direct sum of two distinct irreducible representations of C. Lastly, let us note the character of ρ_1:

$$\chi_1(\iota) = 2, \quad \chi_1(c) = \chi_1(c^2) = -1, \quad \chi_1(r) = \chi_2(rc) = \chi_1(rc^2) = 0,$$
(8.18)

which agrees nicely with the last row in Table **2.2**.

Now let us run through S_3 again, but this time using the subgroup $H = \{\iota, r\}$ and the one-dimensional representation τ specified by $\tau(r) = -1$. The underlying field $\mathbb{F}$ is now arbitrary. The coset decomposition is

$$S_3 = H \cup cH \cup c^2H.$$

Then the induced representation τ_1 has dimension

$$\dim \tau_1 = |S_3/H| \dim \tau = 3.$$

For $\tau_1(c)$ we have

$$\tau_1(c) = \begin{bmatrix} \tau^0(\iota^{-1}c\iota) & \tau^0(\iota^{-1}cc) & \tau^0(\iota^{-1}cc^2) \\ \tau^0(c^{-1}c\iota) & \tau^0(c^{-1}cc) & \tau^0(c^{-1}cc^2) \\ \tau^0(c^{-2}c\iota) & \tau^0(c^{-2}cc) & \tau^0(c^{-2}cc^2) \end{bmatrix}$$

$$= \begin{bmatrix} 0 & 0 & 1 \\ 1 & 0 & 0 \\ 0 & 1 & 0 \end{bmatrix}$$
(8.19)

and for $\tau_1(r)$ we have

$$
\begin{aligned}
\tau_1(r) &= \begin{bmatrix} \tau^0(\iota^{-1}r\iota) & \tau^0(\iota^{-1}rc) & \tau^0(\iota^{-1}rc^2) \\ \tau^0(c^{-1}r\iota) & \tau^0(c^{-1}rc) & \tau^0(c^{-1}rc^2) \\ \tau^0(c^{-2}r\iota) & \tau^0(c^{-2}rc) & \tau^0(c^{-2}rc^2) \end{bmatrix} \\
&= \begin{bmatrix} -1 & 0 & 0 \\ 0 & 0 & -1 \\ 0 & -1 & 0 \end{bmatrix}.
\end{aligned}
\tag{8.20}
$$

The character of τ_1 is given by

$$
\chi_{\tau_1}(\iota) = 3, \quad \chi_{\tau_1}(c) = \chi_{\tau_1}(c^2) = 0, \quad \chi_{\tau_1}(r) = \chi_{\tau_1}(cr) = \chi_{\tau_1}(c^2r) = -1.
\tag{8.21}
$$

Referring back again to the character table for S_3 in Table **2.2**, we see that

$$
\chi_{\tau_1} = \chi_1 + \theta_{+,-}.
\tag{8.22}
$$

The induced representation τ_1 is the direct sum of two irreducible representations, at least when $3 \neq 0$ in $\mathbb{F}$ (in which case χ_1 comes from an irreducible representation; see the solution of Exercise 2.4). In fact,

$$
\mathbb{F}^3 = \mathbb{F}(1, 1, 1) \oplus \{(x_1, x_2, x_3) \in \mathbb{F}^3 : x_1 + x_2 + x_3 = 0\}
$$

decomposes $\mathbb{F}^3$ into a direct sum of irreducible subspaces, provided $3 \neq 0$ in $\mathbb{F}$.

8.4   Universality

At first it might seem that the induced representation is just another clever construction that happened to work out. But there is a certain natural quality to the induced representation, which can be expressed through a "universal property." One way of viewing this universal property is that the induced representation is the "minimal" natural extension of an H-representation to a G-representation.

Theorem 8.2 *Let G be a finite group, H be a subgroup, R be a commutative ring, and E be a left $R[H]$-module. Let $E^G = R[G] \otimes_{R[H]} E$, viewed as a left $R[G]$-module, and $j_E : E \to E^G$ be the map $v \mapsto e \otimes v$, which is linear over*

$R[H]$. *Now suppose F is a left $R[G]$-module and $f : E \to F$ is a map that is linear over $R[H]$. Then there is a unique $R[G]$-linear map*

$$T_f : E^G \to F$$

such that $f = T_f \circ j_E$.

Proof. Pick $g_1, \ldots, g_m \in G$ such that $g_1 H, \ldots, g_m H$ are all the distinct left cosets of H in G. Every $x \in E^G$ has a unique expression as a sum $\sum_i g_i \otimes v_i$ with $v_i \in E$; then define $T_f : E^G \to F$ by setting

$$T_f(x) = g_1 f(v_1) + \cdots + g_m f(v_m).$$

Now consider an element $g \in G$; then $g g_i = g_{i'} h_i$ for a unique $i' \in \{1, \ldots, m\}$ and $h_i \in H$, and so for x as above, we have

$$T_f(gx) = \sum_i T_f(g_{i'} \otimes h_i v_i) = \sum_i g_{i'} f(h_i v_i)$$

$$= \sum_i g_{i'} h_i f(v_i) \tag{8.23}$$

$$= g \sum_i g_i f(v_i) = g T_f(x).$$

So T_f, which is clearly additive as well, is $R[G]$-linear. The relation $f = T_f \circ j_E$ follows immediately from the definition of T_f. Uniqueness of T_f then follows from the fact that the elements $j_E(v) = 1 \otimes v$, with v running over E, span the left $R[G]$-module E^G. $\boxed{\text{QED}}$

8.5 Universal Consequences

Universality is a powerful idea and produces some results with routine automatic proofs. It is often best to think not of E^G by itself, but rather to think of the $R[H]$-linear map

$$j_E : E \to E^G,$$

as a package, as the *induced module*

Let H be a subgroup of a finite group G and let E and F be left $R[H]$-modules, where R is a commutative ring. For any left R-module L, denote by

L^G the left $R[G]$-module $R[G] \otimes_{R[H]} L$ and by j_L the map $L \to L^G : v \mapsto e \otimes v$, where e is the identity in G. Then the map

$$E \oplus F \to E^G \oplus F^G : (v, w) \mapsto \big(j_E(v), j_F(w)\big)$$

is $R[H]$-linear, and so there is a unique $R[G]$-linear map $T : (E \oplus F)^G \to E^G \oplus F^G$ for which

$$Tj(v, w) = \big(j_E(v), j_F(w)\big)$$

for all $v \in E$ and $w \in F$, where $j = j_{E \oplus F}$. In the reverse direction, the $R[H]$-linear mapping

$$E \to (E \oplus F)^G : v \mapsto j(v, 0)$$

gives rise to an $R[G]$-linear map $E^G \to (E \oplus F)^G$, and similarly for F; adding, we obtain an $R[G]$-linear map:

$$S : E^G \oplus F^G \to (E \oplus F)^G : (j_E v, j_F w) \mapsto j(v, 0) + j(0, w) = j(v, w).$$

Then $TS(j_E, j_F) = (j_E, j_F)$ and $STj = j$, which, by the uniqueness in universality, implies that ST and TS are both the identity. To summarize:

Theorem 8.3 *Suppose H is a subgroup of a finite group G, and E and F are left $R[H]$-modules, where R is a commutative ring. Then there is a unique $R[G]$-linear isomorphism*

$$T : (E \oplus F)^G \to E^G \oplus F^G$$

satisfying $Tj_{E \oplus F} = j_E \oplus j_F$, where $j_S : S \to S^G$ denotes the canonical map for the induced representation for any $\mathbb{F}[H]$-module S.

Proof. By Theorem 8.2 there is a unique $R[G]$-linear map $T_f : E^G \to F$ for which

$$T_f j_E = f.$$

Let

$$f^G = j_F T_f.$$

Then

$$f^G j_E = j_F T_d j_E = j_F f.$$

QED

The next such result is *functoriality* of the induced representation; it is an immediate consequence of the universal property of induced modules.

Theorem 8.4 *Suppose H is a subgroup of a finite group G, E and F are left $R[H]$-modules, where R is a commutative ring, and $f : E \to F$ is an $R[H]$-linear map. Let $j_E : E \to E^G$ and $j_F : F \to F^G$ be the induced modules. Then there is a unique $R[G]$-linear map $f^G : E^G \to F^G$ such that $f^G j_E = j_F f$.*

8.6   Reciprocity

The most remarkable consequence of universality is a fundamental "reciprocity" result of Frobenius [33]. As usual, let H be a subgroup of a finite group G, E be a left $R[H]$-module, and F be an $R[G]$-module, where R is a commutative ring.

Recall that, with usual notation, if $f : E \to F$ is $R[H]$-linear, then there is a unique $R[G]$-linear map $T_f : E^G \to F$ for which $T_f j_E = f$. Thus, we have a map

$$\operatorname{Hom}_{R[H]}(E, F_H) \to \operatorname{Hom}_{R[G]}\left(E^G, F\right) : f \mapsto T_f.$$

The domain and codomain here are left R-modules in the obvious way, keeping in mind that R is commutative by assumption. With this preparation, we have a formulation of *Frobenius reciprocity*:

Theorem 8.5 *Let H be a subgroup of a finite group G, E be a left $R[H]$-module, where R is a commutative ring, and F be a left $R[G]$-module. Let F_H denote F viewed as a left $R[H]$-module. Then*

$$\operatorname{Hom}_{R[H]}(E, F_H) \to \operatorname{Hom}_{R[G]}\left(E^G, F\right) : f \mapsto T_f \qquad (8.24)$$

is an isomorphism of R-modules, where T_f is specified by the requirement $T_f j_E = f$.

Proof. If $f \in \operatorname{Hom}_{R[H]}(E, F_H)$, then, by universality, there is a unique $T_f \in \operatorname{Hom}_{R[G]}\left(E^G, F\right)$ such that $T_f \circ j_E = f$. Clearly, $f \mapsto T_f$ is injective. Uniqueness of T_f implies that $T_{f_1+f_2}$ equals $T_{f_1} + T_{f_2}$, because with j_E both produce $f_1 + f_2$ for any $f_1, f_2 \in \operatorname{Hom}_{R[H]}(E, F_H)$. Next, for any $r \in R$, and $f \in \operatorname{Hom}_{R[H]}(E, F_H)$, the map rT_f is in $\operatorname{Hom}_{R[G]}\left(E^G, F\right)$ and satisfies $(rT_f)j_E = rf$, which, again by uniqueness, implies that rT_f is T_{rf}. Now consider any $A \in \operatorname{Hom}_{R[G]}\left(E^G, F\right)$, and let $f = Aj_E$, which is an element of $\operatorname{Hom}_{R[H]}(E, F_H)$. Then uniqueness of T_f implies that $T_f = A$; thus, $f \mapsto T_f$ is surjective. $\boxed{\text{QED}}$

A semisimple module N over a ring A decomposes as a direct sum

$$N = \bigoplus_{i \in I} N_i,$$

where each N_i is a simple A-module. For a simple A-module E, the number of $i \in I$ for which N_i is isomorphic to E, as A-modules, is called the *multiplicity* of E in N. If A is the group algebra $\mathbb{F}[G]$, for a field $\mathbb{F}$ and a finite group G, then the multiplicity is equal to

$$\dim_{\mathbb{F}} \operatorname{Hom}_{\mathbb{F}[G]}(E, N),$$

if $\mathbb{F}$ is algebraically closed (by Schur's lemma).

We bring the reciprocity result (Theorem 8.5) down to earth now by specializing to the case where R is a field $\mathbb{F}$. Then we have the following concrete consequence:

Theorem 8.6 *Let H be a subgroup of a finite group G, E be a simple $\mathbb{F}[H]$-module, where $\mathbb{F}$ is an algebraically closed field in which $|G|1_{\mathbb{F}} \neq 0$, and F be a simple $\mathbb{F}[G]$-module. Let F_H denote F viewed as an $\mathbb{F}[H]$-module. Then the multiplicity of F in E^G is equal to the multiplicity of E in F_H.*

There is one more way to say this. Looking back at Proposition 7.5, we recognize the dimensions of the Hom spaces in (8.24) as the kind of character convolutions that appear in character orthogonality. This at once produces the following Frobenius reciprocity result in terms of characters:

Theorem 8.7 *Let H be a subgroup of a finite group G, E be a representation of H, and F be a representation of G, where E and F are finite-dimensional vector spaces over a field $\mathbb{F}$ in which $|G|1_{\mathbb{F}} \neq 0$. Let F_H denote F viewed as a representation of H, and let E^G be the induced representation of G. Then*

$$\frac{1}{|G|} \sum_{g \in G} \chi_{E^G}(g^{-1})\chi_F(g) = \frac{1}{|H|} \sum_{h \in H} \chi_{F_H}(h^{-1})\chi_E(h). \qquad (8.25)$$

We have seen that on a finite group K there is a useful Hermitian inner product on the vector space of function $K \to \mathbb{C}$ given by

$$\langle f_1, f_2 \rangle_K = \frac{1}{|K|} \sum_{k \in K} \overline{f_1(k)} f_2(k).$$

In this notation, (8.25) reads

$$\langle \chi_{E^G}, \chi_F \rangle_H = \langle \chi_{F_H}, \chi_E \rangle_G. \qquad (8.26)$$

8.7 Afterthoughts: Numbers

In Euclid's *Elements*, ratios of segments are defined by an equivalence class procedure: segments AB, CD, A_1B_1, and C_1D_1 correspond to the same ratio

$$AB : CD = A_1B_1 : C_1D_1$$

if for any positive integers m and n the inequality $m \cdot CD > n \cdot AB$ holds if and only if $m \cdot C_1D_1 > n \cdot A_1B_1$, where whole multiples of segments and the comparison relation $>$ are defined geometrically. Then it is shown, through considerations of similar triangles, that there are well-defined operations of addition and multiplication on ratios of segments. Fast-forwarding through history, and throwing in both 0 and negatives, shows how the axioms of Euclidean geometry lead to number fields. This is also reflected in the traditional ruler and compasses constructions, which show how a number field emerges from the axioms of geometry. A more subtle process leads to constructions of division rings and fields from the sparser axiom set of projective geometry. Turning now to groups, a finite group is, by definition, quite a minimal abstract structure, having just one operation defined on a nonempty set with no other structure. Yet geometric representations of such a group single out certain number fields corresponding to these geometries. Very concretely put, here is a natural question that was addressed from the earliest explorations of group representation theory: for a given finite group, is there a subfield $\mathbb{F}$ of $\mathbb{C}$ such that every irreducible complex representation of G can be realized with matrices having elements all in the subfield $\mathbb{F}$? The following magnificent result of Brauer [7], following up on many intermediate results from the time of Frobenius on, answers this question:

Theorem 8.8 *Let G be a finite group and $m \in \{1, 2, \ldots\}$ be such that $g^m = e$ for all $g \in G$. For any irreducible complex representation ρ of G on a vector space V, there is a basis of V relative to which all entries of the matrix $\rho(g)$ lie in the field $\mathbb{Q}(\zeta_m)$ for all $g \in G$, with $\zeta_m = e^{2\pi i/m}$ being a primitive mth root of unity.*

Here $\mathbb{Q}(\eta_m)$ is the smallest subfield of $\mathbb{C}$ containing the integers and η_m. Weintraub [78] provides a thorough treatment of this result, as well as important other related results. Lang [55] also contains a readable account.

Exercises

8.1. Show that (8.3) is an isomorphism of vector spaces. Work out the representation on E^m which corresponds to $i_H^G \rho$ via this isomorphism.

8.2. For the dihedral group

$$D_4 = \langle c, r : c^4 = r^2 = e, \quad rcr^{-1} = c^{-1} \rangle$$

and the cyclic subgroup $C = \{e, c, c^2, c^3\}$, work out the induced representations for:

(a) The one-dimensional representation ρ of C specified by $\rho(c) = i$

(b) The two dimensional representation τ of C specified by

$$\tau(c) = \begin{bmatrix} 0 & -1 \\ 1 & 0 \end{bmatrix}$$

8.3. Let G be a finite group, H be a subgroup, and R be a commutative ring with 1. Choose $g_1, \ldots, g_m \in G$ such that $g_1 H, \ldots, g_m H$ are all the distinct left cosets of H in G. Show that $g_1, \ldots, g_m \in R[G]$ form a basis of $R[G]$, viewed as a right $R[H]$-module.

Chapter 9

Commutant Duality

Consider an Abelian group E, written additively, and a set S of homomorphisms, addition-preserving mappings, $E \to E$. The *commutant* S_{com} of S is the set of all maps $f : E \to E$ that preserve addition and for which

$$f \circ s = s \circ f \text{ for all } s \in S.$$

We are interested in the case where E is a module over a ring A, and S is the set of all maps $E \to E : x \mapsto ax$ with a running over A. In this case, S_{com} is the ring $C = \mathrm{End}_A(E)$, and E is a module over both ring A and ring C. Our task in this chapter is to study how these two module structures on E interweave with each other.

We return to territory we traveled before in Chap. 5, but on this second pass we have a special focus on the commutant. We pursue three distinct pathways, beginning with a quick, but abstract, approach. The second approach is a more concrete one, in terms of matrices and bases. The third approach focuses more on the relationship between simple left ideals in a ring A and simple C-submodules of an A-module.

9.1 The Commutant

Consider a module E over a ring A. Let us look at what it means for a mapping $f : E \to E$ to be an endomorphism: in addition to the additivity condition

$$f(u + v) = f(u) + f(v) \qquad \text{for all } u, v \in E, \tag{9.1}$$

A.N. Sengupta, *Representing Finite Groups: A Semisimple Introduction*,
DOI 10.1007/978-1-4614-1231-1_9, © Springer Science+Business Media, LLC 2012

the linearity of f means that it *commutes* with the action of A:

$$f(au) = af(u) \qquad \text{for all } a \in A, \text{ and all } u \in E. \tag{9.2}$$

The case of most interest to us is $A = \mathbb{F}[G]$, where G is a finite group and $\mathbb{F}$ is a field, and E is a finite-dimensional vector space over $\mathbb{F}$, with a given representation of G on E. In this case, conditions (9.1) and (9.2) are equivalent to $f \in \text{End}_{\mathbb{F}}(E)$ commuting with all the elements of G represented on E. Thus, $\text{End}_{\mathbb{F}[G]}(E)$ is the commutant for the representation of G on E.

Sometimes the notation

$$\text{End}_G(E)$$

is used instead of $\text{End}_{\mathbb{F}[G]}(E)$, but there is potential for confusion; the minimalist interpretation of $\text{End}_G(E)$ is $\text{End}_{\mathbb{Z}[G]}(E)$, and at the other end it could mean $\text{End}_{\mathbb{F}[G]}(E)$, where $\mathbb{F}$ is some relevant field.

Here is a consequence of Schur's lemma (Lemma 3.1) rephrased in commutant language:

Theorem 9.1 *Let G be a finite group represented on a finite-dimensional vector space V over an algebraically closed field $\mathbb{F}$. Then the commutant of this representation consists of multiples of the identity operator on V if and only if the representation is irreducible.*

(Instant exercise: Check the "only if" part.)

Suppose now that A is a semisimple ring and E is an A-module, decomposing as

$$E = E_1^{n_1} \oplus \ldots \oplus E_r^{n_r}, \tag{9.3}$$

where each E_i is a simple submodule, each $n_i \in \{1, 2, 3, \ldots\}$, and $E_i \not\cong E_j$ as A-modules when $i \neq j$. By Schur's lemma, the only A-linear map $E_i \to E_j$, for $i \neq j$, is 0. Consequently, any element in the commutant $\text{End}_A(E)$ can be displayed as a block-diagonal matrix:

$$\begin{pmatrix} C_1 & 0 & 0 & \ldots & 0 \\ 0 & C_2 & 0 & \ldots & 0 \\ \vdots & \vdots & \vdots & \ldots & 0 \\ 0 & 0 & 0 & \ldots & C_r \end{pmatrix}, \tag{9.4}$$

where each C_i is in $\text{End}_A(E_i^{n_i})$. Moreover, any element of

$$\text{End}_A(E_i^{n_i})$$

is itself an $n_i \times n_i$ matrix, with entries from

$$\mathbb{D}_i = \text{End}_A(E_i), \tag{9.5}$$

which, by Schur's lemma, is a division ring. Conversely, any such matrix clearly specifies an element of the endomorphism ring $\text{End}_A(E_i^{n_i})$. As we saw in Theorem 4.3, the multiplication operation needed on D_i is the opposite of the natural composition, and so we denote this ring by

$$D_i = \mathbb{D}_i^{\text{opp}};$$

note that this is the same as $\mathbb{D}_i$ as a *set*.

To summarize:

Theorem 9.2 *If E is a semisimple module over a ring A and is the direct sum of finitely many simple modules $E_1, \ldots, E_n$,*

$$E \simeq E_1^{m_1} \oplus \ldots \oplus E_n^{m_n},$$

then the ring $\text{End}_A(E)$ is isomorphic to a product of matrix rings:

$$\text{End}_A(E) \simeq \prod_{i=1}^{n} \text{Matr}_{m_i}(D_i), \tag{9.6}$$

where $\text{Matr}_{m_i}(D_i)$ is the ring of $m_i \times m_i$ matrices over the division ring $D_i = \text{End}_A(E_i)^{\text{opp}}$.

9.2 The Double Commutant

Recall that a ring B is *simple* if it is the sum of simple left ideals, all isomorphic to each other as B-modules. In this case any two simple left ideals in B are isomorphic, and B is the internal direct sum of a finite number of simple left ideals.

Consider a left ideal L in a simple ring B, viewed as a B-module. The commutant of the action of B on L is the ring

$$C = \text{End}_B(L).$$

The double commutant is

$$D = \text{End}_C(L).$$

Every element $b \in B$ gives a multiplication map

$$l(b) : L \to L : a \mapsto ba,$$

which, of course, commutes with every $f \in \text{End}_B(L)$. Thus, each $l(b)$ is in $\text{End}_C(L)$. We can now recall Theorem 5.12 in this language:

Theorem 9.3 *Let B be a simple ring, L be a nonzero left ideal in B, and*

$$C = \text{End}_B(L) \qquad \text{and} \qquad D = \text{End}_C(L) \qquad\qquad (9.7)$$

be the commutant and double commutant of the action of B on L. Then the double commutant D is essentially the original ring B, in the sense that the natural map $l : B \to D$, specified by

$$l(b) : L \to L : a \mapsto ba \qquad \text{for all } a \in L \text{ and } b \in B, \qquad (9.8)$$

is an isomorphism.

Stepping up from simplicity, the *Jacobson density theorem* explains how big $l(A)$ is inside D when L is replaced by a semisimple A-module:

Theorem 9.4 *Let E be a semisimple module over a ring A, and let C be the commutant $\text{End}_A(E)$. Then for any $f \in D = \text{End}_C(E)$ and any $x_1, \ldots, x_n \in E$, there exists an $a \in A$ such that*

$$f(x_i) = ax_i \qquad \text{for } i \in [n]. \qquad\qquad (9.9)$$

In particular, if A is an algebra over a field $\mathbb{F}$, and E is finite-dimensional as a vector space over $\mathbb{F}$, then $D = l(A)$; in other words, every element of D is given by multiplication by an element of A.

Proof. View E^n first as a left A-module is the usual way:

$$a(y_1, \ldots, y_n) = (ay_1, \ldots, ay_n)$$

for all $a \in A$, and $(y_1, \ldots, y_n) \in E^n$. Any element of

$$C_n \stackrel{\text{def}}{=} \text{End}_A(E^n)$$

is given by an $n \times n$ matrix with entries in C. To see this in more detail, let ι_j be the inclusion in the jth factor,

$$\iota_j : E \to E^n : y \mapsto (0,\ldots,0,\underbrace{y}_{j\text{th}},0,\ldots,0),$$

and π_j be the projection on the jth factor,

$$\pi_j : E^n \to E : (y_1,\ldots,y_n) \mapsto y_j.$$

Then

$$F\left(\sum_{j=1}^n \iota_j(y_j)\right) = \sum_{j,k=1}^n \pi_k F \iota_j(y_j)$$

$$= \begin{bmatrix} \pi_1 F \iota_1 & \cdots & \pi_1 F \iota_n \\ \vdots & \vdots & \vdots \\ \pi_n F \iota_1 & \cdots & \pi_n F \iota_n \end{bmatrix} \begin{bmatrix} y_1 \\ \vdots \\ y_n \end{bmatrix} \qquad (9.10)$$

shows how to associate with $F \in C_n = \operatorname{End}_A(E^n)$ an $n \times n$ matrix with entries $\pi_i F \iota_j \in C = \operatorname{End}_A(E)$.

Moreover, E^n is also a module over the ring C_n in the natural way. Let $f \in D = \operatorname{End}_C(E)$. The map

$$f_n : E^n \to E^n : (y_1,\ldots,y_n) \mapsto \big(f(y_1),\ldots,f(y_n)\big)$$

is readily checked to be C_n-linear; thus,

$$f_n \in \operatorname{End}_{C_n}(E^n).$$

Now E^n, being semisimple, can be split as

$$E^n = Ax \bigoplus F,$$

where $x = (x_1,\ldots,x_n)$ is any given element of E^n, and F is an A-submodule of E^n. Let

$$p : E^n \to Ax \subset E^n$$

be the corresponding projection. This is, of course, A-linear and is therefore an element of C_n. Consequently, $f_n p = p f_n$, and so

$$f_n\big(p(x)\big) = p\big(f_n(x)\big) \in Ax.$$

Since $p(x) = x$, we have reached our destination: (9.9). $\boxed{\text{QED}}$

9.3   Commutant Decomposition of a Module

Suppose E is a module over a semisimple ring A, L_i is a simple left ideal in A, and D_i is the division ring $\mathrm{End}_A(L_i)$. The elements of D_i are A-linear maps $L_i \to L_i$ and so L_i is, naturally, a left D_i-module. On the other hand, D_i acts naturally on the right on $\mathrm{Hom}_A(L_i, E)$ by taking $(f, d) \in \mathrm{Hom}_A(L_i, E) \times D_i$ to the element $fd = f \circ d \in \mathrm{Hom}_A(L_i, A)$. Thus, $\mathrm{Hom}_A(L_i, E)$ is a right D_i-module. Hence, there is the *balanced tensor product*

$$\mathrm{Hom}_A(L_i, E) \otimes_{D_i} L_i,$$

which, for starters, is just a $\mathbb{Z}$-module. However, the left A-module structure on L_i, which commutes with the D_i-module structure, naturally induces a left A-module structure on $\mathrm{Hom}_A(L_i, E) \otimes_{D_i} L_i$ with multiplications on the second factor. We use this in the following result.

Theorem 9.5 *If E is a module over a semisimple ring A, and $L_1, \ldots, L_r$ is a maximal set of nonisomorphic simple left ideals in A, then the mapping*

$$\bigoplus_{i=1}^{r} \mathrm{Hom}_A(L_i, E) \otimes_{D_i} L_i \to E : (f_1 \otimes x_1, \ldots, f_r \otimes x_r) \mapsto \sum_{i=1}^{r} f_i(x_i) \quad (9.11)$$

is an isomorphism of A-modules. Here D_i is the division ring $\mathrm{End}_A(L_i)$, and the left side of (9.11) has an A-module structure from that on the second factors L_i.

Proof. The module E is a direct sum of simple submodules, each isomorphic to some L_i:

$$E = \bigoplus_{i=1}^{r} \bigoplus_{j \in R_i} E_{ij}, \quad (9.12)$$

where $E_{ij} \simeq L_i$, as A-modules, for each i and $j \in R_i$. In the following we will, as we may, simply assume that $R_i \neq \emptyset$, since $\mathrm{Hom}_A(L_i, E)$ is 0 for all other i. Because L_i is simple, Schur's lemma implies that $\mathrm{Hom}_A(L_i, E_{ij})$ is a one-dimensional (right) vector space over the division ring D_i, and a basis is given by any fixed nonzero element ϕ_{ij}. For any $f_i \in \mathrm{Hom}_A(L_i, E)$ let

$$f_{ij} : L_i \to E_{ij}$$

be the composition of f_i with the projection of E onto E_{ij}. Then

$$f_{ij} = \phi_{ij} d_{ij}$$

for some $d_{ij} \in D_i$. Any element of $\operatorname{Hom}_A(L_i, E) \otimes_{D_i} L_i$ is uniquely of the form

$$\sum_{j \in R_i} \phi_{ij} \otimes x_{ij}$$

with $x_{ij} \in L_i$ (see Theorem 12.10). Consider now the A-linear map

$$J : \bigoplus_{i=1}^{r} \operatorname{Hom}_A(L_i, E) \otimes_{D_i} L_i \to E$$

specified by

$$J\left(\sum_{i=1}^{r} \sum_{j \in R_i} \phi_{ij} \otimes x_{ij}\right) = \sum_{i=1}^{r} \sum_{j \in R_i} \iota_{ij}\big(\phi_{ij}(x_{ij})\big),$$

where $\iota_{ij} : E_{ij} \to E$ is the canonical injection into the direct sum (9.12). If this value is 0, then each $\phi_{ij}(x_{ij}) \in E_{ij}$ is 0 and then, since ϕ_{ij} is an isomorphism, x_{ij} is 0. Thus, J is injective. The decomposition of E into the simple submodules E_{ij} shows that J is also surjective. $\boxed{\text{QED}}$

   Even though $\operatorname{Hom}_A(L_i, E)$ is not, naturally, an A-module, it *is* a left C-module, where

$$C = \operatorname{End}_A(E)$$

is the commutant of the action of A on E: if $c \in C$ and $f \in \operatorname{Hom}_A(L_i, E)$, then

$$cf \stackrel{\text{def}}{=} c \circ f$$

is also in $\operatorname{Hom}_A(L_i, E)$. This makes $\operatorname{Hom}_A(L_i, E)$ a left C-module.

Theorem 9.6 *Let E be a module over a semisimple ring A, and let C be the ring $\operatorname{End}_A(E)$, the commutant of A acting on E. Let L be a simple left ideal in A, and assume that $\operatorname{Hom}_A(L, E) \neq 0$, or, equivalently, that E contains a submodule isomorphic to L. Then the C-module $\operatorname{Hom}_A(L, E)$ is simple.*

Proof. Let $f, h \in \mathrm{Hom}_A(L, E)$, with $h \neq 0$. We will show that $f = ch$ for some $c \in C$. Consequently, any nonzero C-submodule of $\mathrm{Hom}_A(L, E)$ is all of $\mathrm{Hom}_A(L, E)$.

If u is any nonzero element in L, then $L = Au$, and so it will suffice to show that $f(u) = ch(u)$.

We decompose E as the internal direct sum

$$E = F \oplus \bigoplus_{i \in S} E_i,$$

where each E_i is a submodule isomorphic to L, and F is a submodule containing no submodule isomorphic to L. For each $i \in S$ the projection $E \to E_i$, composed with the inclusion $E_i \subset E$, then gives an element

$$p_i \in C.$$

Since $h \neq 0$, and F contains no submodule isomorphic to L, there is some $j \in S$ such that $p_j h(u) \neq 0$. Then $p_j h : L \to E_j$ is an isomorphism. Moreover, for any $i \in S$, the map

$$E_j \to E_i : p_j h(y) \mapsto p_i f(y) \qquad \text{for all } y \in L,$$

is well defined, and extends to an A-linear map

$$c_i : E \to E$$

which is 0 on F and on E_k for $k \neq j$. Note that there are only finitely many i for which $p_i\big(f(u)\big)$ is not 0, and so there are only finitely many i for which c_i is not 0. Let $S' = \{i \in S : c_i \neq 0\}$. Then, piecing together f from its components $p_i f = c_i p_j h$, we have

$$\sum_{i \in S'} c_i p_j h = f.$$

Thus,

$$c = \sum_{i \in S'} c_i p_j$$

is an element of $\mathrm{End}_A(E)$ for which $f = ch$. $\boxed{\text{QED}}$

We have seen that any left ideal L in A is of the form Ay with $y^2 = y$. An element $c \in L$ for which $L = Ac$ is called a *generator* of L.

Here is another interesting observation about $\mathrm{Hom}_A(L, E)$, for a simple left ideal L in A:

Theorem 9.7 *If $L = Ay$ is a left ideal in a semisimple ring A, with y an idempotent, and E is an A-module, then the map*

$$J : \mathrm{Hom}_A(L, E) \to yE : f \mapsto f(y)$$

is an isomorphism of C-modules, where C is the commutant $C = \mathrm{End}_A(E)$. In particular, yE is either 0 or a simple C-module if y is an indecomposable idempotent in A.

Proof. To start with, note that yE is indeed a C-module.

For any $f \in \mathrm{Hom}_A(L, E)$ we have

$$f(y) = f(yy) = yf(y) \in yE.$$

The map

$$J : \mathrm{Hom}_A(L, E) \to yE : f \mapsto f(y) \tag{9.13}$$

is manifestly C-linear.

The kernel of J is clearly 0.

To prove that J is surjective, consider any $v \in yE$; define a map

$$f_v : L \to E : x \mapsto xv.$$

This is clearly A-linear, and $J(f_v) = yv = v$, because $v \in yE$ and $y^2 = y$. Thus, J is surjective.

Finally, if y is an indecomposable idempotent, then $L = Ay$ is a simple left ideal in A and then, by Theorem 9.6, $\mathrm{Hom}_A(E)$, which as we have just proved is C-isomorphic to yE, is either 0 or a simple C-module. $\boxed{\text{QED}}$

The role of the idempotent y in the preceding result is clarified in the following result.

Proposition 9.1 *If idempotents u and v in a ring A generate the same left ideal, and if E is an A-module, then uE and vE are isomorphic C-submodules of E, where $C = \mathrm{End}_A(E)$.*

Proof. Since $Au = Av$, we have

$$u = xv, \quad v = yu \quad \text{for some } x, y \in A.$$

Then the maps

$$f : uE \to vE : w \mapsto yw \quad \text{and} \quad h : vE \to uE : w \mapsto xw$$

act by

$$f(ue) = ve \qquad \text{and} \qquad h(ve) = ue$$

for all $e \in E$. This shows that f and h are inverses to each other. They are also, clearly, both C-linear. $\boxed{\text{QED}}$

Let E be an A-module, where A is a semisimple ring, and $L_1, \ldots, L_r$ are a maximal collection of nonisomorphic simple left ideals in A. Let y_i be a generating idempotent for L_i; thus, $L_i = Ay_i$. We are going to prove that there is an isomorphism

$$\bigoplus_{i=1}^{r} (y_i E \otimes_{D_i} L_i) \simeq E,$$

where both sides have commuting A-module and C-module structures, with C being the commutant $\operatorname{End}_A(E)$ and D_i being the division ring $\operatorname{End}_A(L_i)$. Before looking at a formal statement and proof, let us understand the structures involved here. Easiest is the joint module structure on E: this is simply a consequence of the fact that the actions of A and C on E commute with each other:

$$(a, c)x = a(c(x)) = c(a(x)) \qquad \text{for all } x \in E,\, a \in A,\, c \in C = \operatorname{End}_A(E).$$

Next, consider the action of the division ring D_i on $L_i = Ay_i$:

$$d(ay_i) = d(ay_i y_i) = ay_i d(y_i),$$

which is thus $v \mapsto vd(y_i)$ for all $v \in L_i$. The mapping

$$D_i \to A : d \mapsto d(y_i)$$

is an antihomomorphism:

$$d_1 d_2(y_i) = d_1\big(d_2(y_i)\big) = d_2(y_i)d_1(y_i).$$

The set $y_i E$ is closed under addition and is thus, for starters, just a $\mathbb{Z}$-module. But clearly it is also a C-module, since

$$c(y_i E) = y_i c(E) \subset y_i E.$$

To make matters even more twisted, the mapping $D_i \to A^{\mathrm{opp}} : d \mapsto d(y_i)$ makes $y_i E$ a right module over the division ring D_i with multiplication given by

$$I_\times : y_i E \times D_i \to y_i E : (v, d) \mapsto vd \stackrel{\text{def}}{=} d(y_i)v. \qquad (9.14)$$

Thus, the mapping

$$y_i E \times L_i \to E : (v_i, x_i) \mapsto x_i v_i \tag{9.15}$$

induces first a $\mathbb{Z}$-linear map

$$y_i E \otimes_{\mathbb{Z}} L_i \to E$$

and this quotients to a $\mathbb{Z}$-linear map

$$I : y_i E \otimes_{D_i} L_i \to E \tag{9.16}$$

because

$$I_\times(vd, x) - I_\times(v, dx) = xd(y_i)v - xd(y_i)v = 0.$$

One more thing: $y_i E \otimes_{D_i} L_i$ is both an A-module and a C-module, with commuting module structures, multiplication being given by

$$a \cdot v \otimes x \mapsto v \otimes ax \quad \text{and} \quad c \cdot v \otimes x \mapsto c(v) \otimes x, \tag{9.17}$$

which, as you can check, are well-defined on $y_i E \otimes_{D_i} L_i$ and surely have all the usual necessary properties. This takes us a last step up the spiral: the mapping I is both A-linear and C-linear:

$$\begin{aligned} I(a \cdot v \otimes x) &= I(v \otimes ax) = axv = aI(v \otimes x), \\ I(c \cdot v \otimes x) &= I(c(v) \otimes x) = xc(v) = c(xv) = cI(v \otimes x). \end{aligned} \tag{9.18}$$

At last we are at the end, even if a bit out of breath, of the spiral of tensor product identifications:

Theorem 9.8 *Suppose E is a module over a semisimple ring A, let C be the commutant $\mathrm{End}_A(E)$, and let $L_1 = Ay_1, \ldots, L_r = Ay_r$ be a maximal collection of nonisomorphic simple left ideals in A, with each y_i being an idempotent. Let D_i be the division ring $\mathrm{End}_A(L_i)$. Then the mapping*

$$\bigoplus_{i=1}^{r} y_i E \otimes_{D_i} L_i \to E : \sum_{i=1}^{r} v_i \otimes x_i \mapsto \sum_{i=1}^{r} x_i v_i \tag{9.19}$$

is an isomorphism both for A-modules and for C-modules. Each $y_i E$ is a simple C-module, and, of course, each L_i is a simple A-module.

Proof. On identifying y_iE with $\mathrm{Hom}_A(L_i, E)$ by Theorem 9.7, we find the result becomes equivalent to Theorem 9.5. For a bit more detail, do Exercise 9.7. $\boxed{\mathrm{QED}}$

The awkwardness of phrasing the joint module structures relative to the rings A and C could be eased by bringing in a tensor product ring $A \otimes C$, but let us leave that as an unexplored trail.

Here is another version:

Theorem 9.9 *Let A be a finite-dimensional semisimple algebra over a field $\mathbb{F}$. Suppose E is a module over A and let C be the commutant $\mathrm{End}_A(E)$. Then E, viewed as a C-module, is the direct sum of simple submodules of the form yE, with y running over a set of indecomposable idempotents in A.*

We will explore this in matrix formulation in the next section. But you can also pursue it in Exercise 9.8. The relationship between C-submodules and right ideals in A is explored in greater detail in Exercise 9.6 (which loosely follows Weyl [79]).

9.4   The Matrix Version

In this section we dispel the ethereal elegance of Theorem 9.8 by working through the decomposition in terms of matrices. We will proceed entirely independently of the previous section.

We will work with an algebraically closed field $\mathbb{F}$ of characteristic 0, a finite-dimensional vector space V over $\mathbb{F}$, and a subalgebra A of $\mathrm{End}_{\mathbb{F}}(V)$. Thus, V is an A-module. Let C be the commutant

$$C = \mathrm{End}_A(V).$$

Our objective is to establish Schur's decomposition of V into simple C-modules $e_{ij}V$:

Theorem 9.10 *Let A be a subalgebra of $\mathrm{End}_{\mathbb{F}}(V)$, where $V \neq 0$ is a finite-dimensional vector space over an algebraically closed field $\mathbb{F}$ of characteristic 0. Let*

$$C = \mathrm{End}_A(V)$$

be the commutant of A. Then there exist indecomposable idempotents $\{e_{ij} : 1 \leq i \leq r, 1 \leq j \leq n_i\}$ in A that generate a decomposition of A into simple left ideals

$$A = \bigoplus_{1 \leq i \leq r, 1 \leq j \leq n_i} Ae_{ij}, \tag{9.20}$$

and also decompose V, viewed as a C-module, into a direct sum

$$V = \bigoplus_{1 \leq i \leq r, 1 \leq j \leq n_i} e_{ij}V, \tag{9.21}$$

where each nonzero $e_{ij}V$ is a simple C-submodule of V.

Most of the remainder of this section is devoted to proving this result. We follow Dieudonné and Carrell [22] in examining the detailed matrix structure of A to generate the decomposition of V.

Because A is semisimple and finite-dimensional as a vector space over $\mathbb{F}$, we can decompose it as a direct sum of simple left ideals Ae_j:

$$A = \bigoplus_{j=1}^{N} Ae_j,$$

where the e_j are indecomposable idempotents with

$$e_1 + \cdots + e_N = 1 \quad \text{and} \quad e_i e_j = 0 \quad \text{for all } i \neq j.$$

Then V decomposes as a direct sum:

$$V = e_1 V \oplus \ldots \oplus e_N V. \tag{9.22}$$

(Instant exercise: Why is it a *direct* sum?) The commutant C maps each subspace $e_j V$ into itself. Thus, the $e_j V$ give a decomposition of V as a direct sum of C-submodules. What is, however, not clear is that each nonzero $e_j V$ is a simple C-module; the hard part of Theorem 9.10 provides the simplicity of the submodules in the decomposition (9.21).

We decompose V into a direct sum

$$V = \bigoplus_{i=1}^{r} V^i, \quad \text{with} \quad V^i = V_{i1} \oplus \ldots \oplus V_{in_i}, \tag{9.23}$$

where $V_{i1}, \ldots, V_{in_i}$ are isomorphic simple A-submodules of V, and $V_{i\alpha}$ is *not* isomorphic to $V_{j\beta}$ when $i \neq j$. By Schur's lemma, elements of C map each V^i into itself. To simplify the notation greatly, *we can then just work within a particular V^i*. Thus, let us take for now

$$V = \bigoplus_{j=1}^{n} V_j,$$

where each V_j is a simple A-module and the V_j are isomorphic to each other as A-modules. Let

$$m = \dim_{\mathbb{F}} V_j.$$

Fix a basis

$$u_{11}, \ldots, u_{1m}$$

of the $\mathbb{F}$ vector space V_1 and, using fixed A-linear isomorphisms $V_1 \to V_i$, construct a basis

$$u_{i1}, \ldots, u_{im}$$

in each V_i. Then the matrices of elements in A are block-diagonal, with n blocks, each block being an *arbitrary $m \times m$* matrix T with entries in the field $\mathbb{F}$:

$$\begin{bmatrix} T & & & 0 \\ 0 & T & & \\ & & \ddots & \\ 0 & & & T \end{bmatrix}. \tag{9.24}$$

Thus, the algebra A is isomorphic to the matrix algebra $\mathrm{Matr}_{m \times m}(\mathbb{F})$ by

$$T \mapsto \begin{bmatrix} T & & & 0 \\ 0 & T & & \\ & & \ddots & \\ 0 & & & T \end{bmatrix}. \tag{9.25}$$

(Why "arbitrary" you might wonder; see Exercise 9.10.] The typical matrix in $C = \mathrm{End}_A(V)$ then has the form

$$\begin{bmatrix} s_{11}I & s_{12}I & \cdot & s_{1n}I \\ s_{21}I & s_{22}I & & \cdot \\ \cdot & & \cdot & \cdot \\ s_{n1}I & \cdot & \cdot & s_{nn}I \end{bmatrix}, \tag{9.26}$$

where I is the $m \times m$ identity matrix. Reordering the basis in V as

$$u_{11}, u_{21}, \ldots, u_{n1}, u_{12}, u_{22}, \ldots, u_{n2}, \ldots, u_{1m}, \ldots, u_{nm}$$

displays the matrix (9.26) as the block-diagonal matrix

$$\begin{bmatrix} [s_{ij}] & 0 & \cdot & 0 \\ 0 & [s_{ij}] & \cdot & \\ \cdot & & \cdot & \cdot \\ 0 & \cdot & \cdot & [s_{ij}] \end{bmatrix}, \tag{9.27}$$

where s_{ij} are arbitrary elements of the field $\mathbb{F}$. Thus, C is isomorphic to the algebra of $n \times n$ matrices $[s_{ij}]$ over $\mathbb{F}$. Now the algebra $\mathrm{Matr}_{n \times n}(\mathbb{F})$ is decomposed into a sum of n simple ideals, each consisting of the matrices that have all entries zero except possibly those in one particular column. Thus,

each simple left ideal in C is n-dimensional over $\mathbb{F}$.

Let M_{jh}^i be the matrix for the linear map $V \to V$ which takes u_{ih} to u_{ij} and is 0 on all the other basis vectors. Then, from (9.24), the matrices

$$M_{jh} = M_{jh}^1 + \cdots + M_{jh}^n \tag{9.28}$$

form a basis of A, as a vector space over $\mathbb{F}$. Let

$$e_j = M_{jj}$$

for $1 \leq j \leq m$. This corresponds, in $\mathrm{Matr}_{m \times m}(\mathbb{F})$, to the matrix with 1 at the jj entry and 0 elsewhere. Then A is the direct sum of the simple left ideals Ae_j.

The subspace $e_j V$ has the vectors

$$u_{1j}, u_{2j}, \ldots, u_{nj}$$

as a basis, and so $e_j V$ is n-dimensional. Moreover, $e_j V$ is mapped into itself by C:

$$C(e_j V) = e_j CV \subset e_j V.$$

Consequently, $e_j V$ is a C-module. Since it has the same dimension as any simple C-module, it follows that $e_j V$ cannot have a proper nonzero C-submodule; hence, $e_j V$ is a simple C-module.

We have completed the proof of Theorem 9.10.

Exercises

9.1. Let A be a ring and A^{opp} be the ring formed by the set A with addition the same as for the ring A but multiplication in the opposite order: $a \circ_{\text{opp}} b = ba$ for all $a, b \in A$. For any $a \in A$ let $r_a : A \to A : x \mapsto xa$. Show that $a \mapsto r_a$ gives an isomorphism of A^{opp} with $\text{End}_A(A)$.

9.2. Let A be a semisimple ring. Show that (a) A is also "right semisimple" in the sense that A is the sum of simple right ideals, (b) every right ideal in A has a complementary right ideal, and (c) every right ideal in A is of the form uA, with u an idempotent.

9.3. Let G be a finite group and $\mathbb{F}$ be a field. Denote by $\mathbb{F}[G]_L$ the additive Abelian group $\mathbb{F}[G]$ viewed, in the standard way, as a left $\mathbb{F}[G]$-module. Denote by $\mathbb{F}[G]_R$ the additive Abelian group $\mathbb{F}[G]$ viewed as a left $\mathbb{F}[G]$-module through the multiplication given by

$$x \cdot a = a\hat{x}$$

for $x, a \in \mathbb{F}[G]$, with $\hat{x} = \sum_{g \in G} x(g)g^{-1} \in \mathbb{F}[G]$. Show that the commutant $\text{End}_{\mathbb{F}[G]}\mathbb{F}[G]_L$ is isomorphic to $\mathbb{F}[G]_R$.

9.4. Suppose E is a left module over a semisimple ring A. Then $\hat{E} = \text{Hom}_A(E, A)$ is a right A-module in the natural way via the right-multiplication in A: if $f \in \hat{E}$ and $a \in A$, then $f \cdot a : E \to A : y \mapsto f(y)a$. Show that the map

$$E \to \text{Hom}_A\left(\text{Hom}_A(E, A), A\right) : x \mapsto \text{ev}_x,$$

where $\text{ev}_x(f) = f(x)$ for all $f \in \hat{E}$, is injective.

9.5. Let E be a left A-module, where $A = \mathbb{F}[G]$, with G being a finite group and $\mathbb{F}$ a field. Assume that E is finite-dimensional as a vector space over $\mathbb{F}$. Let $\hat{E} = \text{Hom}_A(E, A)$, E' be the vector space dual $\text{Hom}_{\mathbb{F}}(E, \mathbb{F})$, and $\text{Tr}_e : \mathbb{F}[G] \to \mathbb{F} : x \mapsto x_e$ be the functional which evaluates a general element $x = \sum_{g \in G} x_g g \in A$ at the identity $e \in G$. Show that the mapping

$$I : \hat{E} \to E' : \phi \mapsto \phi_e \stackrel{\text{def}}{=} \text{Tr}_e \circ \phi$$

is an isomorphism of vector spaces over $\mathbb{F}$.

9.6. Let E be a left A-module, where A is a semisimple ring, $C = \mathrm{End}_A(E)$, and $\hat{E} = \mathrm{Hom}_A(E, A)$. We view E as a left C-module in the natural way, and view $\hat{E}$ as a right A-module For any nonempty subset S of E define the subset $S_\#$ of A to be all finite sums of elements $\phi(w)$ with ϕ running over $\hat{E}$ and w over S.

(a) Show that $S_\#$ is a right ideal in A.

(b) Show that $(aE)_\# = aE_\#$ for all $a \in A$.

(c) Show that if W is a C-submodule of E, then $W = W_\# E$.

(d) Suppose U and W are C-submodules of E with $U_\# \subset W_\#$. Show that $U \subset W$. In particular, $U_\# = W_\#$ if and only if $U = W$.

(e) Show that a C-submodule W of E is simple if $W_\#$ is a simple right ideal.

(f) Show that if W is a simple C-submodule of E, and if $E_\# = A$, then $W_\#$ is a simple right ideal in A.

(g) Show that if u is an indecomposable idempotent in A and the right ideal uA lies inside $E_\#$, then uE is a simple C-module.

9.7. With E an A-module, where A is a semisimple ring, and $L = Ay$ a simple left ideal in A with idempotent generator y, use the the map $J : \mathrm{Hom}_A(L, E) \to yE : f \mapsto f(y)$ to transfer the action of the division ring $D = \mathrm{End}_A(L)$ from L to yE.

9.8. Prove Theorem 9.9.

9.9. Prove Burnside's theorem: *If G is a group of endomorphisms of a finite-dimensional vector space E over an algebraically closed field $\mathbb{F}$, and E is simple as a G-module, then $\mathbb{F}G$, the linear span of G inside $\mathrm{End}_\mathbb{F}(E)$, is equal to the whole of $\mathrm{End}_\mathbb{F}(E)$.*

9.10. Prove Wedderburn's theorem: *Let E be a simple module over a ring A, and suppose that it is faithful in the sense that if a is nonzero in A, then the map $l(a) : E \to E : x \mapsto ax$ is also nonzero. If E is finite-dimensional over the division ring $C = \mathrm{End}_A(E)$, then $l : A \to \mathrm{End}_C(E)$ is an isomorphism.* Specialize this to the case where A is a finite-dimensional algebra over an algebraically closed field $\mathbb{F}$.

9.11. Let E be a semisimple module over a ring A.

(a) Show that if the commutant $\mathrm{End}_A(E)$ is a commutative ring, then E is the direct sum of simple sub-A-modules, no two of which are isomorphic.

(b) Suppose E is the direct sum of simple submodules E_α, no two of which are isomorphic to each other, and assume also that each commutant $\mathrm{End}_A(E_\alpha)$ is a field (i.e., it is commutative); show that the ring $\mathrm{End}_A(E)$ is commutative.

(Exercise 5.6 shows that when E is a direct sum of a set of nonisomorphic simple submodules, then every simple submodule of E *is* one of these submodules.) Here is a case which is useful in the Okounkov–Vershik theory for representations of S_n: view S_{n-1} as a subgroup of S_n in the natural way; then it turns out that $\mathbb{C}[S_{n-1}]$ has a commutative centralizer in $\mathbb{C}[S_n]$. This then implies that in the decomposition of a simple $\mathbb{C}[S_n]$-module as a direct sum of simple $\mathbb{C}[S_{n-1}]$-modules, no two of the latter are isomorphic to each other.

Chapter 10

Character Duality

In the chapter we perform a specific implementation of the dual decomposition theory explored in the preceding chapter. The symmetric group S_n has a natural action on $V^{\otimes n}$ for any vector space V, as in (10.1). Our first goal in this chapter is to identify, under some simple conditions, the commutant $\mathrm{End}_{\mathbb{F}[G]} V^{\otimes n}$ as the linear span of the operators $T^{\otimes n}$ on $V^{\otimes n}$ with T running over the group $GL_{\mathbb{F}}(V)$ of all invertible linear endomorphisms of V. The commutant duality theory presented in the previous chapter then produces an interlinking of the representations, and hence also of the characters, of S_n and those of $GL_{\mathbb{F}}(V)$. Following this, we will go through a fast proof of the duality formula connecting characters of S_n and the character of $GL_{\mathbb{F}}(V)$ using the commutant duality theory. In the last section we will prove this duality formula again, but by a more explicit computation.

10.1   The Commutant for S_n on $V^{\otimes n}$

For any vector space V, the permutation group S_n has a natural left action on $V^{\otimes n}$:

$$\sigma \cdot (v_1 \otimes \cdots \otimes v_n) = v_{\sigma^{-1}(1)} \otimes \cdots \otimes v_{\sigma^{-1}(n)}. \tag{10.1}$$

The set of all invertible endomorphisms in $\mathrm{End}_{\mathbb{F}}(V)$ forms the *general linear group*

$$GL_{\mathbb{F}}(V)$$

of the vector space V. Here is a fundamental result from Schur [71]:

A.N. Sengupta, *Representing Finite Groups: A Semisimple Introduction*,
DOI 10.1007/978-1-4614-1231-1_10, © Springer Science+Business Media, LLC 2012

Theorem 10.1 *Suppose V is a finite-dimensional vector space over a field $\mathbb{F}$, and $n \in \{1, 2, \ldots\}$ is such that $n!$ is not divisible by the characteristic of $\mathbb{F}$ and, moreover, the number of elements in $\mathbb{F}$ exceeds $(\dim_{\mathbb{F}} V)^2$. Then the commutant of the action of S_n on $V^{\otimes n}$ is the linear span of all endomorphisms $T^{\otimes n} : V^{\otimes n} \to V^{\otimes n}$, with T running over $GL_{\mathbb{F}}(V)$.*

Proof. Fix a basis $|e_1\rangle, \ldots, |e_d\rangle$ of V, and let $\langle e_1|, \ldots, \langle e_d|$ be the dual basis in V':

$$\langle e_i | e_j \rangle = \delta_{ij}.$$

Any

$$X \in \mathrm{End}_{\mathbb{F}}(V^{\otimes n})$$

is then described in coordinates by the quantities

$$X_{i_1 j_1; \ldots; i_n j_n} = \langle e_{i_1} \otimes \cdots \otimes e_{i_n} | X | e_{j_1} \otimes \cdots \otimes e_{j_n} \rangle. \tag{10.2}$$

Relabel the $m = N^2$ pairs (i, j) with numbers from $1, \ldots, m$. Denote $\{1, \ldots, k\}$ by $[k]$ for all positive integers k; thus, an element a in $[m]^{[n]}$ expands to $(a_1, \ldots, a_n)$ with each $a_i \in \{1, \ldots, m\}$, and encodes an n-tuple of pairs $(i, j) \in \{1, \ldots, N\}^2$.

The condition that X commutes with the action of S_n translates in coordinate language to the condition that the quantities $X_{i_1 j_1; \ldots; i_n j_n}$ in (10.2) remain invariant when $i, j \in [N]^{[n]}$ are replaced by $i \circ \sigma$ and $j \circ \sigma$, respectively, for any $\sigma \in S_n$.

We will show that if $F \in \mathrm{End}_{\mathbb{F}}(V^{\otimes n})$ satisfies

$$\sum_{a \in [m]^{[n]}} F_{a_1 \ldots a_n} (T^{\otimes n})_{a_1 \ldots a_n} = 0 \quad \text{for all } T \in GL_{\mathbb{F}}(V), \tag{10.3}$$

then

$$\sum_{a \in [m]^{[n]}} F_{a_1 \ldots a_n} X_{a_1 \ldots a_n} = 0 \tag{10.4}$$

for all X in the commutant of S_n. This means that any element in the dual of $\mathrm{End}_{\mathbb{F}}(V^{\otimes n})$ that vanishes on the elements $T^{\otimes n}$, with $T \in GL_{\mathbb{F}}(V)$, vanishes on the entire subspace which is the commutant of S_n. This clearly implies that the commutant is spanned by the elements $T^{\otimes n}$.

Consider the polynomial in the $m = N^2$ indeterminates T_a given by

$$p(T) = \left(\sum_{a_1, \ldots, a_n \in \{1, \ldots, m\}} F_{a_1 \ldots a_n} T_{a_1} \ldots T_{a_n} \right) \det[T_{ij}].$$

The hypothesis (10.3) says that this polynomial is equal to 0 for all choices of values of T_a in the field $\mathbb{F}$. If the field $\mathbb{F}$ is not very small, a polynomial $p(T)$ all of whose evaluations are 0 is identically 0 as a polynomial. Let us work through an argument for this. Evaluating the T_k at arbitrary fixed values in $\mathbb{F}$ for all except one $k = k_*$, we find the polynomial $p(T)$ turns into a polynomial $q(T_{k_*})$, of degree $\leq m$, in the one variable T_{k_*}, which vanishes on all the $|\mathbb{F}|$ elements of $\mathbb{F}$; the hypothesis $|\mathbb{F}| > N^2 = m$ then implies that $q(T_{k_*})$ is the zero polynomial. This means that the polynomials in the variables T_a, for $a \neq k_*$, given by the coefficients of powers of T_{k_*} in $p(T)$, evaluate to 0 at all values in $\mathbb{F}$. Reducing the number of variables in this way, we reach all the way to the conclusion that the polynomial $p(T)$ is 0. Since the polynomial $\det[T_{ij}]$ is certainly not 0, it follows that

$$\sum_a F_{a_1 \ldots a_n} T_{a_1} \ldots T_{a_n} = 0 \qquad (10.5)$$

as a polynomial. Keep in mind that

$$F_{a_{\sigma(1)} \ldots a_{\sigma(n)}} = F_{a_1 \ldots a_n}$$

for all $a_1, \ldots, a_n \in \{1, \ldots, m\}$ and $\sigma \in S_n$. Then from (10.5) we see that $n! F_{a_1 \ldots a_n}$ is 0 for all subscripts a_i. Since $n!$ is not 0 on $\mathbb{F}$, it follows that each F_a is 0, and hence we have (10.4). $\boxed{\text{QED}}$

10.2 Schur–Weyl Duality

We can now apply the commutant duality theory of the previous chapter to obtain Schur's decomposition of the representation of S_n on $V^{\otimes n}$. Assume that $\mathbb{F}$ is an algebraically closed field of characteristic 0 (in particular, $\mathbb{F}$ is infinite); then

$$V^{\otimes n} \simeq \bigoplus_{i=1}^{r} L_i \otimes_{\mathbb{F}} y_i V^{\otimes n}, \qquad (10.6)$$

where $L_1, \ldots, L_r$ is a maximal string of simple left ideals in $\mathbb{F}[S_n]$ that are not isomorphic as left $\mathbb{F}[S_n]$-modules, and y_i is a generating idempotent in L_i for each $i \in \{1, \ldots, r\}$. The subspace $y_i V^{\otimes n}$, when nonzero, is a simple C_n-module, where C_n is the commutant $\text{End}_{\mathbb{F}[S_n]}(V^{\otimes n})$. In view of Theorem 10.1, the tensor product representation of $GL_{\mathbb{F}}(V)$ on $V^{\otimes n}$ has a restriction to an irreducible representation on $y_i V^{\otimes n}$ when this is nonzero.

The duality between S_n acting on the n-dimensional space V and the general linear group $GL_{\mathbb{F}}(V)$ is often called Schur–Weyl duality. For much more on commutants and Schur–Weyl duality, see the book by Goodman and Wallach [40].

10.3   Character Duality, the High Road

As before let $\mathbb{F}$ be an algebraically closed field of characteristic 0. If A is a finite-dimensional semisimple algebra over $\mathbb{F}$, E is an A-module with $\dim_{\mathbb{F}} E < \infty$, and C is the commutant $\mathrm{End}_A(E)$, then E decomposes through the map

$$I : \bigoplus_{i=1}^{r} y_i E \otimes_{\mathbb{F}} L_i \to E : \sum_{i=1}^{r} v_i \otimes x_i \mapsto \sum_{i=1}^{r} x_i v_i,$$

which is both A-linear and C-linear, where $y_1, \ldots, y_r$ are idempotents in A such that any simple A-module is isomorphic to $L_i = Ay_i$ for exactly one i. For any $(a, c) \in A \times C$, we have the product ac first as an element of $\mathrm{End}_{\mathbb{F}}(E)$ and then, by I^{-1}, acting on $\bigoplus_{i=1}^{r} y_i E \otimes_{\mathbb{F}} Ay_i$. Comparing traces, we have

$$\mathrm{Tr}(ac) = \sum_{i=1}^{r} \mathrm{Tr}(a|L_i)\mathrm{Tr}(c|y_i E), \tag{10.7}$$

where $a|L_i$ is the element in $\mathrm{End}_{\mathbb{F}}(L_i)$ given by $x \mapsto ax$.

We specialize now to

$$A = \mathbb{F}[S_n]$$

acting on $V^{\otimes n}$, where V is a finite-dimensional vector space over $\mathbb{F}$. Then, as we know, C is spanned by elements of the form $B^{\otimes n}$, with B running over $GL_{\mathbb{F}}(V)$. Nonisomorphic simple left ideals in A correspond to inequivalent irreducible representations of S_n. Let the set $\mathcal{R}$ label these representations; thus, there is a maximal set of nonisomorphic simple left ideals L_α, with α running over $\mathcal{R}$. Then we have, for any $\sigma \in S_n$ and any $B \in GL_{\mathbb{F}}(V)$, the character duality formula

$$\mathrm{Tr}(B^{\otimes n} \cdot \sigma) = \sum_{\alpha \in \mathcal{R}} \chi_\alpha(\sigma)\chi^\alpha(B), \tag{10.8}$$

where χ_α is the characteristic of the representation of S_n on $L_\alpha = \mathbb{F}[S_n]y_\alpha$, and χ^α is that of $GL_{\mathbb{F}}(V)$ on $y_\alpha V^{\otimes n}$.

Recall the character orthogonality relation

$$\frac{1}{n!} \sum_{\sigma \in S_n} \chi_\alpha(\sigma)\chi_\beta(\sigma^{-1}) = \delta_{\alpha\beta} \quad \text{for all } \alpha, \beta \in \mathcal{R}.$$

Using this with (10.8), we have

$$\chi^\alpha(B) = \frac{1}{n!} \sum_{\sigma \in S_n} \chi_\alpha(\sigma^{-1}) s^\sigma(B),$$

where

$$s^\sigma(B) = \text{Tr}(B^{\otimes n} \cdot \sigma). \tag{10.9}$$

Note that s^σ *depends only on the conjugacy class of* σ, rather than on the specific choice of σ. Denoting by K a typical conjugacy class, we then have

$$\chi^\alpha(B) = \sum_{K \in \mathcal{C}} \frac{|K|}{n!} \chi_\alpha(K) s^K(B), \tag{10.10}$$

where $\mathcal{C}$ is the set of all conjugacy classes in S_n, $\chi_\alpha(K)$ is the value of χ_α on any element in K, and s^K is s^σ for any $\sigma \in K$.

In the following section we will prove the character duality formulas (10.8) and (10.10) again, by a more explicit method.

10.4   Character Duality by Calculations

We will now work through a proof of the Schur–Weyl duality formulas by more explicit computations. This section is entirely independent of the preceding section, and the method expounded is similar to that of Weyl [79].

All through this section $\mathbb{F}$ is an algebraically closed field of characteristic 0.

Let $V = \mathbb{F}^N$, on which the group $GL(N, \mathbb{F})$ acts in the natural way. Let

$$e_1, \ldots, e_N$$

be the standard basis of $V = \mathbb{F}^N$.

We know that $V^{\otimes n}$ decomposes as a direct sum of subspaces of the form

$$y_\alpha V^{\otimes n},$$

with y_α running over a set of indecomposable idempotents in $\mathbb{F}[S_n]$, such that the left ideals $\mathbb{F}[S_n]y_\alpha$ form a decomposition of $\mathbb{F}[S_n]$ into simple left submodules.

Let χ^α be the character of the irreducible representation ρ_α of $GL(N,\mathbb{F})$ on the subspace $y_\alpha V^{\otimes n}$, and χ_α be the character of the representation of S_n on $\mathbb{F}[S_n]y_\alpha$. Our goal is to establish the relation between these two characters.

If a matrix $g \in GL(N,\mathbb{F})$ has all eigenvalues distinct, then the corresponding eigenvectors are linearly independent and hence form a basis of V. Changing basis, g is conjugate to a diagonal matrix

$$D(\vec{\lambda}) = D(\lambda_1,\ldots,\lambda_N) = \begin{bmatrix} \lambda_1 & 0 & 0 & \cdots & 0 & 0 \\ 0 & \lambda_2 & 0 & \cdots & 0 & 0 \\ 0 & 0 & \lambda_3 & \cdots & 0 & 0 \\ \vdots & \vdots & \vdots & \vdots & \vdots & \vdots \\ 0 & 0 & 0 & \cdots & 0 & \lambda_N \end{bmatrix}.$$

Then $\chi^\alpha(g)$ equals $\chi^\alpha\big(D(\vec{\lambda})\big)$. We will evaluate the latter.

The tensor product

$$e_{i_1} \otimes \cdots \otimes e_{i_n}$$

is an eigenvector of $D(\vec{\lambda})$ with eigenvalue $\lambda_{i_1}\ldots\lambda_{i_N}$, and these form a basis of $\mathbb{F}^N$ as $(i_1,\ldots,i_n)$ runs over $[N]^{[n]}$. Hence, every eigenvalue of $D(\vec{\lambda})$ is of the form $\lambda_{i_1}\ldots\lambda_{i_N}$. Moreover, the eigensubspace for $\lambda_{i_1}\ldots\lambda_{i_N}$ is the same for all $\vec{\lambda} \in \mathbb{F}^N$.

Fix a partition of n given by

$$\vec{f} = (f_1,\ldots,f_N) \in \mathbb{Z}_{\geq 0}^N$$

with

$$|\vec{f}| = f_1 + \cdots + f_N = n,$$

and let

$$\vec{\lambda}^{\vec{f}} = \prod_{j=1}^N \lambda_j^{f_j}$$

and

$$V(\vec{f}) = \{v \in V^{\otimes n} : D(\vec{\lambda})v = \vec{\lambda}^{\vec{f}}v \quad \text{for all } \vec{\lambda} \in \mathbb{F}^N\}.$$

Thus, every eigenvalue of $D(\vec{\lambda})$ is of the form $\vec{\lambda}^{\vec{f}}$. From the observation in the previous paragraph, it follows that $\mathbb{F}^N$ is the direct sum of the subspaces $V(\vec{f})$, with $\vec{f}$ running over all partitions of n.

Since the action of $GL(N, \mathbb{F})$ on $V^{\otimes n}$ commutes with that of S_n, the action of $D(\vec{\lambda})$ on the vector

$$y_\alpha(e_{i_1} \otimes \cdots \otimes e_{i_n})$$

is also multiplication by $\lambda_{i_1} \ldots \lambda_{i_N}$. The subspaces $y_\alpha V(\vec{f})$, for fixed $\vec{f}$ and y_α running over the string of indecomposable idempotents adding up to 1, directly sum to $V(\vec{f})$. Consequently,

$$\chi^\alpha(D(\vec{\lambda})) = \sum_{\vec{f} \in \mathbb{Z}_{\geq 0}^N} \vec{\lambda}^{\vec{f}} \dim(y_\alpha V(\vec{f})). \tag{10.11}$$

The space $V(\vec{f})$ has a basis given by the *set*

$$\{\sigma \cdot e_1^{\otimes f_1} \otimes \cdots \otimes e_N^{\otimes f_N} : \sigma \in S_n\}.$$

Note that

$$\vec{e}^{\otimes \vec{f}} = e_1^{\otimes f_1} \otimes \cdots \otimes e_N^{\otimes f_N}$$

is indeed in $V^{\otimes n}$, because $|\vec{f}| = n$.

The dimension of $y_\alpha V(\vec{f})$ is

$$\dim(y_\alpha V(\vec{f})) = \frac{1}{f_1! \ldots f_N!} \sum_{\sigma \in S_n(\vec{f})} \chi_\alpha(\sigma), \tag{10.12}$$

where

$$S_n(\vec{f})$$

is the subgroup of S_n consisting of elements that preserve the sets

$$\{1, \ldots, f_1\}, \{f_1 + 1, \ldots, f_2\}, \ldots, \{f_{N-1} + 1, \ldots, f_N\}$$

and we have used the fact that χ_α equals the character of the representation of S_n on $\mathbb{F}[S_n]y_\alpha$. (If you have a short proof of (10.12), write it in the margins here, or else work through Exercise 10.2.)

Thus,

$$\chi^\alpha(D(\vec{\lambda})) = \sum_{\vec{f} \in \mathbb{Z}_{\geq 0}^N} \vec{\lambda}^{\vec{f}} \frac{1}{f_1! \ldots f_N!} \sum_{\sigma \in S_n(\vec{f})} \chi_\alpha(\sigma). \tag{10.13}$$

The character χ_α is constant on conjugacy classes. So the second sum on the right here should be reduced to a sum over conjugacy classes. Note that, with obvious notation,

$$S_n(\vec{f}) \simeq S_{f_1} \times \cdots \times S_{f_N}.$$

The conjugacy class of a permutation is completely determined by its cycle structure: i_1 1-cycles, i_2 2-cycles,.... For a given sequence

$$\vec{i} = (i_1, i_2, \ldots, i_m) \in \mathbb{Z}_{\geq 0}^m,$$

the number of such permutations in S_m is

$$\frac{m!}{(i_1!1^{i_1})(i_2!2^{i_2})(i_3!3^{i_3}) \ldots (i_m!m^{i_m})} \tag{10.14}$$

because, in distributing $1, \ldots, m$ among such cycles, the i_k k-cycles can be arranged in $i_k!$ ways and each such k-cycle can be expressed in k ways. Alternatively, the denominator in (10.14) is the size of the isotropy group of any element of the conjugacy class.

The cycle structure of an element of

$$(\sigma_1, \ldots, \sigma_N) \in S_{f_1} \times \cdots \times S_{f_N}$$

is described by a sequence

$$[\vec{i}_1, \ldots, \vec{i}_N] = (\underbrace{i_{11}, i_{12}, \ldots, i_{1f_1}}_{\vec{i}_1}, \ldots, \underbrace{i_{N1}, \ldots, i_{Nf_N}}_{\vec{i}_N}),$$

with i_{jk} being the number of k-cycles in the permutation σ_j. Let us denote by

$$\chi_\alpha([\vec{i}_1, \ldots, \vec{i}_N])$$

the value of χ_α on the corresponding conjugacy class in S_n. Then

$$\sum_{\sigma \in S_n(\vec{f})} \chi_\alpha(\sigma) = \sum_{[\vec{i}_1, \ldots, \vec{i}_N] \in [\vec{f}]} \chi_\alpha([\vec{i}_1, \ldots, \vec{i}_N]) \prod_{j=1}^N \frac{f_j!}{(i_{j1}!1^{i_{j1}})(i_{j2}!2^{i_{j2}}) \ldots}.$$

Here the sum is over the set $[\vec{f}]$ of all $[\vec{i}_1, \ldots, \vec{i}_N]$ for which

$$i_{j1} + 2i_{j2} + \cdots + ni_{jn} = f_j \quad \text{for all } j \in \{1, \ldots, N\}.$$

(Of course, i_{jn} is 0 when $n > f_j$.)

Returning to the expression for χ^α in (10.13), we have

$$\chi^\alpha(D(\vec{\lambda})) = \sum_{\vec{f} \in \mathbb{Z}_{\geq 0}^N} \vec{\lambda}^{\vec{f}} \sum_{[\vec{i}_1,\dots,\vec{i}_N] \in [\vec{f}]} \chi_\alpha([\vec{i}_1,\dots,\vec{i}_N]) \prod_{j=1}^N \frac{1}{(i_{j1}!1^{i_{j1}})(i_{j2}!2^{i_{j2}}) \dots (i_{jn}!n^{i_{jn}})}$$

$$= \sum_{\vec{f} \in \mathbb{Z}_{\geq 0}^N} \vec{\lambda}^{\vec{f}} \sum_{[\vec{i}_1,\dots,\vec{i}_N] \in [\vec{f}]} \chi_\alpha([\vec{i}_1,\dots,\vec{i}_N]) \prod_{1 \leq j \leq N,\, 1 \leq k \leq n} \frac{1}{i_{jk}!\, k^{i_{jk}}}.$$

Now χ_α is constant on conjugacy classes in S_n. The conjugacy class in $S_{f_1} \times \cdots \times S_{f_N}$ specified by the cycle structure

$$[\vec{i}_1,\dots,\vec{i}_N]$$

corresponds to the conjugacy class in S_n specified by the cycle structure

$$\vec{i} = (i_1,\dots,i_n)$$

with

$$\sum_{j=1}^N i_{jk} = i_k \quad \text{for all } k \in \{1,\dots,n\}. \tag{10.15}$$

Recall again that

$$\sum_{k=1}^n k i_{jk} = f_j. \tag{10.16}$$

Note that

$$\vec{\lambda}^{\vec{f}} = \prod_{k=1}^n (\lambda_1^{k i_{1k}} \dots \lambda_N^{k i_{Nk}}).$$

Combining these observations we have

$$\chi^\alpha(D(\vec{\lambda})) = \sum_{\vec{i} \in \mathbb{Z}_{\geq 0}^N} \chi_\alpha(\vec{i}) \frac{1}{1^{i_1} 2^{i_2} \dots n^{i_n}} \sum_{i_{jk}} \prod_{k=1}^n \frac{\lambda_1^{k i_{1k}} \dots \lambda_N^{k i_{Nk}}}{i_{1k}! i_{2k}! \dots i_{Nk}!}, \tag{10.17}$$

where the inner sum on the right is over all $[\vec{i}_1,\dots,\vec{i}_N]$ corresponding to the cycle structure $\vec{i} = (i_1,\dots,i_n)$ in S_n, hence satisfying (10.15). We observe now that this sum simplifies to

$$\sum_{i_{jk}} \prod_{k=1}^n \frac{\lambda_1^{k i_{1k}} \dots \lambda_N^{k i_{Nk}}}{i_{1k}! i_{2k}! \dots i_{Nk}!} = \frac{1}{i_1! \dots i_n!} \prod_{k=1}^n (\lambda_1^k + \cdots + \lambda_N^k)^{i_k}. \tag{10.18}$$

This produces

$$\chi^\alpha(D(\vec{\lambda})) = \sum_{\vec{i} \in \mathbb{Z}_{\geq 0}^N} \chi_\alpha(\vec{i}) \frac{1}{(i_1!1^{i_1})(i_2!2^{i_2})\ldots(i_n!n^{i_n})} \prod_{k=1}^n s_k(\vec{\lambda})^{i_k}, \qquad (10.19)$$

where $s_1, \ldots, s_n$ are the power sums given by

$$s_m(X_1, \ldots, X_n) = X_1^m + \cdots + X_n^m. \qquad (10.20)$$

We can also conveniently define

$$s_m(B) = \mathrm{Tr}(B^m). \qquad (10.21)$$

Then

$$\chi^\alpha(B) = \sum_{\vec{i} \in \mathbb{Z}_{\geq 0}^N} \chi_\alpha(\vec{i}) \frac{1}{(i_1!1^{i_1})(i_2!2^{i_2})\ldots(i_n!n^{i_n})} \prod_{k=1}^n s_k(B)^{i_k} \qquad (10.22)$$

for all $B \in GL(N, \mathbb{F})$ with distinct eigenvalues, and hence for all $B \in GL(N, \mathbb{F})$. (So there is a leap of logic which you should explore.) The beautiful formula (10.22) for the character χ^α of the $GL(V)$ in terms of characters of S_n was obtained by Schur [71].

The sum on the right in (10.22) is over all conjugacy classes in S_n, each labeled by its cycle structure

$$\vec{i} = (i_1, \ldots, i_n).$$

Note that the number of elements in this conjugacy class is exactly $n!$ divided by the denominator which appears on the right inside the sum. Thus, we can also write the *Schur–Weyl duality* formula as

$$\chi^\alpha(B) = \sum_{K \in \mathcal{C}} \frac{|K|}{n!} \chi_\alpha(K) s^K(B), \qquad (10.23)$$

where $\mathcal{C}$ is the set of all conjugacy classes in S_n, and

$$s^K \overset{\mathrm{def}}{=} \prod_{m=1}^n s_m^{i_m} \qquad (10.24)$$

if K has the cycle structure $\vec{i} = (i_1, \ldots, i_n)$.

Up to this point we have *not needed to assume that α labels an irreducible representation of S_n*. We have merely used the character χ_α corresponding to some left ideal $\mathbb{F}[S_n]y_\alpha$ in $\mathbb{F}[S_n]$, and the corresponding $GL(n, \mathbb{F})$-module $y_\alpha V^{\otimes n}$.

We will now assume that χ_α indeed labels the irreducible characters of S_n. Then we have the Schur orthogonality relations

$$\frac{1}{n!} \sum_{\sigma \in S_n} \chi_\alpha(\sigma)\chi_\beta(\sigma^{-1}) = \delta_{\alpha\beta}.$$

These can be rewritten as

$$\sum_{K \in \mathcal{C}} \chi_\alpha(K)\frac{|K|}{n!}\chi_\beta(K^{-1}) = \delta_{\alpha\beta}. \tag{10.25}$$

Thus, the $|\mathcal{C}| \times |\mathcal{C}|$ square matrix $[\chi_\alpha(K^{-1})]$ has the inverse $\frac{1}{n!}[|K|\chi_\alpha(K)]$. Therefore, also

$$\sum_{\alpha \in \mathcal{R}} \chi_\alpha(K^{-1})\frac{|K'|}{n!}\chi_\alpha(K') = \delta_{KK'}, \tag{10.26}$$

where $\mathcal{R}$ labels a maximal set of inequivalent irreducible representations of S_n. Consequently, multiplying (10.23) by $\chi_\alpha(K^{-1})$ and summing over α, we obtain

$$\sum_{\alpha \in \mathcal{R}} \chi^\alpha(B)\chi_\alpha(K) = s^K(B) \tag{10.27}$$

for every conjugacy class K in S_n, where we used the fact that $K^{-1} = K$.

Observe that

$$s^K(B) = \text{Tr}(B^{\otimes n} \cdot \sigma), \tag{10.28}$$

where σ, any element of the conjugacy class K, appears on the right here by its representation as an endomorphism of $V^{\otimes n}$. The identity (10.28) is readily checked (Exercise 10.3) if σ is the cycle $(12 \ldots n)$, and then the general case follows by observing (and verifying in Exercise 10.3) that

$$\text{Tr}(B^{\otimes j} \otimes B^{\otimes l} \cdot \phi\theta) = \text{Tr}(B^{\otimes j}\phi)\text{Tr}(B^{\otimes l}\theta) \tag{10.29}$$

if ϕ and θ are the disjoint cycles $(12 \ldots j)$ and $(j+1 \ldots n)$.

Thus, the duality formula (10.27) coincides exactly with the formula (10.8) we proved in the previous section.

Exercises

10.1. Let E be a module over a ring A, $\vec{e}$ be an element of E, and N be the left ideal in A consisting of all $n \in A$ for which $n\vec{e} = 0$. Assume that A decomposes as $N \oplus N_c$, where N_c is also a left ideal, and let $P_c : A \to A$ be the projection map onto N_c; thus, every $a \in A$ splits as $a = a_N + P_c(a)$, with $a_N \in N$ and $P_c(a) \in N_c$. Show that for any right ideal R in A:

(a) $P_c(R) \subset R$.

(b) There is a well-defined map given by

$$f : R\vec{e} \to P_c(R) : x\vec{e} \mapsto P_c x.$$

(c) The map

$$P_c(R) \to R\vec{e} : x \mapsto x\vec{e}$$

is the inverse of f.

10.2. Let G be a finite group, represented on a finite-dimensional vector space E over a field $\mathbb{F}$ characteristic 0. Suppose $\vec{e} \in E$ is such that the *set* $G\vec{e}$ is a basis of E. Denote by H the isotropy subgroup $\{h \in G : h\vec{e} = \vec{e}\}$, and $N = \{n \in \mathbb{F}[G] : n\vec{e} = 0\}$.

(a) Show that
$$\mathbb{F}[G] = N \oplus \mathbb{F}[G/H],$$

where $\mathbb{F}[G/H]$ is the left ideal in $\mathbb{F}[G]$ consisting of all x for which $xh = x$ for every $h \in H$, and that the projection map onto $\mathbb{F}[G/H]$ is given by

$$\mathbb{F}[G] \to \mathbb{F}[G] : x \mapsto \frac{1}{|H|} \sum_{h \in H} xh.$$

(b) Let y be an idempotent and $L = \mathbb{F}[G]y$. Show that

$$\hat{L}\vec{e} = \hat{y}E, \tag{10.30}$$

where $\mathbb{F}[G] \to \mathbb{F}[G] : x \mapsto \hat{x}$ is the $\mathbb{F}$-linear map carrying g to g^{-1} for every $g \in G \subset \mathbb{F}[G]$. Then, using Exercise 10.1, obtain the dimension formula

$$\dim_{\mathbb{F}}(\hat{y}E) = \frac{1}{|H|} \sum_{h \in H} \chi_L(h), \qquad (10.31)$$

where $\chi_L(a)$ is the trace of the map $L \to L : y \mapsto ay$.

10.3. Verify the identity (10.28) in the case σ is the cycle $(12 \ldots n)$. Next verify the identity (10.29).

Chapter 11

Representations of $U(N)$

The unitary group $U(N)$ consists of all $N \times N$ complex matrices U that satisfy the unitarity condition

$$U^*U = I.$$

It is a group under matrix multiplication, and, being a subset of the linear space of all $N \times N$ complex matrices, it is a topological space as well. Multiplication of matrices is, clearly, continuous. The inversion map $U \mapsto U^{-1} = U^*$ is continuous as well. This makes $U(N)$ a *topological group*. It has much more structure, but we will need no more.

By a *representation* ρ of $U(N)$ we will mean a continuous mapping

$$\rho : U(N) \to \mathrm{End}_{\mathbb{C}}(V)$$

for some finite-dimensional complex vector space V. Notice the additional condition of continuity required of ρ. The *character* of ρ is the function

$$\chi_\rho : U(N) \to \mathbb{C} : U \mapsto \mathrm{tr}\,(\rho(U)). \tag{11.1}$$

The representation ρ is said to be *irreducible* if the only subspaces of V invariant under the action of $U(N)$ are 0 and V, and $V \neq 0$.

Representations ρ_1 and ρ_2 of $U(N)$, on finite-dimensional vector spaces V_1 and V_2, respectively, are said to be *equivalent* if there is a linear isomorphism

$$\Theta : V_1 \to V_2$$

that *intertwines* ρ_1 and ρ_2 in the sense that

$$\Theta\rho_1(U)\Theta^{-1} = \rho_2(U) \quad \text{for all } U \in U(N).$$

A.N. Sengupta, *Representing Finite Groups: A Semisimple Introduction*,
DOI 10.1007/978-1-4614-1231-1_11, © Springer Science+Business Media, LLC 2012

If there is no such Θ, then the representations are *inequivalent*. As for finite groups (Proposition 1.2), if ρ_1 and ρ_2 are equivalent, then they have the same character.

In this chapter we will explore the representations of $U(N)$. Although $U(N)$ is definitely not a finite group, Schur–Weyl duality interweaves the representation theories of $U(N)$ and of the permutation group S_n, making the exploration of $U(N)$ a natural digression from our main journey through finite groups. For an interesting application of this duality, and duality between other compact groups and discrete groups, see the article by Lévy [56].

11.1 The Haar Integral

For our exploration of $U(N)$ there is one essential piece of equipment we cannot do without: the Haar integral. Its construction would take us far off the main route, and so we will accept its existence and one basic formula that we will see in the next section. A readable exposition of the construction of the Haar integral on a general topological group is given by Cohn [13, Chap. 9]; an account specific to compact Lie groups, such as $U(N)$, is given in the book by Bröcker and tom Dieck [8].

On the space of complex-valued continuous functions on $U(N)$ there is a unique linear functional, the normalized *Haar integral*

$$f \mapsto \langle f \rangle = \int_{U(N)} f(U)\,\mathrm{d}U$$

satisfying the following conditions:

- It is nonnegative in the sense that

$$\langle f \rangle \geq 0 \quad \text{if } f \geq 0,$$

and, moreover, $\langle f \rangle$ is 0 if and only if f is 0.

- It is invariant under left and right translations in the sense that

$$\int_{U(N)} f(xUy)\,\mathrm{d}U = \int_{U(N)} f(U)\,\mathrm{d}U \quad \text{for all } x, y \in U(N)$$

and all continuous functions f on $U(N)$.

- The integral is normalized:

$$\langle 1 \rangle = 1.$$

In more standard notation, the Haar integral of f is denoted

$$\int_{U(N)} f(g) \, dg.$$

Let T denote the subgroup of $U(N)$ consisting of all diagonal matrices. A typical element of T has the form

$$D(\lambda_1, \ldots, \lambda_N) \stackrel{\text{def}}{=} \begin{bmatrix} \lambda_1 & 0 & 0 & \cdots & 0 & 0 \\ 0 & \lambda_2 & 0 & \cdots & 0 & 0 \\ 0 & 0 & \lambda_3 & \cdots & 0 & 0 \\ \vdots & \vdots & \vdots & \vdots & \vdots & \vdots \\ 0 & 0 & 0 & \cdots & 0 & \lambda_N \end{bmatrix},$$

where $\lambda_1, \ldots, \lambda_N$ are complex numbers of unit modulus.

Thus, T is the product of N copies of the circle group $U(1)$ of unit modulus complex numbers:

$$T \simeq U(1)^N.$$

This makes it, geometrically, a torus, and hence the choice of notation. There is a natural Haar integral over T, specified by

$$\int_T h(t) \, dt = (2\pi)^{-N} \int_0^{2\pi} \cdots \int_0^{2\pi} h\big(D(e^{i\theta_1}, \ldots, e^{i\theta_N})\big) \, d\theta_1 \ldots d\theta_N \qquad (11.2)$$

for any continuous function h on T.

11.2   The Weyl Integration Formula

Recall that a function f on a group is *central* if

$$f(xyx^{-1}) = f(y)$$

for all elements x and y of the group.

For every continuous central function f on $U(N)$ the following integration formula (Weyl [79, Sect. 17]) holds:

$$\int_{U(N)} f(U)\, dU = \frac{1}{N!} \int_T f(t)|\Delta(t)|^2\, dt,$$ (11.3)

where

$$\Delta\big(D(\lambda_1,\ldots,\lambda_N)\big) = \det \begin{bmatrix} \lambda_1^{N-1} & \lambda_2^{N-1} & \cdots & \lambda_{N-1}^{N-1} & \lambda_N^{N-1} \\ \lambda_1^{N-2} & \lambda_2^{N-2} & \cdots & \lambda_{N-1}^{N-2} & \lambda_N^{N-2} \\ \vdots & \vdots & \vdots\vdots\vdots & \vdots & \vdots \\ \lambda_1 & \lambda_2 & \cdots & \lambda_{N-1} & \lambda_N \\ 1 & 1 & \cdots & 1 & 1 \end{bmatrix}$$ (11.4)

$$= \prod_{1 \le j < k \le N} (\lambda_j - \lambda_k),$$

the last step being a famed identity. This *Vandermonde determinant*, written out as an alternating sum, is

$$\Delta\big(D(\lambda_1,\ldots,\lambda_N)\big) = \sum_{\sigma \in S_N} \operatorname{sgn}(\sigma)\lambda_1^{N-\sigma(1)} \ldots \lambda_N^{N-\sigma(N)}.$$ (11.5)

The diagonal term is

$$\lambda_1^{N-1}\lambda_2^{N-2} \ldots \lambda_{N-1}^1 \lambda_N^0.$$

Observe that among all the monomial terms $\lambda_1^{w_1} \ldots \lambda_N^{w_N}$, where $\vec{w} = (w_1,\ldots, w_N) \in \mathbb{Z}^N$, which appear in the determinant, this is the "highest" in the sense that all such $\vec{w}$ are $\le (N-1, N-2,\ldots,0)$ in lexicographic order (check the dominance in the first component, then the second, and so on).

11.3   Character Orthogonality

As with finite groups, every representation is a direct sum of irreducible representations. Hence, every character is a sum of irreducible representation characters with positive integer coefficients. (The details of this are farmed out to Exercise 11.1.)

Just as for finite groups, the character orthogonality relations hold for representations of $U(N)$. If ρ_1 and ρ_2 are inequivalent irreducible representations of $U(N)$, then

$$\int_{U(N)} \chi_{\rho_1}(U^{-1})\chi_{\rho_2}(U)\, dU = 0$$ (11.6)

and

$$\int_{U(N)} \chi_\rho(U^{-1}) \chi_\rho(U) \, dU - 1, \tag{11.7}$$

for any irreducible representation ρ. (You can work through the proofs in Exercise 11.3.)

Analogously to the case of finite groups, each $\rho(U)$ is diagonal in some basis, with diagonal entries being of unit modulus.

It follows that

$$\chi_\rho(U^{-1}) = \overline{\chi_\rho(U)}. \tag{11.8}$$

The Haar integral specifies a Hermitian inner product on the space of continuous functions on $U(N)$ by

$$\langle f, h \rangle = \int_{U(N)} \overline{f(U)} h(U) \, dU. \tag{11.9}$$

In terms of this inner product, the character orthogonality relations say that the characters χ_ρ of irreducible representations form an orthonormal set of functions on $U(N)$.

11.4 Weights

Consider an irreducible representation ρ of $U(N)$ on a finite-dimensional vector space V.

The linear maps

$$\rho(t) : V \to V,$$

with t running over the Abelian subgroup T, commute with each other,

$$\rho(t)\rho(t') = \rho(tt') = \rho(t't) = \rho(t')\rho(t),$$

and so there is a basis $\{v_j\}_{1 \leq j \leq d_V}$ of V with respect to which the matrices of $\rho(t)$, for all $t \in T$, are diagonal:

$$\rho(t) = \begin{bmatrix} \rho_1(t) & 0 & \cdots & 0 \\ 0 & \rho_2(t) & \cdots & 0 \\ \vdots & \vdots & \vdots & 0 \\ 0 & 0 & \cdots & \rho_{d_V}(t) \end{bmatrix},$$

where

$$\rho_r : T \to U(1) \subset \mathbb{C}$$

are continuous homomorphisms. Thus,

$$\rho_r\big(D(\lambda_1, \ldots, \lambda_N)\big) = \rho_{r1}(\lambda_1) \ldots \rho_{rN}(\lambda_N),$$

where $\rho_{rk}(\lambda)$ is ρ_r evaluated on the diagonal matrix that has λ at the kth diagonal entry and all other diagonal entries are 1. Since each ρ_{rk} is a continuous homomorphism

$$U(1) \to U(1),$$

they necessarily have the form

$$\rho_{rk}(\lambda) = \lambda^{w_{rk}} \qquad\qquad (11.10)$$

for some integer w_{rk}. We will refer to

$$\vec{w}_r = (w_{r1}, \ldots, w_{rN}) \in \mathbb{Z}^N$$

as a *weight* for the representation ρ.

11.5   Characters of $U(N)$

Continuing with the framework as above, we have

$$\rho_r\big(D(\lambda_1, \ldots, \lambda_N)\big) = \lambda_1^{w_{r1}} \ldots \lambda_N^{w_{rN}}.$$

Thus,

$$\chi_\rho\big(D(\lambda_1, \ldots, \lambda_N)\big) = \sum_{r=1}^{d_V} \lambda_1^{w_{r1}} \ldots \lambda_N^{w_{rN}}. \qquad (11.11)$$

It will be convenient to write

$$\vec{\lambda} = (\lambda_1, \ldots, \lambda_N)$$

and analogously for $\vec{w}$.

Two diagonal matrices in $U(N)$ whose diagonal entries are permutations of each other are conjugate within $U(N)$ (permutation of the basis vectors implements the conjugation transformation). Consequently, a character will have the same value on two such diagonal matrices. Thus,

$$\chi_\rho\big(D(\lambda_1, \ldots, \lambda_N)\big) \text{ is invariant under permutations of the } \lambda_j.$$

Then, by gathering similar terms, we can rewrite the character as a sum of symmetric sums

$$\sum_{\sigma \in S_N} \lambda_{\sigma(1)}^{w_1} \cdots \lambda_{\sigma(N)}^{w_N} \tag{11.12}$$

with $\vec{w} = (w_1, \ldots, w_N)$ running over a certain set of elements in $\mathbb{Z}^N$. (If "gathering similar terms" bothers you, wade through Theorem 12.7.)

Thus, we can express each character as a Fourier sum (with only finitely many nonzero terms):

$$\chi_\rho(D(\vec{\lambda})) = \sum_{\vec{w} \in \mathbb{Z}_\downarrow^N} c_{\vec{w}} s_{\vec{w}}(\vec{\lambda}), \tag{11.13}$$

where each coefficient $c_{\vec{w}}$ is a nonnegative integer, and $s_{\vec{w}}$ is the symmetric function given by

$$s_{\vec{w}}(\vec{\lambda}) = \sum_{\sigma \in S_N} \prod_{j=1}^{N} \lambda_{\sigma(j)}^{w_j}. \tag{11.14}$$

The subscript $\downarrow$ in $\mathbb{Z}_\downarrow^N$ signifies that it consists of integer strings

$$w_1 \geq w_2 \geq \cdots \geq w_N.$$

Now ρ is irreducible if and only if

$$\int_{U(N)} |\chi_\rho(U)|^2 \, dU = 1. \tag{11.15}$$

(Verify this as Exercise 11.4.) Using the Weyl integration formula, and our expression for χ_ρ, we find this is equivalent to

$$\int_{U(1)^N} \left| \chi_\rho(\vec{\lambda}) \Delta(\vec{\lambda}) \right|^2 d\lambda_1 \ldots d\lambda_N = N! \tag{11.16}$$

Now the product

$$\chi_\rho(\vec{\lambda}) \Delta(\vec{\lambda})$$

is *skew-symmetric* in $\lambda_1, \ldots, \lambda_N$, and is an integer linear combination of terms of the form

$$\lambda_1^{m_1} \ldots \lambda_N^{m_N}.$$

So, collecting similar terms together, we can express $\chi_\rho(\vec{\lambda})\Delta(\vec{\lambda})$ as an integer linear combination of the elementary skew-symmetric sums

$$a_{\vec{f}}(\vec{\lambda}) = \sum_{\sigma \in S_N} \text{sgn}(\sigma) \lambda_{\sigma(1)}^{f_1} \cdots \lambda_{\sigma(N)}^{f_N}$$

$$= \sum_{\sigma \in S_N} \text{sgn}(\sigma) \lambda_1^{f_{\sigma(1)}} \cdots \lambda_N^{f_{\sigma(N)}}$$

$$= \det \begin{bmatrix} \lambda_1^{f_1} & \lambda_2^{f_1} & \cdots & \lambda_N^{f_1} \\ \lambda_1^{f_2} & \lambda_2^{f_2} & \cdots & \lambda_N^{f_2} \\ \vdots & \vdots & \vdots & \vdots \\ \lambda_1^{f_N} & \lambda_2^{f_N} & \cdots & \lambda_N^{f_N} \end{bmatrix}, \tag{11.17}$$

with $\vec{f} = (f_1, \ldots, f_N) \in \mathbb{Z}^N$. (Again, the "collecting terms" argument is set on more serious foundations by Theorem 12.7.) Therefore,

$$\int_{U(1)^N} \left| \chi_\rho(\vec{\lambda})\Delta(\vec{\lambda}) \right|^2 d\lambda_1 \ldots d\lambda_N$$

is an integer linear combination of inner products

$$\int_{U(1)^N} \overline{a_{\vec{f}}(\vec{\lambda})} a_{\vec{f}}(\vec{\lambda}) \, d\lambda_1 \ldots d\lambda_N. \tag{11.18}$$

Now we use the simple, yet crucial, fact that on $U(1)$ there is the orthogonality relation

$$\int_{U(1)} \overline{\lambda^m} \lambda^n \, d\lambda = \delta_{nm}.$$

Consequently, distinct monomials such as $\lambda_1^{a_1} \ldots \lambda_N^{a_N}$, with $\vec{a} \in \mathbb{Z}^N$, are orthonormal. Hence, if $f_1 > f_2 > \cdots > f_N$, then the first two expressions in (11.17) for $a_{\vec{f}}(\vec{\lambda})$ *are sums of orthogonal terms, each of norm 1.*

If $\vec{f}$ and $\vec{f}'$ are *distinct* elements of $\mathbb{Z}_\downarrow^N$, each a strictly decreasing sequence, then no permutation of the entries of $\vec{f}$ could be equal to $\vec{f}'$, and so

$$\int_{U(1)^N} \overline{a_{\vec{f}'}(\vec{\lambda})} a_{\vec{f}}(\vec{\lambda}) \, d\lambda_1 \ldots d\lambda_N = 0. \tag{11.19}$$

On the other hand,

$$\int_{U(1)^N} \overline{a_{\vec{f}}(\vec{\lambda})} a_{\vec{f}}(\vec{\lambda}) \, d\lambda_1 \ldots d\lambda_N = N! \tag{11.20}$$

because $a_{\vec{f}}(\vec{\lambda})$ is a sum of $N!$ orthogonal terms each of norm 1.

Putting all these observations, especially the norms (11.16) and (11.20), together, we see that an expression of $\chi_\rho(\vec{\lambda})\Delta(\vec{\lambda})$ as an integer linear combination of the elementary skew-symmetric functions $a_{\vec{f}}$ will involve exactly one of the latter, and with coefficient ± 1:

$$\chi_\rho(\vec{\lambda})\Delta(\vec{\lambda}) = \pm a_{\vec{h}}(\vec{\lambda}) \tag{11.21}$$

for some $\vec{h} \in \mathbb{Z}_\downarrow^N$. To determine the sign here, it is useful to use the lexicographic ordering on $\mathbb{Z}^N$, with $v \in \mathbb{Z}^N$ being $>$ than $v' \in \mathbb{Z}^N$ if the first nonzero entry in $v - v'$ is positive. With this ordering, let $\vec{w}$ be the highest (maximal) of the weights.

Then the "highest" term in $\chi_\rho(\vec{\lambda})$ is

$$\lambda_1^{w_1}\ldots\lambda_N^{w_N},$$

appearing with some positive integer coefficient, and the "highest" term in $\Delta(\vec{\lambda})$ is the diagonal term

$$\lambda_1^{N-1}\ldots\lambda_N^0.$$

Thus, the highest term in the product $\chi_\rho(\vec{\lambda})\Delta(\vec{\lambda})$ is

$$\lambda_1^{w_1+N-1}\ldots\lambda_{N-1}^{w_{N-1}+1}\lambda_N^{w_N},$$

appearing with coefficient $+1$.

We conclude that

$$\chi_\rho(\vec{\lambda})\Delta(\vec{\lambda}) = a_{(w_1+N-1,\ldots,w_{N-1}+1,w_N)}(\vec{\lambda}), \tag{11.22}$$

and also that the highest weight term

$$\lambda_1^{w_1}\ldots\lambda_N^{w_N}$$

appears with coefficient 1 in the expression for $\chi_\rho(D(\vec{\lambda}))$. This gives a remarkable consequence:

Theorem 11.1 *In the decomposition of the representation of T given by ρ on V, the representation corresponding to the highest weight appears exactly once.*

The orthogonality relations (11.19) imply that

$$\int_{U(1)^N} \overline{\chi_{\rho'}(\vec{\lambda})}\chi_\rho(\vec{\lambda})|\Delta(\vec{\lambda})|^2\, d\lambda_1\ldots d\lambda_N = 0 \qquad (11.23)$$

for irreducible representations ρ and ρ' corresponding to *distinct* highest weights $\vec{w}$ and $\vec{w}'$.

Thus, we have the following theorem:

Theorem 11.2 *Representations corresponding to different highest weights are inequivalent.*

Finally, we also have an explicit expression, Weyl's formula [79, (16.9)], for the character χ_ρ of an irreducible representation ρ as a ratio of determinants:

Theorem 11.3 *The character χ_ρ of an irreducible representation ρ of $U(N)$ is the unique central function on $U(N)$ whose value on diagonal matrices is given by*

$$\chi_\rho\big(D(\vec{\lambda})\big) = \frac{a_{(w_1+N-1,\ldots,w_{N-1}+1,w_N)}(\vec{\lambda})}{a_{(N-1,\ldots,1,0)}(\vec{\lambda})}, \qquad (11.24)$$

where $(w_1,\ldots,w_N)$ is the highest weight for ρ. The division on the right in (11.24) is to be understood as division of polynomials, treating the $\lambda_j^{\pm 1}$ as indeterminates.

Note that in (11.24) the denominator is $\Delta(\vec{\lambda})$ from (11.4).

11.6 Weyl Dimension Formula

The *dimension* of the representation ρ is equal to $\chi_\rho(I)$, but (11.24) reads $0/0$ on putting $\vec{\lambda} = (1,1,\ldots,1)$ into the numerator and the denominator. L'Hôpital's rule may be applied, but it is simplified by a trick borrowed from Weyl. Take an indeterminate t, and evaluate the ratio in (11.24) at

$$\vec{\lambda} = (t^{N-1}, t^{N-2}, \ldots, t, 1).$$

Then $a_{\vec{h}}(\vec{\lambda})$ becomes a Vandermonde determinant:

$$a_{(h_1,\ldots,h_N)}(t^{N-1},\ldots,t,1) = \det \begin{bmatrix} t^{h_1(N-1)} & t^{h_1(N-2)} & \cdots & t^{h_1} & 1 \\ t^{h_2(N-1)} & t^{h_2(N-2)} & \cdots & t^{h_2} & 1 \\ \vdots & \vdots & \vdots & \vdots & \vdots \\ t^{h_N(N-1)} & t^{h_N(N-2)} & \cdots & t^{h_N} & 1 \end{bmatrix}$$

$$= \prod_{1 \le j < k \le N} \left(t^{h_j} - t^{h_k} \right).$$

Consequently,

$$\frac{a_{(h_1,\ldots,h_N)}(t^{N-1},\ldots,t,1)}{a_{(h'_1,\ldots,h'_N)}(t^{N-1},\ldots,t,1)} = \prod_{1 \le j < k \le N} \frac{t^{h_j} - t^{h_k}}{t^{h'_j} - t^{h'_k}}.$$

Evaluating of the rational function in t on the right at $t = 1$ gives us

$$\prod_{1 \le j < k \le N} \frac{h_j - h_k}{h'_j - h'_k} = \frac{\mathrm{VD}(h_1,\ldots,h_N)}{\mathrm{VD}(h'_1,\ldots,h'_N)},$$

where VD denotes the Vandermonde determinant.

Applying this to the Weyl character formula yields the wonderful *Weyl dimension formula*:

Theorem 11.4 *If ρ is an irreducible representation of $U(N)$, then the dimension of the corresponding representation space is*

$$\dim(\rho) = \prod_{1 \le j < k \le N} \frac{w_j - w_k + k - j}{k - j}, \tag{11.25}$$

where $(w_1,\ldots,w_N)$ is the highest weight for ρ.

11.7   From Weights to Representations

Our next goal is to construct an irreducible representation of $U(N)$ with a given weight $\vec{w} \in \mathbb{Z}^N_{\downarrow}$. We will produce such a representation inside a tensor product of exterior powers of $\mathbb{C}^N$.

It will be convenient to work first with a vector $\vec{f} \in \mathbb{Z}^N_{\downarrow}$, all of whose components are greater than or equal to 0. We can take $\vec{f}$ to be simply $\vec{w}$ in the case when all w_j are nonnegative. If, on the other hand, some $w_i < 0$, then we set

$$f_j = w_j - w_N \quad \text{for all } j \in \{1,\ldots,N\}.$$

Display $\vec{f}$ as a tableau of empty boxes, with the first row having f_1 boxes, followed beneath by a row of f_2 boxes, and so on, with the Nth row containing f_N boxes. (We ignore the trivial case where all f_j are 0.) For example,

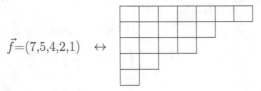

$$\vec{f} = (7,5,4,2,1) \quad \leftrightarrow$$

Let f_1' be the number of boxes in column 1; this is the largest i for which $f_i \geq 1$. In this way, let f_j' be the number of boxes in column j (the largest i for which $f_i \geq j$). Now consider

$$V_{\vec{f}} = \overset{f_1'}{\bigwedge} \mathbb{C}^N \otimes \overset{f_2'}{\bigwedge} \mathbb{C}^N \otimes \cdots \otimes \overset{f_N'}{\bigwedge} \mathbb{C}^N, \tag{11.26}$$

where the zeroth exterior power is, by definition, just $\mathbb{C}$, and thus effectively dropped. (If $\vec{f} = 0$, then $V_{\vec{f}} = \mathbb{C}$.)

The group $U(N)$ acts on $V_{\vec{f}}$ in the obvious way through tensor powers, and we have thus a representation ρ of $U(N)$. The appropriate tensor products of the standard basis vectors $e_1, \ldots, e_N$ of $\mathbb{C}^N$ form a basis of $V_{\vec{f}}$, and these basis vectors are eigenvectors of the diagonal matrix

$$D(\vec{\lambda}) \in T$$

acting on $V_{\vec{f}}$. Indeed, a basis is formed by the vectors

$$e_a = \bigotimes_{j=1}^{N} (e_{a_{1,j}} \wedge \ldots \wedge e_{a_{f_j',j}}),$$

with each string $a_{1,j}, \ldots, a_{f_j',j}$ being strictly increasing and drawn from $[N]$. We can visualize e_a as being obtained by placing the number $a_{i,j}$ in the box in the ith row in the jth column, and then taking the wedge-product of the vectors $e_{a_{i,j}}$ down each column and then taking the tensor product across all the columns. For example,

1	3	4	8
2	4	6
5	7
3

$$\leftrightarrow \quad (e_1 \wedge e_2 \wedge e_5 \wedge e_3) \otimes (e_3 \wedge e_4 \wedge e_7) \otimes (e_4 \wedge e_6) \otimes e_8.$$

Clearly,

$$\rho(D(\vec{\lambda}))e_a = \left(\prod_{i,j} \lambda_{a_{i,j}}\right) e_a. \tag{11.27}$$

The highest-weight term corresponds to precisely e_{a^*}, where a^* has the entry 1 in all boxes in row 1, then the entry 2 in all boxes in row 2, and so on. The eigenvalue corresponding to e_{a^*} is

$$\lambda_1^{f_1} \dots \lambda_N^{f_N}.$$

The corresponding subspace inside $V_{\vec{f}}$ is one-dimensional, spanned by e_{a^*}. Decomposing $V_{\vec{f}}$ into a direct sum of irreducible subspaces under the representation ρ, we find that e_{a^*} lies inside (exactly) one of these subspaces. This subspace $V_{\vec{f}}$ must then be the irreducible representation of $U(N)$ corresponding to the highest weight $\vec{f}$.

We took $\vec{f} = \vec{w}$ if $w_N \geq 0$, and so we are finished with that case. Now suppose $w_N < 0$. We have to make an adjustment to $V_{\vec{f}}$ to produce an irreducible representation corresponding to the original highest weight $\vec{w} \in \mathbb{Z}_\downarrow^N$.

Consider then

$$V(\vec{w}) = V_{\vec{f}} \otimes \left(\bigwedge^{-N}(\mathbb{C}^N)\right)^{\otimes |w_N|}, \tag{11.28}$$

where a negative exterior power is defined as a dual

$$\bigwedge^{-m} V = \left(\bigwedge^m V\right)' \quad \text{for } m \geq 1.$$

The representation of $U(N)$ on $\bigwedge^{-N}(\mathbb{C}^N)$ is given by

$$U \cdot \phi = (\det U)^{-1}\phi \quad \text{for all } U \in U(N) \text{ and } \phi \in \bigwedge^{-N}(\mathbb{C}^N).$$

This is a one-dimensional representation with weight $(-1, \dots, -1)$, because the diagonal matrix $D(\vec{\lambda})$ acts by multiplication by $\lambda_1^{-1} \dots \lambda_N^{-1}$.

For the representation of $U(N)$ on $V_{\vec{w}}$, we have a basis of $V_{\vec{w}}$ consisting of eigenvectors of $\rho(D(\vec{\lambda}))$; the highest weight is

$$\vec{f} + (-w_N)(-1, \dots, -1) = (f_1 + w_N, \dots, f_N + w_N) = (w_1, \dots, w_N),$$

by our choice of $\vec{f}$. Thus, $V(\vec{w})$ contains an irreducible representation with highest weight $\vec{w}$. But

$$\dim V(\vec{w}) = \dim V_{\vec{f}},$$

and, on using Weyl's dimension formula, we find this is equal to the dimension of the irreducible representation of highest weight $\vec{w}$. Thus, $V(\vec{w})$ is the irreducible representation with highest weight $\vec{w}$.

11.8 Characters of S_n from Characters of $U(N)$

We will now see how Schur–Weyl duality leads to a way of determining the characters of S_n from the characters of $U(N)$.

Let $N, n \in \{1, 2, \ldots\}$, and consider the vector space $(\mathbb{C}^N)^{\otimes n}$. The permutation group S_n acts on this by

$$\sigma \cdot (v_1 \otimes \cdots \otimes v_n) = v_{\sigma^{-1}(1)} \otimes \cdots \otimes v_{\sigma^{-1}(n)}. \tag{11.29}$$

The group $GL(N, \mathbb{C})$ of invertible linear maps on $\mathbb{C}^N$ also acts on $(\mathbb{C}^N)^{\otimes n}$ in the natural way:

$$B \cdot (v_1 \otimes \cdots \otimes v_n) = B^{\otimes n}(v_1 \otimes \cdots \otimes v_n) = Bv_1 \otimes \ldots \otimes Bv_n.$$

In Theorem 10.1, these actions are dual in the sense that the commutant of the action of $\mathbb{C}[S_n]$ on $(\mathbb{C}^N)^{\otimes n}$ is the linear span of the operators $B^{\otimes n}$ with B running over $GL(N, \mathbb{C})$. We can leverage this to the following duality for the unitary group:

Theorem 11.5 *Let $N, n \in \{1, 2, \ldots\}$, and consider $(\mathbb{C}^N)^{\otimes n}$ as a $\mathbb{C}[S_n]$-module by means of the multiplication specified in (11.29). Then the commutant $\mathrm{End}_{\mathbb{C}[S_n]}(\mathbb{C}^N)^{\otimes n}$ is spanned by the elements $U^{\otimes n}$, with U running over $U(N)$.*

For a complex vector space W let us, for our purposes here only, declare the elements $A, B \in \mathrm{End}(W)$ to be *orthogonal* if $\mathrm{Tr}\,(AB) = 0$. For any subspace $L \subset \mathrm{End}(W)$ let $L^{\perp}$ be the set of all $A \in \mathrm{End}(W)$ orthogonal to all elements of L. We will use the fact that $L \mapsto L^{\perp}$ is injective. Note also that if A and UBU^{-1} are orthogonal, then $U^{-1}AU$ and B are orthogonal for any $U \in \mathrm{End}(W)$. You can work these out as Exercise 11.5.

Proof. In Theorem 10.1 we showed that $\mathrm{End}_{\mathbb{C}[S_n]}(\mathbb{C}^N)^{\otimes n}$ is the linear span of the operators $B^{\otimes n}$ with B running over $GL(N, \mathbb{C})$. Suppose now that $S \in \mathrm{End}_{\mathbb{C}}(\mathbb{C}^N)^{\otimes n}$ is orthogonal to $D^{\otimes n}$ for all $D \in U(N)$. Then for any fixed

$T \in U(N)$, the element $S_1 = T^{\otimes n} S (T^{-1})^{\otimes n}$ is also orthogonal to $D^{\otimes n}$ for all $D \in U(N)$. From this it follows that S_1 is orthogonal to $D^{\otimes n}$ for all *diagonal* matrices $D \in GL(N, \mathbb{C})$, because $\mathrm{Tr}\,(S_1 D^{\otimes n})$, viewed as a polynomial in every particular diagonal entry of D, is zero on the infinite set $U(1) \subset \mathbb{C}$ and hence is 0 on all elements of $\mathbb{C}$. Now for any $N \times N$ Hermitian matrix H there is a unitary matrix $T_1 \in U(N)$ such that $T_1^{-1} H T_1 = D$ is a diagonal matrix. Hence, S is orthogonal to $H^{\otimes n}$ for every Hermitian matrix H. If H_1 and H_2 are Hermitian, then

$$\mathrm{Tr}\,\left(S(H_1 + t H_2)^{\otimes n} \right) = 0 \tag{11.30}$$

for all *real* t, and hence the left side of (11.30), viewed as a *polynomial* in the variable t, is identically 0. Therefore, (11.30) holds for all $t \in \mathbb{C}$. Now for a general $B \in GL(N, \mathbb{C})$ we have $B = H_1 + i H_2$, where H_1 and H_2 are Hermitian. Hence, S is orthogonal to $B^{\otimes n}$ for all $N \times N$ matrices $B \in GL(N, \mathbb{C})$. Thus, the linear span of $\{U^{\otimes n} : U \in U(N)\}$ is equal to the linear span of $\{B^{\otimes n} : B \in GL(N, \mathbb{C})\}$. $\boxed{\text{QED}}$

From the Schur–Weyl duality formula it follows that

$$\mathrm{Tr}(B^{\otimes n} \cdot \sigma) = \sum_{\alpha \in \mathcal{R}} \chi_\alpha(\sigma) \chi^\alpha(B), \tag{11.31}$$

where, on the left, σ represents the action of $\sigma \in S_n$ on $(\mathbb{C}^N)^{\otimes n}$, and $B \in U(N)$, and, on the right, $\mathcal{R}$ is a maximal set of inequivalent irreducible representations of S_n. For the representation α of S_n given by the regular representation restricted on a simple left ideal L_α in $\mathbb{C}[S_n]$, χ^α is the character of the representation of $U(N)$ on

$$y_\alpha (\mathbb{C}^N)^{\otimes n}, \tag{11.32}$$

where y_α is a nonzero idempotent in L_α.

Now the simple left ideals in $\mathbb{C}[S_n]$ correspond to

$$\vec{f} = (f_1, \dots, f_n) \in \mathbb{Z}_{\geq 0, \downarrow}^n \tag{11.33}$$

(the subscript $\downarrow$ signifying that $f_1 \geq \cdots \geq f_n$), which are partitions of n:

$$f_1 + f_2 + \cdots + f_n = n.$$

Recall that associated with this partition we have a Young tableau $T_{\vec{f}}$ of the numbers $1,\ldots,n$ in r rows of boxes:

1	2	$\cdots$	$\cdots$	$\cdots$	$\cdots$	f_1
$1+f_1$	$\cdots$	$\cdots$	$\cdots$	$\cdots$	f_2+f_1	
$\cdots$	$\cdots$	$\cdots$	$\cdots$			
$\cdots$	$\cdots$	$\cdots$	$\cdots$			
$\cdots$	$\cdots$	n				

If $r < n$, then $f_j = 0$ for $r < j \le n$. Associated with $T_{\vec{f}}$ there is the idempotent

$$y_{\vec{f}} = \sum_{q\in C_{T_{\vec{f}}},\, p\in R_{T_{\vec{f}}}} (-1)^{\mathrm{sgn}\,(q)} qp, \qquad (11.34)$$

where $C_{T_{\vec{f}}}$ is the subgroup of S_n which, acting on the tableau $T_{\vec{f}}$, map the entries of each column into the same column, and $R_{T_{\vec{f}}}$ preserves rows. Let

$$a_{ij} \in \{1,\ldots,n\}$$

be the entry in the box in row i and column j in the tableau $T_{\vec{f}}$. For example,

$$a_{23} = f_1 + 3.$$

Let $e_1,\ldots,e_N$ be the standard basis of $\mathbb{C}^N$. Place e_1 in each of the boxes in the first row, then e_2 in each of the boxes in the second row, and so on till the rth row. Let

$$e^{\otimes \vec{f}} = e_1^{\otimes f_1} \otimes \cdots \otimes e_n^{\otimes f_n}$$

be the tensor product of these vectors (recall that if $r < j \le n$, then $f_j = 0$ and the corresponding terms are simply absent from $e^{\otimes \vec{f}}$). Then

$$y_{\vec{f}} e^{\otimes \vec{f}}$$

is a positive integral multiple of

$$\sum_{q \in C_{T_{\vec{f}}}} (-1)^{\text{sgn}(q)} q e^{\otimes \vec{f}}.$$

Let θ be the permutation that rearranges the entries in the tableau such that as one reads the new tableau book-style (row 1 left to right, then row 2 left to right, and so on) the numbers are as in $T_{\vec{f}}$ read down column 1 first, then down column 2, and so on:

$$\theta : a_{ij} \mapsto a_{ji}.$$

Then $y_{\vec{f}} e^{\otimes \vec{f}}$ is a nonzero multiple of θ applied to

$$\otimes_{j \geq 1} \wedge_{i \geq 1} e_{a_{ij}}.$$

In particular,

$$y_{\vec{f}}(\mathbb{C}^N)^{\otimes n} \neq 0$$

if the columns in the tableau $T_{\vec{f}}$ have at most N entries each.

Under the action of a diagonal matrix

$$D(\vec{\lambda}) \in U(N)$$

with diagonal entries given by

$$\vec{\lambda} = (\lambda_1, \ldots, \lambda_N),$$

on $(\mathbb{C}^N)^{\otimes n}$, the vector $y_{\vec{f}} e^{\otimes \vec{f}}$ is an eigenvector with eigenvalue

$$\lambda_1^{f_1} \ldots \lambda_N^{f_N}.$$

Clearly, the highest weight for the representation of $U(N)$ on $y_{\vec{f}}(\mathbb{C}^N)^{\otimes n}$ is $\vec{f}$.

Returning to the Schur–Weyl character duality formula and using in it the character formula for $U(N)$, we have

$$\text{Tr}\left(D(\vec{\lambda})^{\otimes n} \cdot \sigma\right) = \sum_{\vec{w}} \chi_{\vec{w}}(\sigma) \frac{a_{(w_1+N-1,\ldots,w_{N-1}+1,w_N)}(\vec{\lambda})}{a_{(N-1,\ldots,1,0)}(\vec{\lambda})}, \tag{11.35}$$

where the sum is over all $\vec{w} \in \mathbb{Z}_{\geq 0, \downarrow}^N$ satisfying $|\vec{w}| = n$.

Multiplying through in (11.35) by the Vandermonde determinant in the denominator on the right, we have

$$\text{Tr}\left(D(\vec{\lambda})^{\otimes n}\cdot\sigma\right)a_{(N-1,\ldots,1,0)}(\vec{\lambda}) = \sum_{\vec{w}\in\mathbb{Z}_{\geq 0,\downarrow}^N,|\vec{w}|=n}\chi_{\vec{w}}(\sigma)a_{(w_1+N-1,\ldots,w_{N-1}+1,w_N)}(\vec{\lambda}).$$

$$(11.36)$$

To obtain the character value $\chi_{\vec{w}}(\sigma)$, view

$$\text{Tr}\left(D(\vec{\lambda})^{\otimes n}\cdot\sigma\right)a_{(N-1,\ldots,1,0)}(\vec{\lambda}) \qquad\qquad (11.37)$$

as a polynomial in $\lambda_1,\ldots,\lambda_N$. Examining the right side of (11.36), we see that

$$w_1+N-1 > w_2+N-2 > \cdots > w_{N-1}+1 > w_N$$

and the coefficient of

$$\lambda_1^{w_1+N-1}\ldots\lambda_N^{w_N}$$

is precisely $\chi_{\vec{w}}(\sigma)$. This provides a way of reading off the character value $\chi_{\vec{w}}(\sigma)$ as a coefficient in $\text{Tr}\left(D(\vec{\lambda})^{\otimes n}\cdot\sigma\right)a_{(N-1,\ldots,1,0)}(\vec{\lambda})$, treated as a polynomial in $\lambda_1,\ldots,\lambda_N$.

We can work out the trace in (11.37) by using the identity (10.28) taking σ to be a product of cycles of lengths $l_1,\ldots,l_m$; this leads to

$$\text{Tr}\left(D(\vec{\lambda})^{\otimes n}\cdot\sigma\right) = \prod_{j=1}^m(\lambda_1^{l_j}+\cdots+\lambda_N^{l_j}). \qquad (11.38)$$

In (11.4) we saw that

$$a_{(N-1,\ldots,1,0)}(\vec{\lambda}) = \prod_{1\leq j<k\leq N}(\lambda_j-\lambda_k).$$

Thus, for the partition $\vec{w}=(w_1,\ldots,w_N)$ of n, the value of the character $\chi_{\vec{w}}$ on a permutation with cycle structure given by the partition $(l_1,\ldots,l_m)$ of n is the coefficient of $\lambda_1^{w_1+N-1}\ldots\lambda_N^{w_N}$ in

$$\prod_{j=1}^m(\lambda_1^{l_j}+\cdots+\lambda_N^{l_j})\prod_{1\leq j<k\leq N}(\lambda_j-\lambda_k). \qquad (11.39)$$

Even though it is not explicit, this formula, due to Frobenius, is a wonderful concrete specification of the irreducible characters of the symmetric group.

Exercises

11.1. Prove that any finite-dimensional representation of $U(N)$ is a direct sum of irreducible representations. Conclude that every character of $U(N)$ is a linear combination, with nonnegative integer coefficients, of irreducible characters. [Hint: If $\rho : U(N) \to \text{End}_{\mathbb{C}}(V)$ is a representation, consider $\rho(U(N))$ as a subset of the algebra $\text{End}_{\mathbb{C}}(V)$.]

11.2. Prove Schur's lemma for $U(N)$: if $\rho_j : U(N) \to \text{End}_{\mathbb{C}}(V_j)$, for $j \in \{1, 2\}$, are irreducible representations of $U(N)$, then the vector space $\text{Hom}_{U(N)}(V_1, V_2)$ of all linear maps $T : V_1 \to V_2$ that satisfy $T\rho_1(g) = \rho_2(g)T$ for all $g \in U(N)$ is $\{0\}$ if ρ_1 is not equivalent to ρ_2, and is one-dimensional if ρ_1 is equivalent to ρ_2. [Hint: As with the case of finite groups, see what irreducibility implies for the kernel and range of any $T \in \text{Hom}_{U(N)}(V_1, V_2)$.]

11.3. For continuous functions f_1 and f_2 on $U(N)$, the convolution $f_1 \times f_2$ is defined to be the function on $U(N)$ whose value at any $g \in U(N)$ is given by

$$(f_1 \times f_2)(g) = \int_{U(N)} f_2(gh)f_1(h^{-1}) \, dh. \qquad (11.40)$$

(More honestly, this is $f_2 \times f_1$ by standard convention.) Let $\rho_1 : U(N) \to \text{End}_{\mathbb{C}}(V_1)$ and $\rho_2 : U(N) \to \text{End}_{\mathbb{C}}(V_2)$ be irreducible representations of $U(N)$. Show first that

$$\chi_{\rho_1} \times \chi_{\rho_2} = \begin{cases} \frac{1}{\dim_{\mathbb{C}} V_1} \chi_{\rho_1} & \text{if } \rho_1 \text{ and } \rho_2 \text{ are equivalent,} \\ 0 & \text{if } \rho_1 \text{ and } \rho_2 \text{ are not equivalent.} \end{cases} \qquad (11.41)$$

Then deduce the character orthogonality relation

$$\int_{U(N)} \chi_{\rho_1}(g^{-1})\chi_{\rho_2}(g) \, dg = \dim_{\mathbb{C}} \text{Hom}_{U(N)}(V_1, V_2), \qquad (11.42)$$

holding for any finite-dimensional representations ρ_1 and ρ_2 on spaces V_1 and V_2, respectively. (Hint: Imitate the case of finite groups, replacing the average over the group with the Haar integral.)

11.4. Show that a representation ρ of $U(N)$ is irreducible if and only if

$$\int_{U(N)} |\chi_\rho(U)|^2 \, dU = 1.$$

(Hint: Use Exercise 11.3.)

11.5. Let V be a finite-dimensional vector space over a field $\mathbb{F}$, and for $A, B \in E = \mathrm{End}_{\mathbb{F}}(V)$ define

$$(A, B)_{\mathrm{Tr}} = \phi_A(B) = \mathrm{Tr}\,(AB).$$

(a) Show that the map $\phi_A : E \to E'$, where E' is the dual of E, is an isomorphism.

(b) For L any subspace of E, let $L^{\perp} = \cap_{A \in L}\phi_A$. Show that $(L^{\perp})^{\perp} = L$.

(c) For any $A, B, T \in E$, with T invertible, show that $(A, TBT^{-1})_{\mathrm{Tr}} = (T^{-1}AT, B)_{\mathrm{Tr}}$.

11.6. Let ρ_i, with i running over an indexing set $\mathcal{R}$, be all the inequivalent complex representations of $U(N)$, where N is a positive integer. Let E_i denote the representation space of ρ_i. For any positive integer n and any $\vec{i} = (i_1, \ldots, i_n) \in \mathcal{R}^n$, work out a formula, in terms of the characters χ_{ρ_i}, for the dimension of the space

$$E_{\vec{i}} = (E_{i_1} \otimes \cdots \otimes E_{i_n})^{U(N)}$$

consisting of all elements of $E_{i_1} \otimes \cdots \otimes E_{i_n}$ fixed under $\rho_{i_1}(g) \otimes \ldots \otimes \rho_{i_n}(g)$ for all $g \in U(N)$.

Chapter 12

Postscript: Algebra

This lengthy postscript summarizes definitions, results, and proofs from algebra, some of it used earlier in the book and some providing a broader cultural background. The self-contained account here is strongly steered towards uses we make in representation theory. This chapter may be read on its own, but the reader would have to be patient with the long lists of definitions. The choice of topics and results becomes most meaningful in the context of the rest of this book. We have left Galois theory as a field too vast, *ein zu weites Feld*, for us to explore [28].

12.1 Groups and Less

A *group* is a set G along with an operation

$$G \times G \to G : (a, b) \mapsto a \cdot b$$

satisfying the following conditions:

1. The operation is associative:

$$a \cdot (b \cdot c) = (a \cdot b) \cdot c \quad \text{for all } a, b, c \in G.$$

2. There is an element $e \in G$, called the *identity* element, for which

$$a \cdot e = e \cdot a = a \quad \text{for all } a \in G. \tag{12.1}$$

A.N. Sengupta, *Representing Finite Groups: A Semisimple Introduction*,
DOI 10.1007/978-1-4614-1231-1_12, © Springer Science+Business Media, LLC 2012

3. For each element $a \in G$ there is an element $a^{-1} \in G$, called the inverse of a, for which

$$a \cdot a^{-1} = a^{-1} \cdot a = e. \tag{12.2}$$

If $e' \in G$ is an element with the same property (12.1) as e, then

$$e' = e \cdot e' = e',$$

and so the identity element is unique. If $a, a_L \in G$ are such that $a_L \cdot a$ is e, then

$$a_L = a_L \cdot e = a_L \cdot (a \cdot a^{-1}) = (a_L \cdot a) \cdot a^{-1} = e \cdot a^{-1} = a^{-1},$$

and, similarly, if $a \cdot a_R$ is e, then a_R is equal to a^{-1}. Thus, the inverse of an element is unique.

Usually, we drop the $\cdot$ in the operation and simply write ab for $a \cdot b$:

$$ab = a \cdot b.$$

If $ab = ba$, we say that a and b *commute*. The number of elements in G is called the *order* of G and is denoted $|G|$. The *order* of an element $g \in G$ is $\min\{n \geq 1 : g^n = e\}$.

If G_1 and G_2 are groups and $f : G_1 \to G_2$ a mapping satisfying

$$f(ab) = f(a)f(b) \quad \text{for all } a, b \in G_1, \tag{12.3}$$

then f is a *homomorphism* of groups. Such a homomorphism carries the identity of G_1 to the identity of G_2, and $f(a^{-1}) = f(a)^{-1}$ for all $a \in G_1$. A homomorphism that is a bijection is an *isomorphism*. The identity map $G \to G$, for any group G, is clearly an isomorphism. The composite of homomorphisms is a homomorphism, and the inverse of an isomorphism is an isomorphism.

The *symmetric group* S_n is the set of all bijections $[n] \to [n]$, under the operation of composition. A *cycle* in S_n is a permutation c for which there exist distinct $i_1, \ldots, i_k \in [n]$, with $k \geq 2$ such that $c(i_j) = i_{j+1}$ for $j \in [k-1]$, $c(i_k) = i_1$, and $c(m) = m$ for all other $m \in [n]$; the set $\{i_1, \ldots, i_k\}$ is the *support* of the cycle, and we say that two cycles are *disjoint* if their supports are disjoint. For a cycle c with k elements in its support we define its *length* $l(c)$ to be $k - 1$:

$$l(c) = |\text{support of } c| - 1.$$

[Warning: a much more common practice is to define the length to be the order of c, which is 1 more than $l(c)$.] For example, the length of $(1\,2\,3)$ is 2, and the length of any *transposition* $t = (a\,b)$ is 1. Every permutation other than the identity ι can be decomposed into a product of a unique set of disjoint cycles. The sum of the lengths of the disjoint cycles whose product is a given permutation s is the *length* $l(s)$ of s. Multiplying a permutation s by a transposition $t = (a\,b)$ either splits a cycle into a product of two disjoint cycles or combines two disjoint cycles into one; in either case

$$l(st) = l(s) \pm 1. \tag{12.4}$$

The *signature* map

$$\epsilon : S_n \to \{+1, -1\} : s \mapsto \epsilon(s) \overset{\text{def}}{=} (-1)^{l(s)} \tag{12.5}$$

is then a homomorphism, viewing $\{+1, -1\}$ as a group under multiplication. If $\epsilon(s) = 1$, we say s is *even* and if $\epsilon(s) = -1$, we say s is odd.

A *subgroup* of a group G is a nonempty subset H for which $ab \in H$ and $a^{-1} \in H$ for all $a, b \in H$; this means that H is a group when the group operation of G is restricted to H. A *left coset* of H in G is a subset of the form $xH = \{xh : h \in H\}$ for some $x \in G$. The set of all left cosets forms the *quotient* G/H:

$$G/H = \{xH : x \in G\}. \tag{12.6}$$

The fact that H is a subgroup ensures that distinct cosets are disjoint, and this implies

$$|H| \text{ is a divisor of } |G|, \tag{12.7}$$

an observation Lagrange made (for the symmetric groups S_n). A subgroup H of G is *normal* if $gH = Hg$ for all $g \in G$; for a normal subgroup H, there is a natural operation on G/H given by

$$(aH)(bH) = (ab)H \quad \text{for all } a, b \in G, \tag{12.8}$$

which is well defined and makes G/H also a group. In this case the natural projection map $G \to G/H : g \mapsto gH$ is a homomorphism.

The subset of even permutations in S_n is a subgroup, called the *alternating group* and denoted A_n.

Elements a, b in a group are *conjugate* if $b = gag^{-1}$ for some $g \in G$. Conjugacy is an equivalence relation and partitions G into a union of disjoint *conjugacy classes*. The conjugacy class of a is the set $\{gag^{-1} : g \in G\}$.

The *center* Z_G of a group G is the set of all elements $c \in G$ that commute with all elements of G:

$$Z_G = \{c \in G : cg = gc \quad \text{for all } g \in G\}. \tag{12.9}$$

An *action* of a group G on a nonempty set S is a mapping

$$G \times S \to S : (g, s) \mapsto gs,$$

such that $es = s$ for all $s \in S$, where e is the identity element of G, and

$$(gh)s = g(hs) \quad \text{for all } g, h \in G \text{ and all } s \in S.$$

The set $Gs = \{gs : g \in G\}$ is called the *orbit* of $s \in S$, and

$$\text{Stab}(s) = \{g \in G : gs = s\} \tag{12.10}$$

is a subgroup of G called the *stabilizer* or *isotropy* subgroup for $s \in S$. The map

$$G \to Gs : g \mapsto gs$$

is surjective, and the preimage of any gs is the coset $g\text{Stab}(s)$ whose cardinality is

$$|G|/|\text{Stab}(s)|$$

if G is finite. Since S is the union of all the distinct (and disjoint) orbits, we have

$$|S| = \sum_{j=1}^{m} \frac{|G|}{|\text{Stab}(s_j)|}, \tag{12.11}$$

where $s_1, \ldots, s_m \in S$ are such that $Gs_1, \ldots, Gs_m$ are all the distinct orbits. As a typical application of this formula, suppose $|G| = p^n$, where p is prime and n is a positive integer, and $|S|$ is divisible by p; then (12.11) implies that the number of j for which $Gs_j = \{s_j\}$ is divisible by p and hence this number is greater than 1 if it is positive. The solution of Exercise 4.13 uses this.

If $f : G_1 \to G_2$ is a homomorphism, then the *kernel*

$$\ker f = \{g \in G_1 : f(g) = e_2\}, \tag{12.12}$$

where e_2 is the identity in G_2, is a subgroup of G_1; moreover, the *image* $\text{Im}(f) = f(G_1)$ is a subgroup of G_2. Writing K for $\ker f$, we have a well-defined induced mapping

$$\bar{f} : G_1/K \to G_2 : gK \mapsto f(g), \tag{12.13}$$

which is an injective homomorphism.

A group A is *Abelian* or *commutative* if

$$ab = ba \quad \text{for all } a, b \in A.$$

For many Abelian groups, the group operation is written additively

$$G \times G \to G : (a, b) \mapsto a + b,$$

the identity element being denoted 0, and the inverse of a then being denoted $-a$.

A group C is *cyclic* if there is an element $c \in C$ such that C consists precisely of all the powers c^n with n running over $\mathbb{Z}$. Such an element c is called a *generator* of C.

A *semigroup* is a nonempty set T with a binary operation $T \times T \to T$: $(a, b) \mapsto ab$ which is associative. A *monoid* is a semigroup with an identity element; as with groups, this element is necessarily unique.

Let I be a nonempty set and for each $i \in I$ suppose we have a set Y_i. Let $U = \cup_{i \in I} Y_i$; then there is the Cartesian product set

$$\prod_{i \in I} Y_i \overset{\text{def}}{=} \{y \in U^I : y(i) \in Y_i \quad \text{for every } i \in I\} \tag{12.14}$$

and a projection map

$$\pi_k : \prod_{i \in I} Y_i \to Y_k : y \mapsto m_k = y(k) \tag{12.15}$$

for each $k \in I$. For $y \in \prod_{i \in I} Y_i$, the element $\pi_k(y)$ is the kth *component* of y.

If S is a nonempty set, and $n \in \{0, 1, 2, \ldots\}$, we have the set $S^n = S^{\{1,\ldots,n\}}$ of all maps $\{1, \ldots, n\} \to S$, where S^0 is taken to be the one-element set $1 = \{\emptyset\}$. Display an element $x \in S^n$, for now, as a string $x_1 \ldots x_n$, where $x_j = x(j)$ for each j. Then let

$$\langle S \rangle = \cup_{n \geq 0} S^n,$$

and define the product of $x, y \in \langle S \rangle$ to be

$$xy = x_1 \ldots x_n y_1 \ldots y_m$$

if $x \in S^n$ and $y \in S^m$. This makes $\langle S \rangle$ a semigroup, with $1 \in S^0$ as an identity element. This is the *free monoid* over the set S. If $S = \emptyset$, we take $\langle S \rangle$ to be the one-element group $\{1\}$.

12.2 Rings and More

A *ring* A is a set with two operations,

$$A \times A \to A : (a, b) \mapsto a + b \quad \text{(addition)},$$
$$A \times A \to A : (a, b) \mapsto ab \quad \text{(multiplication)},$$

such that addition makes A an Abelian group, multiplication is associative, multiplication distributes over addition,

$$a(b + c) = ab + ac,$$
$$(b + c)a = ba + ca, \tag{12.16}$$

and A contains a multiplicative identity element 1_A (or, simply, 1). Since not everyone requires a ring to have 1, we will often restate the existence of 1 explicitly when discussing a ring.

If A is a ring, then on the set A we can define addition as for A but reverse multiplication to

$$a \circ_{\text{opp}} b = ba,$$

for all $a, b \in A$. These operations make the set A again a ring, called the *opposite ring* of A and denoted A^{opp}.

The set $\mathbb{Z}$ of all integers, with usual addition and multiplication, is a ring.

A *division ring* is a ring in which $1 \neq 0$ and every nonzero element has a multiplicative inverse. A *field* is a division ring in which multiplication is commutative.

A *left ideal* L in a ring A is a nonempty subset of A for which

$$al \in L \quad \text{for all } a \in A \text{ and } l \in L.$$

A *right ideal* J is a nonempty subset of A for which $ja \in J$ for all $j \in J$ and $a \in A$. A subset of A is a *two-sided ideal* if it is both a left ideal and a right ideal.

A left (or right) ideal in A is *principal* if it is of the form Ac (or cA) for some $c \in A$. Note that $Ax \subset Ay$ is equivalent to y being a right *divisor* of x in the sense that $x = ay$ for some $a \in A$.

In $\mathbb{Z}$ every ideal is principal and has a unique nonnegative generator.

Proof. If I is a nonzero ideal in $\mathbb{Z}$, choose $m \in I$ for which $|m|$ is least; then for any $a \in I$, dividing by m produces a quotient $q \in \mathbb{Z}$ and a remainder $r \in \{0, \ldots, |m| - 1\}$, and then $a - qm = r$ is a nonnegative element of I which is less than $|m|$ and is therefore 0, and so $a = qm \in m\mathbb{Z}$; thus, $I \subset m\mathbb{Z} \subset I$ and so $I = m\mathbb{Z}$. If m and m_1 both generate, I then each is a divisor of the other and so $m = \pm m_1$, and nonnegativity picks out a unique generator.

If A is a ring and I is a two-sided ideal in A, then the quotient

$$A/I \stackrel{\text{def}}{=} \{x + I : x \in A\} \tag{12.17}$$

is a ring under the operations

$$(x + I) + (y + I) = (x + y) + I, \quad (x + I)(y + I) = xy + I.$$

The multiplicative identity in A/I is $1 + A$ (which is 0 if and only if $I = A$).

If S is a subset of a ring A, then the set of all finite sums of elements of the form xsy, with x and y running over A, is a two-sided ideal; clearly, it is the smallest two-sided ideal of A containing S as a subset, and is called the two-sided ideal *generated* by S.

If $a \in A$ and $m \in \{1, 2, 3, \ldots\}$, the sum of m copies of a is denoted ma; more officially, define inductively

$$1a = a \quad \text{and} \quad (m+1)a = ma + a.$$

Further, setting

$$0a = 0,$$

wherein 0 on the left is the integer 0, and for $m \in \{1, 2, \ldots\}$, setting

$$(-m)a = m(-a),$$

gives a map

$$\mathbb{Z} \times A \to A : (n, a) \mapsto na$$

that is additive in n and in a, and also satisfies

$$m(na) = (mn)a \quad \text{for all } m, n \in \mathbb{Z} \text{ and } a \in A.$$

The nonnegative generator of the ideal $I_A = \{m \in \mathbb{Z} : mA = 0\}$ in $\mathbb{Z}$ is the *characteristic* of A. The term is generally used only when A is a field. Suppose $1 \neq 0$ in A and also that whenever $ab = 0$, with $a, b \in A$, a or b is 0; then the characteristic p of A is either 0 or prime.

Proof. If m and n are integers such that mn is divisible by p, then $mn \in I_A$, that is, $mn1_A = 0$, and so $m1_A n1_A = 0$, which then implies $m \in I_A$ or $n \in I_A$, so m or n is divisible by p.

Theorem 12.1 *Let A be a ring, p be any positive integer, and C be the two-sided ideal generated by the set of elements of the form $ab - ba$ with a and b running over A. Then the map $\phi_p : x \mapsto x^p$ maps C into itself. Assume now that p is prime and $pa = 0$ for all $a \in A$. Then*

$$\overline{\phi}_p : A/C \to A/C : x + C \mapsto \phi_p(x) + C \tag{12.18}$$

is a well-defined map and is a homomorphism of rings. Equivalently,

$$\begin{aligned} \phi_p(x + y) - \phi_p(x) - \phi_p(y) &\in C, \\ \phi_p(xy) - \phi_p(x)\phi_p(y) &\in C \end{aligned} \tag{12.19}$$

for all $x, y \in A$.

The map ϕ_p is called the *Frobenius map* [31].
Proof. Observe that for any $x_j, y_j, a_j, b_j \in A$, for $j \in [n]$ with n any positive integer,

$$\left(\sum_{j=1}^{n} x_j(a_j b_j - b_j a_j) y_j \right)^p$$

is a sum of n terms each of the form $x(ab - ba)y$ for some $x, a, b \in A$. This means ϕ_p maps C into itself. The definition of C implies that $abcd - acbd \in C$ for all $a, b, c, d \in A$. Then, by the binomial theorem, for any $x, y \in A$ and any positive integer q we have

$$(x + y)^q - \sum_{j=0}^{q} \binom{q}{j} x^j y^{q-j} \in C.$$

If p is prime, then $\binom{p}{j} = p!/[j!(p-j)!]$ is divisible by p when $j \in \{1, \dots, p-1\}$, because the denominator $j!(p-j)!$ contains no factor p, whereas the numerator $p!$ does. Thus, if $pA = 0$, then all terms in $\sum_{j=0}^{p} \binom{p}{j} x^j y^{p-j}$ are 0 except the terms for $j \in \{0, p\}$; so

$$(x + y)^p - x^p - y^p \in C. \tag{12.20}$$

In particular,

$$(x - y)^p - (x^p - y^p) \in C$$

for all $x, y \in A$, which is clear from (12.20) if p is odd, whereas if $p = 2$, then $-a = a$ for all $a \in A$ and so again we are back at (12.20). Thus, if $x + C = y + C$, which means $x - y \in C$, then

$$\phi_p(x) - \phi_p(y) \in \phi_p(x - y) + C \subset C.$$

Hence, the mapping $\overline{\phi}_p : A/C \to A/C$ in (12.18) is well defined. From (12.20) it follows that $\overline{\phi}_p$ preserves addition. Next, $(xy)^p - x^p y^p \in C$ because, as noted above, every time we commute two elements in A their difference is in C. Hence, $\overline{\phi}_p$ also preserves multiplication. Lastly, $\overline{\phi}_p$ maps 1 to 1, because so does ϕ_p. $\boxed{\text{QED}}$

Suppose A_1 and A_2 are rings and $f : A_1 \to A_2$ is a mapping for which

$$\begin{aligned} f(a + b) &= f(a) + f(b), \\ f(ab) &= f(a)f(b) \end{aligned} \tag{12.21}$$

for all $a, b \in A_1$, and f maps the multiplicative identity in A_1 to that in A_2. Then we say that f is a *homomorphism*, or simply *morphism*, of rings. A morphism that is a bijection is an *isomorphism*. The identity map $A \to A$, for any ring A, is clearly an isomorphism. The composite of morphisms is a morphism, and the inverse of an isomorphism is an isomorphism.

A *subring* of a ring A is a nonempty subset B for which $x + y \in B$ and $xy \in B$ for all $x, y \in B$, and B contains a multiplicative identity; this means that B is a ring when the ring operations of A are restricted to B. Note that 1_A might not be in B, in which case, of course, $1_B \neq 1_A$. The terminology here is a bit awkward.

If $f : A_1 \to A_2$ preserves addition and multiplication, then the *kernel*

$$\ker f = f^{-1}(0)$$

is a two-sided ideal in A_1. The *image* $\text{Im}(f) = f(A_1)$ is a subring of A_2. Writing J for $\ker f$, we have a well-defined induced mapping

$$\overline{f} : A_1/J \to A_2 : a + J \mapsto f(a) \tag{12.22}$$

that is injective, preserves addition and multiplication, and is a morphism if f is a morphism of rings.

Now let A_i be a ring for each i in a nonempty set $\mathcal{I}$. On the product set

$$P = \prod_{i \in \mathcal{I}} A_i$$

define addition and multiplication componentwise:

$$(x + y)_i = x_i + y_i,$$
$$(xy)_i = x_i y_i \tag{12.23}$$

for all $i \in \mathcal{I}$. This makes P a ring, called the *product* of the family of rings A_i. For each i, the projection map $P \to A_i : x \mapsto x_i$ is a morphism of rings, and this property completely specifies the product ring structure on P.

For each $i \in \mathcal{I}$ we have an injective mapping $j_i : A_i \to P$ where, for any $a \in A_i$, the element $j_i(a)$ has the ith component equal to a and all other components are 0. Note that j_i preserves addition and multiplication, but does not generally carry 1 to 1. Identifying A_i with $j_i(A_i)$, we can view A_i as a subring of R.

If A is a ring and m and n are positive integers, an $m \times n$ *matrix* M with entries in A is a mapping

$$M : [m] \times [n] \to A : (i, j) \mapsto M_{ij}.$$

This is usually denoted $[M_{ij}]$ or, more precisely, $[M_{ij}]_{i \in [m], j \in [n]}$, and is displayed as

$$M = [M_{ij}] = \begin{bmatrix} M_{11} & M_{12} & \dots & M_{1n} \\ \vdots & \vdots & \vdots & \vdots \\ M_{m1} & M_{m2} & \dots & M_{mn} \end{bmatrix}.$$

The value M_{ij} is the (i, j)th *entry* of M, and is a *diagonal* entry of $i = j$. The *transpose* M^{t}, or M^{tr}, is the $n \times m$ matrix with entries specified by

$$(M^{\mathrm{tr}})_{ij} = (M^{\mathrm{t}})_{ij} = M_{ji}$$

for all $i \in [n]$, $j \in [m]$. The sum of $m \times n$ matrices M and N is defined pointwise:

$$(M + N)_{ij} = M_{ij} + N_{ij} \quad \text{for all } i \in [m], j \in [n]. \tag{12.24}$$

If M is an $m \times n$ matrix and N is an $n \times r$ matrix, then MN is the $m \times r$ matrix with entries specified by

$$(MN)_{ij} = \sum_{k=1}^{n} M_{ik} N_{kj} \tag{12.25}$$

for all $i \in [m]$ and $j \in [r]$. The set of all $m \times m$ matrices is a ring, denoted $\mathrm{Matr}_{m \times m}(A)$ under this multiplication, with the multiplicative identity being the matrix I whose diagonal entries are all 1 and all other entries are 0.

More generally, we may have matrices indexed by more general sets. Let $\mathcal{R}$ and $\mathcal{C}$ be nonempty sets; a matrix with rows labeled by $\mathcal{R}$ and with columns labeled by $\mathcal{C}$, and entries in a set S is a map

$$M : \mathcal{R} \times \mathcal{C} \to S : (i, j) \mapsto M_{ij}.$$

In cases of algebraic interest, the entry M_{ij} lies in an additive Abelian group A_{ij}, and then addition of matrices is defined pointwise as in (12.24). If $\mathcal{R} = \mathcal{C}$, then multiplication, as defined by (12.25) modified with the sum being over $k \in \mathcal{R}$, is meaningful as long as the sums are meaningful (all terms 0 except for finitely many).

A *commutative ring* is a ring in which multiplication is commutative.

An element a in a commutative ring R is a *divisor* of $b \in R$ if $b = ac$, for some $c \in R$. A divisor of 1 is called a *unit*.

An ideal I in a commutative ring R is a *prime ideal* if it is not R and has the property that if $a, b \in R$ have their product ab in I, then a or b is in I. In the ring $\mathbb{Z}$ a nonzero ideal is prime if and only if it consists of all multiples of some prime number.

An ideal I in a commutative ring R is *maximal* if $I \neq R$, and if J is any ideal containing I, then either $J = R$ or $J = I$. Applying Zorn's lemma to increasing chains of ideals not containing 1 shows that every commutative ring with $1 \neq 0$ has a maximal ideal. (In the annoying distraction $R = \{0\}$ there is, of course, no maximal ideal.)

Every maximal ideal in a commutative ring with 1 is prime.

Proof. If $x, y \in R$ have product xy lying in a maximal ideal M, and $y \notin M$, then $M + Ry$, being an ideal properly containing M, is all of R and hence contains 1, which is then of the form $m + ry$; multiplying by x shows that $x = xm + rxy$, which is in the ideal M.

A commutative ring R with multiplicative identity $1 \neq 0$ is an *integral domain* if whenever $ab = 0$ for some $a, b \in R$ at least one of a and b is 0. Thus, an ideal I in a commutative ring R with 1 is prime if and only if $R \neq I$ and R/I is an integral domain. The most basic example of an integral domain is $\mathbb{Z}$.

A narrower generalization of $\mathbb{Z}$ is the notion of a *principal ideal domain*: this is an integral domain in which every ideal is principal.

In a principal ideal domain every nonzero prime ideal is maximal.

Proof. Suppose $pR \neq 0$ is prime and cR is an ideal properly containing pR; then $p = ac$ for some $a \in R$ and so $a \in pR$ or $c \in pR$; proper containment rules out $c \in pR$, and we have $a = pu$ for some $u \in R$. Then $p = pcu$ and then, since $p \neq 0$ and R is an integral domain, we conclude that $cu = 1$, which implies $1 \in cR$, and hence $cR = R$. Hence, pR is maximal.

The argument above also shows that a generating element p of a nonzero prime ideal in a principal ideal domain is *irreducible*: p is not a unit and its only divisors are units and multiples of itself by units. In a principal ideal domain R an element p is irreducible if and only if it is *prime*, in the sense that $p \neq 0$ and if p is a divisor of ab, for some $a, b \in R$, then p is a divisor of a or of b.

The essential idea of the following result on greatest common divisors goes back to Euclid's *Elements*:

Theorem 12.2 *If $a_1, \ldots, a_n \in R$, where R is a principal ideal domain, then there is a $c \in R$ of the form $c = a_1 b_1 + \cdots + a_n b_n$, with $b_1, \ldots, b_n \in R$, such that $d \in R$ is a common divisor of $a_1, \ldots, a_n$ if and only if it is a divisor of c. If $a_1, \ldots, a_n$ are coprime in the sense that their only common divisors are the units in R, then $a_1 d_1 + \cdots + a_n d_n = 1$ for some $d_1, \ldots, d_n \in R$.*

Proof. Let c be a generator of the ideal $\sum_{i=1}^{n} Ra_i$, hence of the form $\sum_{i=1}^{n} a_i b_i$ for some $b_i \in R$. Now $d \in R$ is a common divisor of the a_i if and only if $a_1, \ldots, a_n \in Rd$, and this holds if and only if $Rc \subset Rd$, which is equivalent to d being a divisor of c. If $a_1, \ldots, a_n$ are coprime, then c, being a common divisor, is a unit; multiplying $c = \sum_i a_i b_i$ by an inverse of c produces $1 = \sum_i a_i d_i$ for some $d_i \in R$. $\boxed{\text{QED}}$

Returning to general rings, here is a useful little stepping stone:

Proposition 12.1 *Let $A_1, \ldots, A_n$ be two-sided ideals in a ring A, with $n \geq 2$, such that $A_i + A_j = A$ for all pairs i, j with $i \neq j$. Let B_k be the intersection of the A_i's except for $i = k$:*

$$B_k = \underbrace{A_1 \cap \ldots \cap A_n}_{\text{drop } k\text{th term}} = \cap_{m \in [n] - \{k\}} A_m \quad \text{for all } i \in [n],$$

with $[n]$ being $\{1, \ldots, n\}$. Then

$$A_k + B_k = A \qquad \text{for all } k \in [n] \tag{12.26}$$

and

$$B_1 + \cdots + B_n = A. \tag{12.27}$$

Proof. Fix any $k \in [n]$, and for $j \neq k$ pick $a_j \in A_j$ and $a'_j \in A_k$ such that $1 = a_j + a'_j$. Then

$$1 = \underbrace{(a_1 + a'_1) \ldots (a_n + a'_n)}_{\text{drop } k\text{th term}} = \text{terms involving } a'_j + \underbrace{a_1 \ldots a_n}_{\text{drop } k\text{th term}} \in A_k + B_k,$$

because each A_j is a two-sided ideal. Hence , $A_k + B_k = A$. We prove (12.27) inductively. It is clearly true when n is 2. Assuming its validity for smaller values of $n > 2$, let B'_i be defined as B_i except for the collection $A_1, \ldots, A_{n-1}$. Then

$$B'_1 + \cdots + B'_{n-1} = A.$$

Picking $b'_i \in B'_i$, summing up to 1, and $a_n \in A_n$, $b_n \in B_n$ adding to 1, we have

$$\begin{aligned} 1 &= (b'_1 + \cdots + b'_{n-1})(a_n + b_n) \\ &= \underbrace{b'_1 a_n}_{\in B_1} + \cdots + \underbrace{b'_{n-1} a_n}_{\in B_n} + \underbrace{1 \cdot b_n}_{\in B_n}, \end{aligned} \tag{12.28}$$

which is just (12.27). $\boxed{\text{QED}}$

This brings us to the ever-useful Chinese remainder theorem:

Theorem 12.3 *Suppose $A_1, \ldots, A_n$ are two-sided ideals in a ring A such that $A_j + A_k = A$ for every $j, k \in [n] = \{1, \ldots, n\}$ with $j \neq k$, and let $C = A_1 \cap \ldots \cap A_n$. Then, for any $y_1, \ldots, y_n \in A$ there exists an element $y \in A$ such that $y \in y_j + A_j$ for all $j \in [n]$. More precisely, the mapping*

$$f : A/C \to \prod_{j=1}^n A/A_j : a + C \mapsto (a + A_j)_{j \in [m]} \tag{12.29}$$

is a well-defined isomorphism of rings.

For variations on this using only the lattice structure of sets of ideals in A, see Exercise 5.19.

Proof. The map f is well defined and injective since $a + C = b + C$ is equivalent to $a - b \in C \subset A_j$ for each j, and this is equivalent to $a + A_j = b + A_j$ for

all $j \in [n]$. Clearly f preserves addition and multiplication, and maps 1 to
1. Surjectivity will be proved by induction. To start off the induction, take
$n = 2$; since $y_1 - y_2 \in A = A_1 + A_2$, we have $y_1 - y_2 = b_1 - b_2$ for some
$b_1 \in A_1$ and $b_2 \in A_2$, and so $y = y_1 - b_1 = y_2 - b_2$ satisfies $y + A_1 = y_1 + A_1$
and $y + A_2 = y_2 + A_2$. Next, assuming $n > 2$, let $B = A_1 \cap \ldots \cap A_{n-1}$.
By Proposition 12.1, $A_n + B = A$. Let $y_1, \ldots, y_n \in A$; inductively we can
assume that there exists $x \in A$ such that

$$x + A_j = y_j + A_j \qquad (12.30)$$

for all $j \in [n - 1]$. Then by the case of two ideals, it follows that there exists
$y \in A$ such that $y + A_n = y_n + A_n$ and $y + B = x + B$, with the latter being
equivalent to $y + A_j = x + A_j$ for all $j \in [n - 1]$. Together with (12.30), this
shows that there exists $y \in A$ for which $f(y) = (y_1 + A_1, \ldots, y_n + A_n)$. $\boxed{\text{QED}}$

12.3   Fields

Recall that a field is a ring, with $1 \neq 0$, in which multiplication is commutative and every nonzero element has a multiplicative inverse. Thus, in a field,
the nonzero elements form a group under multiplication.

Suppose R is a commutative ring with a multiplicative identity element
$1 \neq 0$; then an ideal M in R is maximal if and only if the quotient ring R/M
is a field.

Proof. Suppose M is maximal; if $x \in R \setminus M$, then $M + Rx$, being an ideal
containing M, is all of R, which implies that $1 = m + yx$ for some $y \in R$,
and so $(y + M)(x + M) = 1 + M$, thus producing a multiplicative inverse
for $x + M$ in R/M. Conversely, if R/M is a field, then, first $M \neq R$, and
if $x \in J \setminus M$, where J is an ideal containing M, then there is $y \in R$ with
$xy \in 1 + M$, and so $1 = xy - m$ for some $m \in M$, which implies $1 \in J$ and
so $J = R$.

Applying the construction above to the ring $\mathbb{Z}$ and a prime number p
produces the finite field

$$\mathbb{Z}_p = \mathbb{Z}/p\mathbb{Z}. \qquad (12.31)$$

Let R be an integral domain and S be a nonempty subset of $R - \{0\}$. On
the set $S \times R$ define the relation $\simeq$ by

$$(s_1, r_1) \simeq (s_2, r_2) \text{ if and only if } s_2 r_1 = s_1 r_2.$$

You can easily check that this is an equivalence relation. The Set of equivalence classes is denoted $S^{-1}R$ and the image of (s, r) in $S^{-1}R$ denoted by r/s.

Assume now that (1) if $s_1, s_2 \in S$, then $s_1 s_2 \in S$; and (2) $1 \in S$. Then $S^{-1}R$ is a ring with operations

$$r_1/s_1 + r_2/s_2 = (r_1 s_2 + r_2 s_1)/(s_1 s_2), \quad (r_1/s_1)(r_2/s_2) = r_1 r_2/(s_1 s_2),$$

with $0/1$ as a zero element, and $1 = 1/1$ as a multiplicative identity, which is $\neq 0$. Inside $S^{-1}R$ we have a copy of R sitting in through the elements $a/1$. A crucial fact is that each element s of S is a unit element in $S^{-1}R$ because $s/1$ clearly has $1/s$ as a multiplicative inverse. Elements r/s are called *fractions* and $S^{-1}R$ is the *ring of fractions* of R by S. A special case of great use is when $S = R - \{0\}$; in this case, the relation $(r/s)(s/r) = 1/1$ for $r, s \in S$ implies that $S^{-1}R$ is in fact a field, the *field of fractions* of R.

Suppose $\mathbb{F}_1$ is a field and $\mathbb{F} \subset \mathbb{F}_1$ is a subset that is a field under the operations inherited from $\mathbb{F}_1$. Then $\mathbb{F}_1$ is called an *extension* of $\mathbb{F}$.

12.4 Modules over Rings

In this section A is a ring with a multiplicative identity element 1_A. A *left A-module* M is a set M that is an Abelian group under an addition operation $+$, and there is an operation of scalar multiplication

$$A \times M \to M : (a, v) \mapsto av$$

for which the following hold:

$$(a + b)v = av + bv,$$
$$a(v + w) = av + aw,$$
$$a(bv) = (ab)v,$$
$$1_A v = v$$

for all $v, w \in M$, and all $a, b \in A$. Note that $0 = 0 + 0$ in A implies, on multiplying by v,

$$0v = 0 \quad \text{for all } v \in M,$$

where 0 on the left is the 0 in A, and 0 on the right is the 0 in M.

A *right A-module* is defined analogously, except that the multiplication by scalars is on the right:

$$M \times A \to M : (v, a) \mapsto va,$$

and so the "associative law" reads

$$(va)b = v(ab).$$

By leftist bias, the party line rule is that *an A-module means a left A-module.*

A *vector space* over a division ring is a module over the division ring.

Any Abelian group A is automatically a $\mathbb{Z}$-module, using the multiplication

$$\mathbb{Z} \times A \to A : (n, a) \mapsto na.$$

If M and N are A-modules, a map $f : M \to N$ is *linear* if

$$\begin{aligned} f(v + w) &= f(v) + f(w), \\ f(av) &= af(v) \end{aligned} \qquad (12.32)$$

for all $v, w \in M$ and all $a \in A$. The set of all linear maps $M \to N$ is denoted

$$\mathrm{Hom}_A(M, N)$$

and is an Abelian group under addition. An invertible linear map is an *isomorphism* of modules. When $M = N$, we use the notation

$$\mathrm{End}_A(M)$$

for $\mathrm{Hom}_A(M, M)$, and the elements of $\mathrm{End}_A(M)$ are *endomorphisms* of M. If M and N are modules over a commutative ring R, then $\mathrm{Hom}_R(M, N)$ is an R-module, with multiplication of an element $f \in \mathrm{Hom}_R(M, N)$ by a scalar $r \in R$ defined to be the map

$$rf : M \to N : v \mapsto rf(v).$$

Note that rf is linear only on using the commutativity of R.

The set $\mathrm{Matr}_{m \times n}(A)$ of $m \times n$ matrices over the ring A is both a left A-module and a right A-module under the natural multiplications:

$$a[M_{ij}] = [aM_{ij}] \quad \text{and} \quad [M_{ij}]a = [M_{ij}a]. \qquad (12.33)$$

A subset $N \subset M$ of an A-module M is a *submodule* of M if it is a module under the restrictions of addition and scalar multiplication, or, equivalently, if $N + N \subset N$ and $AN \subset N$. In this case, the quotient

$$M/N = \{v + N : v \in M\}$$

is an A-module with the natural operations

$$(v + N) + (w + N) \overset{\text{def}}{=} (v + w) + N \quad \text{and} \quad a(v + N) \overset{\text{def}}{=} av + N$$

for all $v, w \in M$ and $a \in A$. Thus, it is the unique A-module structure on M/N that makes the quotient map

$$M \to M/N : v \mapsto v + N$$

linear.

Let I be a nonempty set and for each $i \in I$ suppose we have an A-module M_i. Let $U = \cup_{i \in I} M_i$; then the product set

$$\prod_{i \in I} M_i = \{m \in U^I : m(i) \in M_i \quad \text{for every } i \in I\} \tag{12.34}$$

has a unique A-module structure which makes the projection map

$$\pi_k : \prod_{i \in I} M_i \to M_k : m \mapsto m_k = m(k) \tag{12.35}$$

A-linear for each $k \in I$. This module, along with these *canonical projection* maps, is called the *product* of the family of modules $\{M_i\}_{i \in I}$. Inside it consider the subset $\oplus_{i \in I} M_i$ consisting of all m for which $\{i \in I : \pi_i(m) \neq 0\}$ is a finite set. For each $k \in I$ and any $x \in M_k$, there is a unique element $\iota_k(x) \in \oplus_{i \in I} M_i$ for which the kth component is x and all other components are 0. Then $\oplus_{i \in I} M_i$ is a submodule of $\prod_{i \in I} M_i$, and, along with the A-linear *canonical injections*

$$\iota_k : M_k \to \oplus_{i \in I} M_i, \tag{12.36}$$

is called the *direct sum* of the family of modules $\{M_i\}_{i \in I}$. For the moment let us write M for the direct sum $\sum_{i \in I} M_i$. The linear maps

$$p_k = \iota_k \circ \pi_k | \oplus_{i \in I} M_i : M \to M \tag{12.37}$$

are projections onto the subspaces $\iota_k(M_k)$ of M and are *orthogonal idempotents*:

$$p_i^2 = p_i \qquad p_i p_k = 0 \quad \text{if } i, k \in I \text{ and } i \neq k,$$

$$\sum_{i \in I} p_i(x) = x \quad \text{for all } x \in M, \tag{12.38}$$

on observing that in the sum above only finitely many $p_i(x)$ are nonzero. Conversely, if M is an A-module and $\{p_i\}_{i \in I}$ is any family of elements in $\text{End}_A(M)$ satisfying (12.38), then M is isomorphic to the direct sum of the subspaces $p_i(M)$ via the addition map

$$\bigoplus_{i \in I} p_i(M) \to M : x \mapsto \sum_{i \in I} p_i(x).$$

The following Chinese-remainder-flavored result will be useful later in establishing the uniqueness of the Jordan decomposition:

Proposition 12.2 *Let $A_1, \ldots, A_n$ be two-sided ideals in a ring A such that $A_j + A_k = A$ for all pairs $j \neq k$. Suppose E is an A-module such that $CE = 0$, where $C = A_1 \cap \ldots \cap A_n$. Then E is the direct sum of the submodules $E_j = \{v \in E : A_j v = 0\}$. Moreover, if $c_1, \ldots, c_n \in A$, then there exists $s \in A$ such that $sv = c_j v$ for all $v \in E_j$ and $j \in [n]$.*

Proof. Let B_i be the intersection of all A_j except for $j = i$. Then by Proposition 12.1 there exist $b_1 \in B_1, \ldots, b_n \in B_n$, for which $b_1 + \cdots + b_n = 1$. So then for any $v \in E$,
$$v = b_1 v + \cdots + b_n v$$
and $A_j b_j v \subset Cv = 0$, because $A_j b_j \subset A_j \cap B_j = C$, and so each $b_j v$ lies in E_j. Next, suppose

$$w_1 + \cdots + w_n = 0, \tag{12.39}$$

where $w_j \in E_j$ for each $j \in [n]$. By Proposition 12.1, there exist $a_j \in A_j$ and $b'_j \in B_j$ such that $a_j + b'_j = 1$ for each $j \in [n]$. Then, since $a_j w_j = 0$, we have

$$w_j = 1 w_j = a_j w_j + b'_j w_j = b'_j w_j,$$

and, for $i \neq j$ we have

$$b'_j w_i \in B_j w_i \subset A_i w_i = 0 \quad \text{if } i \neq j.$$

Thus, multiplying (12.39) by b'_j produces $w_j = 0$. Thus, E is the direct sum of the E_j. Note that E_j is indeed a submodule, because if $y \in E_j$ and $a \in A$, then $A_j a y \subset A_j y = \{0\}$ and so $ay \in E_j$. Finally, consider $c_1, \ldots, c_n \in A$. By the Chinese remainder theorem (Theorem 12.3) there exists $s \in A$ such that $s - c_j \in A_j$ for each $j \in [n]$, and so $sv = (c_j + s - c_j)v = c_j v$ for all $v \in E_j$. $\boxed{\text{QED}}$

An *algebra* A over a ring R is an R-module equipped with a binary operation of "multiplication"

$$A \times A \to A : (a, b) \mapsto ab,$$

which is bilinear:

$$(ra)b = r(ab) = a(rb)$$

for all $r \in R$ and all $a, b \in A$. Then

$$(rs - sr)(ab) = (ra)(sb) - (ra)(sb) = 0 \quad \text{for any } r, s \in R \text{ and } a, b \in A,$$

and we work only with algebras over commutative rings. If A_1 and A_2 are algebras, a mapping $f : A_1 \to A_2$ is a *morphism* of algebras if f preserves both addition and multiplication: $f(a+b) = f(a) + f(b)$ and $f(ab) = f(a)f(b)$ for all $a, b \in A_1$. In this book we use only algebras for which multiplication is associative. If we are working with algebras which have multiplicative identities, a morphism is required to take the identity for A_1 to that for A_2. A morphism of algebras that is a bijection is an *isomorphism* of algebras. The identity map $A_1 \to A_1$ is clearly an isomorphism. The composition of morphisms is a morphism and the inverse of an isomorphism is an isomorphism.

Subalgebras and products of algebras are defined exactly as for rings, except that we note that subalgebras and product algebras also have R-module structures.

12.5 Free Modules and Bases

For a module M over a ring A, the *span* of a subset T of an A-module is the set of all elements of M that are linear combinations of elements of T; this is, of course, a submodule of M. The module M is said to be *finitely generated* if it is the span of a finite subset. (Take the span of the empty set to be $\{0\}$.)

A set $I \subset M$ is *linearly independent* if for any $n \in \{1, 2, \ldots\}$, $v_1, \ldots, v_n \in I$ and $a_1, \ldots, a_n \in A$ with $a_1 v_1 + \cdots + a_n v_n = 0$ the elements $a_1, \ldots, a_n$ are all 0. A *basis* of M is a linearly subset of M whose span is M. If M has a basis it is said to be a *free* module. (The zero module is free if you accept the empty set as its basis.)

From the general results of Theorems 5.3 and 5.6 it follows that any vector space V over a division ring D has a basis whose cardinality is uniquely determined.   The cardinality of a basis of V is called the *dimension* of V and is denoted $\dim_D V$. Theorem 5.3 also shows that if I is a linearly independent subset of V and S is a subset of V that spans V, then there is a basis of V consisting of all the vectors in I and some of the vectors in S.

Theorem 12.4 *Let R be a principal ideal domain. Any submodule of a finitely generated R-module is finitely generated. Any submodule of a finitely generated free R-module is again a finitely generated free R-module. Any two bases of a free R-module have the same cardinality.*

Proof. Leaving aside the trivial case of zero modules, let M be an R-module which is the linear span of a set $S = \{a_1, \ldots, a_n\}$ of n elements, and let N be a submodule of M. To produce a spanning set for N, the only immediate idea is to somehow pick a "smallest" element among the linear combinations $r_1 a_1 + \cdots + r_n a_n$ that lie in N; a reasonable first step in making this precise is to pick the one for which the coefficient r_1 is the "least." To fill this out to something sensible, observe that the set I_1 consisting of all $r_1 \in R$ for which $r_1 a_1 + \cdots + r_n a_n \in N$ for some $r_2, \ldots, r_n \in R$ is an ideal in R and hence is of the form $r_1^* R$ for some $r_1^* \in R$; in particular, there is an element of N of the form $b_1 = r_1^* a_1 + \cdots + r_n^* a_n$ for some $r_2^*, \ldots, r_n^* \in R$. Then every element of N can be expressed as an R-multiple of b_1 plus an element of N that is a linear combination of $a_2, \ldots, a_n$. Working our way down the induction ladder with n being the rung-count, we touch the ground level $n = 0$, where the claimed result is obviously valid. Thus, N is the linear span of a subset containing at most n elements.

Next we turn to the case of free modules and assume that the spanning set S is a basis of M. Let b_1 be constructed as above. Inductively, we can assume that there exists a basis B' of the submodule N' of N spanned by $a_2, \ldots, a_n$:

$$N' = N \cap \sum_{j=2}^{n} R a_j.$$

If $b_1 \in N'$, then $N' = N$ and $B = B'$ is a basis of N. If $b_1 \notin N'$ and $t_1 b_1$, with $t_1 \in R$, and an element in the span of B' is 0, then, expressing everything in terms of the linearly independent a_i, it follows that $t_1 r_1^* = 0$ and so, since $r_1^* \neq 0$ as $b_1 \notin N$, we have $t_1 = 0$ and this, coupled with the linear independence of B', implies that $B = \{b_1\} \cup B'$ is linearly independent.

Finally, consider a free R-module $M \neq 0$, and let B be a basis of M and J be a maximal ideal in R. There is the quotient map $M \to M/JM : x \mapsto$

$\overline{x} = x + JM$, and M/JM is a vector space over the field R/J. If $b_1, \ldots, b_n$ are distinct elements in the basis B, then for any $r_1, \ldots, r_n \in R$ for which the linear combination $r_1 b_1 + \cdots + r_n b_n$ is in JM, the fact that B is a basis implies that $r_1, \ldots, r_n$ are in J. Thus $b \mapsto \overline{b}$ is an injection on B and the image $\overline{B}$ is a basis for the vector space M/JM. The uniqueness of dimension for vector spaces then implies that the cardinality of B is $\dim_{R/J} M/JM$, independent of the choice of B. $\boxed{\text{QED}}$

An element m in an R-module M is a *torsion* element if it is not 0 and if $rm = 0$ for some nonzero $r \in R$. The module M is said to be *torsion-free* if it contains no torsion elements. Thus, M is torsion-free if for each nonzero $r \in R$, the mapping $M \to M : m \mapsto rm$ is injective.

A set $B \subset M$ is a basis of M if and only if M is the direct sum of the submodules Rb, with b running over B, and the mapping $R \to Rb : r \mapsto rb$ is injective.

Theorem 12.5 *A finitely generated torsion-free module over a principal ideal domain is free.*

Notice that $\mathbb{Q}$, as a $\mathbb{Z}$-module, is torsion-free but is not free because no subset of $\mathbb{Q}$ containing at least two elements is linearly independent and nor is any one-element set a basis of $\mathbb{Q}$ over $\mathbb{Z}$.

Proof. Let M be a torsion-free module over a principal ideal domain R, and, focusing on $M \neq \{0\}$, let $b_1, \ldots, b_r$ span M. Assume, without loss of generality, that $b_1, \ldots, b_k$ are linearly independent for some $k \leq r$, and every b_i, with $k + 1 \leq i \leq r$, has a nonzero multiple, say, $t_i b_i$, in the span of $b_1, \ldots, b_k$. Hence, with t being the product of these nonzero t_i, we have $t b_i \in N \stackrel{\text{def}}{=} R b_1 + \cdots + R b_k$ for all $i \in \{k + 1, \ldots, r\}$ (it holds automatically for $i \in [k]$). Thus, the mapping $M \to M : x \mapsto tx$ has an image in N, and so, since M is torsion-free, $\lambda_t : M \to N : x \mapsto tx$ is an isomorphism. Being isomorphic to the free module N (which has $b_1, \ldots, b_k$ as a basis), M is also free. $\boxed{\text{QED}}$

If S is a nonempty set and R is a ring with identity 1_R, then the set $R[S]$ of all maps $f : S \to R$ for which $f^{-1}(R - \{0\})$ is finite is an R-module with the natural operations of addition and multiplication induced from R:

$$(f + g)(x) = f(x) + g(x), \quad (rf)(x) = rf(x)$$

for all $x \in S$, $r \in R$, and $f, g \in R[S]$. The R-module $R[S]$ is called the *free R-module over S*. It is convenient to write an element $f \in R[S]$ in the form

$$f = \sum_{x \in S} f(x)x.$$

For $x \in S$, let $j(x)$ be the element of $R[S]$ equal to 1_R on x and 0 elsewhere. Then $j : S \to R[S]$ is an injection that can be used to identify S with the subset $j(S)$ of $R[S]$. Note that $j(S)$ is a *basis* of $R[S]$; that is, every element of $R[S]$ can be expressed in a unique way as a linear combination of the elements of $j(S)$:

$$f = \sum_{x \in S} f(x)j(x),$$

wherein all but finitely many elements are 0. If M is an R-module and $\phi : S \to M$ is a map, then $\phi = \phi_1 \circ j$, where $\phi_1 : R[S] \to M$ is uniquely specified by requiring that it be linear and equal to $\phi(x)$ on $j(x)$. (For $S = \emptyset$ take $R[S] = \{0\}$.)

Let A be a ring and E and F be free A-modules with an n-element basis $b_1, \ldots, b_n$ of E and an m-element basis $c_1, \ldots, c_m$ of F. Then for any $f \in \mathrm{Hom}_A(E, F)$ we have

$$f\left(\sum_{j=1}^n a_j b_j\right) = \sum_{j=1}^n a_j f(b_j) = \sum_{i=1}^m \left(\sum_{j=1}^n a_j f_{ij}\right) c_i, \qquad (12.40)$$

with f_{ij} being the c_ith component of $f(b_j)$. This relation is best displayed in matrix form:

$$[a_1, \ldots, a_n] \mapsto [a_1, \ldots, a_n] \begin{bmatrix} f_{11} & f_{21} & \cdots & f_{m1} \\ \vdots & \vdots & \vdots\vdots\vdots & \vdots \\ f_{1n} & f_{2n} & \cdots & f_{mn} \end{bmatrix}. \qquad (12.41)$$

Note that in the absence of commutativity of A, the matrix operation appears more naturally on the right, and clearly the matrix on the right here is not $[f_{ij}]$ itself but is the transpose $[f_{ij}]^t$. A further significance of (12.41) is that, if we work with one fixed basis of E, for $f, g \in \mathrm{End}_A(E)$,

$$(gf)_{ik} = \sum_{j=1}^m f_{jk} g_{ij} = \sum_{j=1}^m g_{ij} \circ_{\mathrm{opp}} f_{jk},$$

so that the mapping

$$\mathrm{End}_A(E) \to \mathrm{Matr}_{m \times m}(A^{\mathrm{opp}}) : f \mapsto [f_{ij}]^{\mathrm{t}}, \tag{12.42}$$

is an isomorphism of rings, where A^{opp} is the opposite ring.

The method used above to associate a matrix with a linear mapping between free modules works even when the modules are not finitely generated. If E and F are free A-modules, $\mathcal{C}$ is a basis of E, and $\mathcal{R}$ is a basis of F, then with any A-linear map $f : E \to F$ we associate the matrix $[f_{rc}]_{r \in \mathcal{R}, c \in \mathcal{C}}$, where f_{rc} is the coefficient of the basis element $r \in \mathcal{R}$ in the expression of $f(c)$ as a linear combination of the elements of $\mathcal{R}$ for all $c \in \mathcal{C}$.

12.6 Power Series and Polynomials

In this section R is a commutative ring with multiplicative identity 1, and $\mathbb{F}$ is a field.

A power series in a variable X with coefficients in R is, formally, an expression of the form

$$a_0 + a_1 X + a_2 X^2 + \cdots,$$

where the coefficients a_j are all drawn from R.

For an official definition, consider an abstract element X, called a *variable* or *indeterminate*, and let, $\langle X \rangle$ be the free monoid over $\{X\}$. Then let $R[[X]]$ be the set of all maps

$$a : \langle X \rangle \to R.$$

Denote by a_j the image of X^j under a. Define addition in $R[[X]]$ pointwise:

$$(a + b)_j = a_j + b_j \quad \text{for all } j \in \{0, 1, 2, \dots\}.$$

Define multiplication by

$$(ab)_n = \sum_{j=0}^{n} a_j b_{n-j} \quad \text{for all } j \in \{0, 1, 2, \dots\}.$$

These operations make $R[[X]]$ a ring, called the *ring of power series in X with coefficients in R*. An element $a \in R[[X]]$ is best written in the form

$$a(X) = \sum_j a_j X^j,$$

with the understanding that j runs over $\{0, 1, 2, \ldots\}$. With this notation, both multiplication and addition make notational sense; for example, the product of the power series rX^j and the power series sX^k is indeed the power series rsX^{j+k}, and

$$\left(\sum_j a_j X^j\right) \left(\sum_j b_j X^j\right) = \sum_j c_j X^j,$$

where

$$c_j = \sum_{k=0}^{j} a_k b_{j-k} \quad \text{for all } j \in \{0, 1, 2, \ldots\}.$$

If $1 \neq 0$ in R, then $1 \neq 0$ in $R[[X]]$ as well.

More generally, if S is a nonempty set, then we have first the set $R[[S]]_{\mathrm{nc}}$ of power series in noncommuting indeterminates $X \in S$, defined to be the set of all maps

$$a : \langle S \rangle \to R,$$

where $\langle S \rangle$ is the free monoid over S. Such a map is more conveniently displayed as

$$a = \sum_{f \in \langle S \rangle} a_f f.$$

An element a for which $a_f = 0$ except for exactly one $f \in S^n$, for some $n \in \{1, 2, \ldots\}$, is a *monomial*. Addition is defined on $R[[S]]_{\mathrm{nc}}$ pointwise and multiplication is defined by

$$ab = \sum_{f \in \langle S \rangle} \left(\sum_{h, k \in \langle S \rangle, hk = f} a_h b_k \right) f, \tag{12.43}$$

where the inner sum on the right is necessarily a sum of a finite number of terms. This makes $R[[S]]_{\mathrm{nc}}$ a ring.

Quotienting by the two-sided ideal generated by all elements of the form $XY - YX$ with $X, Y \in S$ produces the ring $R[[S]]$ of *power series* in the set S of *variables* or *indeterminates*, with coefficients in R. If S consists of the distinct variables $X_1, \ldots, X_n$, then $R[[S]]$ is written as $R[[X_1, \ldots, X_n]]$.

Inside the ring $R[[X_1, \ldots, X_n]]$ is the *polynomial ring* $R[X_1, \ldots, X_n]$ consisting of all elements $\sum_j a_j X_1^{j_1} \ldots X_n^{j_n}$, with j running over $\{0, 1, \ldots\}^n$, for which the set $\{j : a_j \neq 0\}$ is finite. Thus, the *monomials* $X_1^{j_1} \ldots X_n^{j_n}$ form a basis of the free R-module $R[X_1, \ldots, X_n]$.

Quotienting $R[X_1, Y_1, \ldots, X_n, Y_n]$ by the ideal generated by the elements $X_1 Y_1 - 1, \ldots, X_n Y_n - 1$ produces a ring which we will denote

$$R[X_1, X_1^{-1}, \ldots, X_n, X_n^{-1}]. \tag{12.44}$$

This is a free R-module with basis $\{X_1^{j_1} \ldots X_n^{j_n} : j_1, \ldots, j_n \in \mathbb{Z}\}$, with X^0 being 1. An element of this ring is called a *Laurent polynomial*.

For a nonzero polynomial $p(X) \in R[X]$, the largest j for which the coefficient of X^j is not zero is called the *degree* of the polynomial. We take the degree of 0 to be 0 by convention.

A polynomial $p(X) \in R[X]$ is *monic* if it is of the form $\sum_{j=0}^{n} p_j X^j$ with $p_n = 1$ and $n \geq 1$.

If $a(X), b(X) \in \mathbb{F}[X]$, and the degree of $b(X)$ is ≥ 1 or greater, then there are polynomials $q(X), r(X) \in \mathbb{F}[X]$, with the degree of $r(X)$ being less than the degree of $b(X)$, such that

$$a(X) = q(X)b(X) + r(X).$$

This is the *division algorithm* in $\mathbb{F}[X]$.

Inductive proof: If $a(X)$ has degree less than the degree of $b(X)$. simply set $q(X) = 0$ and $r(X) = a(X)$. If $a(X)$ has degree $n \geq m$, the degree of $b(X)$, then $a(X) - (a_n b_m^{-1}) X^{n-m} b(X)$ has degree less than n and so by induction there exist $q_1(X), r_1(X) \in \mathbb{F}[X]$, with the degree of $r_1(X)$ being less than the degree $b(X)$, such that

$$a(X) - (a_n b_m^{-1}) X^{n-m} b(X) = q_1(X)b(X) + r_1(X),$$

and so we obtain the desired result with $q(X) = q_1(X) + (a_n b_m^{-1}) X^{n-m}$.

The polynomial ring $\mathbb{F}[X]$, for any field $\mathbb{F}$, is clearly an integral domain; it is, moreover, a principal ideal domain.

Proof. For an ideal I that is neither 0 nor $\mathbb{F}[X]$, let $b(X)$ be a nonzero element of lowest degree; then for any $p(X) \in I$, we have $p(X) = q(X)b(X) + r(X)$ with $r(X)$ of lower degree than $b(X)$, but, on the other hand $r(X) = p(X) - q(X)b(X) \in I$ and so $r(X)$ must be 0, and hence $I = b(X)\mathbb{F}[X]$.

If $q(X) \in \mathbb{F}[X]$ has no polynomial divisors other than constants (elements of $\mathbb{F}$) and constant multiples of $q(X)$, then $q(X)$ is said to be *irreducible*. The ideal $q(X)\mathbb{F}[X]$ is maximal if and only if $q(X)$ is irreducible. Thus, $q(X)$ is irreducible if and only if $\mathbb{F}[X]/q(X)\mathbb{F}[X]$ is a field.

For any commutative ring R, the *derivative map*

$$D : R[X] \to R[X] : \sum_{j=0}^{m} a_j X^j \mapsto \sum_{j=1}^{m} j a_j X^{j-1} \qquad (12.45)$$

is a *derivation* on the ring $R[X]$ in the sense that it satisfies the following two conditions:

$$\begin{aligned} D(p+q) &= Dp + Dq, \\ D(pq) &= (Dp)q + pDq \end{aligned} \qquad (12.46)$$

for all $p, q \in R[X]$. These conditions are readily verified.

If $p(X) = \sum_{j=1}^{d} a_j X^j \in R[X]$, where R is a commutative ring, and $\alpha \in R$, then the *evaluation* of $p(X)$ at (or on) α is

$$p(\alpha) = \sum_{j=1}^{d} a_j \alpha^j \in R.$$

The element α is called a *root* of $p(X)$ if $p(\alpha)$ is 0.

For a field $\mathbb{F}$ and polynomial $p(X) \in \mathbb{F}[X]$ of positive degree, let $p_1(X)$ be a divisor of $p(X)$ of positive degree and $\mathbb{F}_1$ be the field $\mathbb{F}[X]/p_1(X)\mathbb{F}[X]$. Since $p_1(X)$ is of positive degree, the map $c \mapsto c + p_1(X)\mathbb{F}[X]$ maps $\mathbb{F}$ injectively into $\mathbb{F}_1$, and so we can view $\mathbb{F}$ as being a subset of $\mathbb{F}_1$. Let

$$\alpha = X + p_1(X)\mathbb{F}[X] \in \mathbb{F}_1;$$

then $p_1(\alpha) = 0$, and so $p(\alpha)$ is also 0. Thus, in the field $\mathbb{F}_1$ the polynomial $p(X)$ has a root.

A field $\mathbb{F}$ is *algebraically closed* if each polynomial $p(X) \in \mathbb{F}$ of degree 1 or greater has a root in $\mathbb{F}$. In this case, a polynomial $p(X)$ of degree $d \geq 1$ splits into a product of d terms each of the form $X - \alpha$, for $\alpha \in \mathbb{F}$, and a constant.

Theorem 12.6 *Let $\mathbb{F}$ be a field and n be a positive integer. Then $\mathbb{F}$ has an extension that contains n distinct nth roots of unity if and only if $n 1_{\mathbb{F}} \neq 0$ in $\mathbb{F}$.*

Proof. Assume first that $n 1_{\mathbb{F}} \neq 0$. Let $\mathbb{F}_1$ be an extension of $\mathbb{F}$ in which $X^n - 1$ splits as a product of linear terms: $X^n - 1 = \prod_{j=1}^{n} (X - \alpha_j)$ (we write

1 for $1_\mathbb{F}$). Suppose that α_k and α_l are equal to some common value α for some distinct $k, l \in [n]$. Thus, $X^n - 1 = (X - \alpha)^2 q(X)$ for a polynomial $q(X) \in \mathbb{F}_1[X]$. Applying the derivative D to this factorization of $X^n - 1$ produces

$$nX^{n-1} = 2(X - \alpha)q(X) + (X - \alpha)^2 Dq(X) = (X - \alpha)h(X),$$

where $h(X) \in \mathbb{F}_1[X]$. But this contradicts the fact that $X^n - 1$ and nX^{n-1} are coprime,

$$\begin{aligned} n1_\mathbb{F} &= XnX^{n-1} - n(X^n - 1) \\ &= X(X - \alpha)h(X) - n(X - \alpha)^2 q(X), \end{aligned} \tag{12.47}$$

which is impossible since $X - \alpha$ is not a divisor of $n1_\mathbb{F} \neq 0$. Thus, the nth roots of 1 are distinct in $\mathbb{F}_1$.

For the converse, assume that $n1_\mathbb{F} = 0$, and let p be the characteristic of $\mathbb{F}$. Then

$$X^p - 1 = (X - 1)^p$$

because the intermediate binomial coefficients are all divisible by p (see Theorem 12.1). Since p divides n, we have $n = pk$, for a positive integer k, and $X^n - 1 = (X^p)^k - 1$, of which $X^p - 1 = (X - 1)^p$ is a factor, thus showing that not all nth roots of 1 are distinct in this case. $\boxed{\text{QED}}$

An *algebraic closure* of a field $\mathbb{F}$ is an algebraically closed field $\overline{\mathbb{F}}$ that contains a subfield isomorphic to $\mathbb{F}$. Every field has an algebraic closure (for a proof, see Lang [55]).

Let $\mathbb{Z}^n_\downarrow$ be the subset of $\mathbb{Z}^n$ consisting of all strings $(j_1, \ldots, j_n)$ with $j_1 \geq \ldots \geq j_n$. Inside $\mathbb{Z}^n_\downarrow$ is the subset $\mathbb{Z}^n_{\downarrow\downarrow}$ of all strictly decreasing sequences.

Let R be a commutative ring with $1 \neq 0$. Denote a typical element of $R[X_1, X_1^{-1}, \ldots, X_n, X_n^{-1}]$ as $f(X_1, \ldots, X_n)$, or simply f. It can be expressed uniquely as a linear combination of monomials $X^{\vec{j}} = X_1^{j_1} \ldots X_n^{j_n}$, where $\vec{j} = (j_1, \ldots, j_n) \in \mathbb{Z}^n$, with coefficients $f_{\vec{j}} \in R$, all but finitely many of which are 0. If R_1 is any commutative R-algebra and $a_1, \ldots, a_n \in R_1$, then denote by $f(a_1, \ldots, a_n)$ the *evaluation* of f at $X_1 = a_1, \ldots, X_n = a_n$:

$$f(a_1, \ldots, a_n) = \sum_{\vec{j} \in \mathbb{Z}^n} f_{\vec{j}} a_1^{j_1} \ldots a_n^{j_n}, \tag{12.48}$$

whenever meaningful (i.e., noninvertible a_j are not raised to negative powers). Note that, in particular, the a_i could be drawn from the algebra

$R[X_1, X_1^{-1}, \ldots, X_n, X_n^{-1}]$ itself. If $\sigma \in S_n$, denote by $f_\sigma(X_1, \ldots, X_n)$ the element $f(X_{\sigma(1)}, \ldots, X_{\sigma(n)})$.

For the following result we say that f is *symmetric* if $f_\sigma = f$ for all $\sigma \in S_n$. The set of all such symmetric f forms a subring $R_{\text{sym}}[X_1, \ldots, X_n]$ of $R[X_1, X_1^{-1}, \ldots, X_n, X_n^{-1}]$. We say that f is *alternating* if $f(Y_1, \ldots, Y_n) = 0$ whenever $\{Y_1, \ldots, Y_n\}$ is a strictly proper subset of $\{X_1, \ldots, X_n\}$.

Theorem 12.7 *Let $\mathbb{F}$ be a field that contains m distinct mth roots of 1 for every $m \in \{1, 2, \ldots\}$ and R be a subring of $\mathbb{F}$.*

1. *If $f \in R[X_1, X_1^{-1}, \ldots, X_n, X_n^{-1}]$ is such that $f(\lambda_1, \ldots, \lambda_n) = 0$ for all roots of unity $\lambda_1, \ldots, \lambda_n \subset \mathbb{F}$, then $f = 0$.*

2. *$R_{\text{sym}}[X_1, X_1^{-1}, \ldots, X_n, X_n^{-1}]$ is a free R-module with basis given by the symmetric sums*

$$s(\vec{w}) = \sum_{\sigma \in S_n} X_{\sigma(1)}^{w_1} \cdots X_{\sigma(n)}^{w_n} \tag{12.49}$$

with $\vec{w} = (w_1, \ldots, w_n)$ running over $\mathbb{Z}_\downarrow^n$ and $s_{\vec{0}}$ defined to be 1.

3. *$R_{\text{alt}}[X_1, X_1^{-1}, \ldots, X_n, X_n^{-1}]$ is a free R-module with basis given by the alternating sums*

$$a(\vec{w}) = \sum_{\sigma \in S_n} (-1)^\sigma X_{\sigma(1)}^{w_1} \cdots X_{\sigma(n)}^{w_n} \tag{12.50}$$

with $\vec{w} = (w_1, \ldots, w_n)$ running over $\mathbb{Z}_{\downarrow\downarrow}^n$.

Restricting $\vec{w}$ to the indexing set $\mathbb{Z}_\downarrow^n$ avoids repeating basis elements; restricting it further to $\mathbb{Z}_{\downarrow\downarrow}^n$ avoids including both an element and its negative in a basis.

Proof. 1. First suppose $n = 1$, and $\phi \in R[X, X^{-1}]$ is 0 when X is evaluated at any root of unity in $\mathbb{F}$. Suppose $\phi = \sum_{k \in \mathbb{Z}} \phi_k X^k$, with $\phi_k = 0$ for k not between integers l and u, with $l < u$, and let $a = \max\{0, -l\}$. Then $X^a \phi(X)$ is a polynomial that vanishes on infinitely many elements (all roots of unity) in the field $\mathbb{F}$ and so $X^a \phi(X) = 0$, whence $\phi = 0$. Next, consider $n \geq 2$, and suppose $f \in R[X_1, X_1^{-1}, \ldots, X_n, X_n^{-1}]$ satisfies the condition given. Write f as an element of $R[X_2, X_2^{-1}, \ldots, X_n, X_n^{-1}][X_1, X_1^{-1}]$, with X_1^j having coefficient $f_j \in R[X_2, X_2^{-1}, \ldots, X_n, X_n^{-1}]$. Then by the $n = 1$ case, each $f_j(\lambda_2, \ldots, \lambda_n) = 0$ for each j and all roots λ_k of unity. Then, inductively, each f_j is 0.

2. Consider a nonzero $f \in R[X_1, X_1^{-1}, \ldots, X_n, X_n^{-1}]$, let W_f be the finite set $\{\vec{w} \in \mathbb{Z}_\downarrow^n : f_{\vec{w}} \neq 0\}$, and let $\vec{W}_f = \max W_f$ in the lexicographic order. Then

$$g = f - f_{\vec{W}} s_{\vec{W}}$$

is symmetric and if it is not 0 then $\vec{W}_g < \vec{W}$; working down the induction ladder of the finite set W_f, we see that the symmetric sums span $R[X_1, X_1^{-1}, \ldots, X_n, X_n^{-1}]$. The linear independence follows from observing that if $\vec{w}, \vec{w}'$ are distinct elements of $\mathbb{Z}_\downarrow^n$, then $s_{\vec{w}}$ and $s_{\vec{w}'}$ are sums over disjoint sets of monomials.

3. The argument is virtually the same as in part 2 of the proof except one substitutes $a_{\vec{W}}$ for $s_{\vec{W}}$. $\boxed{\text{QED}}$

12.7 Algebraic Integers

If R is a subring of a commutative ring R_1 with multiplicative identity $1 \neq 0$ lying in R, then an element $a \in R_1$ is said to be *integral* over R if $p(a) = 0$ for some monic polynomial $p(X) \in R[X]$. All elements r of R are integral over R (think $X - r$).

With R and R_1 as above, if $b_1, \ldots, b_m \in R_1$, then by $R[b_1, \ldots, b_m]$ is meant the subring of R_1 consisting of all elements of the form $p(b_1, \ldots, b_m)$ with $p(X_1, \ldots, X_m)$ running over all elements of the polynomial ring $R[X_1, \ldots, X_m]$. Note that $R[b_1, \ldots, b_m]$ is a subalgebra of R_1 when both are also equipped with the obvious R-module structures.

Theorem 12.8 *Suppose R is a subring of a commutative ring R_1 with $1 \neq 0$ lying in R, and assume that R is a principal ideal domain. Then an element $a \in R_1$ is integral over R if and only if the R-module $R[a]$ is finitely generated. If $a, b \in R_1$ are integral over R, then so are $a + b$ and ab. Thus, the subset of R_1 consisting of all elements integral over R is a subring of R_1.*

Proof. Suppose a is an integral over R. Then $a^n + p_{n-1} a^{n-1} + \cdots + p_1 a + p_0 = 0$ for some positive integer n and $p_0, \ldots, p_{n-1} \in R$. Thus, a^n lies in the R-linear span of $1, a, \ldots, a^{n-1}$, and hence by an induction argument all powers of a lie in the R-linear span of $1, \ldots, a^{n-1}$. Consequently, the R-module $R[a]$ is finitely generated. Conversely, suppose $R[a]$ is finitely generated as an R-module. Then there exist polynomials $q_1(X), \ldots, q_m(X) \in R[X]$ such that the R-linear span of $q_1(a), \ldots, q_m(a)$ is all of $R[a]$. Let n be 1 more than the degree of $q_1(X) \ldots q_m(X)$; then a^n is an R-linear combination of $q_1(a), \ldots, q_m(a)$, and so this produces a monic polynomial, of degree n, which vanishes on a.

Suppose $a, b \in R_1$ are integral over R. Then, by the first part, the R-modules $R[a]$ and $R[b]$ are finitely generated, and then $R[a] + R[b]$ and $R[a]R[b]$ (consisting of all sums of products of elements from $R[a]$ and $R[b]$) are also finitely generated. Since $R[a+b] \subset R[a] + R[b]$ and $R[ab] \subset R[a]R[b]$, it follows from Theorem 12.4 that these are also finitely generated and so, by the first part, $a + b$ and ab are integral over R. $\boxed{\text{QED}}$

Elements of $\mathbb{C}$ (or, if you prefer, $\overline{\mathbb{Q}}$) that are integral over $\mathbb{Z}$ are called *algebraic integers*. Firmly setting aside the temptation to explore the vast and deep terrain of algebraic number theory, let us mention only one simple observation:

Proposition 12.3 *If $a, b \in \mathbb{Z}$ are such that a/b is an algebraic integer, then $a/b \in \mathbb{Z}$.*

Proof. Let $p(X) = \sum_{j=0}^{n} p_j X^j \in \mathbb{Z}[X]$ be a monic polynomial that vanishes on a/b. Assume, without loss of generality, that a and b are coprime. From $p(a/b) = 0$ and $p_n = 1$ we have $a^n = -\sum_{j=0}^{n-1} p_j b^{n-j} a^j$, but the latter is clearly divisible by b, which, since a and b are coprime, implies that $b = \pm 1$. $\boxed{\text{QED}}$

12.8   Linear Algebra

Let V be a vector space over a field $\mathbb{F}$. In this section we will prove some useful results in linear algebra on decompositions of elements of $\mathrm{End}_{\mathbb{F}}(V)$ into convenient standard forms.

An *eigenvalue* of an endomorphism $T \in \mathrm{End}_{\mathbb{F}}(V)$ is an element $\lambda \in \mathbb{F}$ for which there exists a nonzero $v \in V$ satisfying

$$Tv = \lambda v. \tag{12.51}$$

We will say that a linear map $S : V \to V$ is *semisimple* if there is a basis of V with respect to which the matrix of S is diagonal and there are only finitely many distinct diagonal entries. For such S there is then a nonzero polynomial $p(X)$ for which $p(S) = 0$. Compare this with the definition of a semisimple element in the algebra $\mathrm{End}_{\mathbb{F}}(V)$ given in Exercise 5.13.

An $n \times n$ matrix M is said to be *upper-triangular* if $M_{ij} = 0$ whenever $i > j$. It is *strictly* upper-triangular if $M_{ij} = 0$ whenever $i \geq j$.

An element $N \in \mathrm{End}_{\mathbb{F}}(V)$ is *nilpotent* if $N^k = 0$ for some positive integer k. Clearly, a nilpotent that is also semisimple is 0. Moreover, the sum of two commuting nilpotents is nilpotent.

Here is a concrete picture of nilpotent elements in terms of ordered bases:

Proposition 12.4 *Let $V \neq 0$ be a finite-dimensional vector space and $\mathcal{N}$ be a nonempty set of commuting nilpotent elements in $\mathrm{End}_{\mathbb{F}}(V)$. Then V has a basis relative to which all matrices in $\mathcal{N}$ are strictly upper-triangular.*

Proof. First we show that there is a nonzero vector on which all $N \in \mathcal{N}$ vanish. Choose $N_1, \ldots, N_r$ in $\mathcal{N}$, which span the linear span of $\mathcal{N}$. We show, by induction on r, that there is a nonzero $b \in \cap_{i=1}^{r} \ker N_i$. Observe that if ν is the smallest positive integer for which $N_1^{\nu_1} = 0$, then there is a vector b_1 for which

$$N_1^{\nu_1 - 1} b_1 \neq 0 \quad \text{and} \quad N_1^{\nu_1} b_1 = 0.$$

So $N_1^{\nu_1 - 1} b_1$ is a nonzero vector in $\ker N_1$. Since N_j commutes with N_1 for $j \in \{2, \ldots, r\}$, we have

$$N_j(\ker N_1) \subset \ker N_1.$$

Hence, inductively, focusing on the subspace $\ker N_1$ and the restrictions of N_2, $\ldots$, N_r to $\ker N_1$, we find there is a nonzero $v \in \ker N_1$ on which $N_2, \ldots, N_r$ vanish. Hence, $b_1 \in \cap_{j=1}^{r} \ker N_j$.

Now we use induction on $n = \dim_{\mathbb{F}} V > 1$. The result that there is a basis making all $N \in \mathcal{N}$ strictly upper-triangular is valid in a trivial way for one-dimensional spaces because in this case 0 is the only nilpotent endomorphism. Assume that $n > 1$ and that the result holds for dimension $< n$. Pick nonzero $b_1 \in \cap_{j=1}^{r} \ker N_j$. Let

$$\overline{V} = V/\mathbb{F}b_1$$

and

$$\overline{N}_j \in \mathrm{End}_{\mathbb{F}}(V_1),$$

the map given by

$$w + \mathbb{F}b_1 \mapsto N_j w + \mathbb{F}b_1.$$

Note that $\dim_{\mathbb{F}} \overline{V} = n - 1 < n$, and each $\overline{N}_i$ is nilpotent. So, by the induction hypothesis, $\overline{V}$ has a basis $\overline{b}_2, \ldots, \overline{b}_n$ such that

$$\overline{N}_j \overline{b}_k = \sum_{2 \leq l < k} (\overline{N}_j)_{lk} \overline{b}_l$$

for some $(\overline{N}_j)_{lk} \in \mathbb{F}$, and all $j \in [r]$ and $k \in \{2, \ldots, n\}$. Then the matrix for each N_j relative to the basis $b_1, \ldots, b_n$ is strictly upper-triangular. $\boxed{\text{QED}}$

The ladder of consequences of the Chinese remainder theorem we have built is tall enough to pluck a pleasant prize, the Chevalley–Jordan decomposition:

Theorem 12.9 *Let V be a vector space over a field $\mathbb{F}$ and $T \in \mathrm{End}_{\mathbb{F}}(V)$ satisfy $p(T) = 0$, where $p(X) \in \mathbb{F}[X]$ is of the form*

$$p(X) = \prod_{j=1}^{m} (X - c_j)^{\nu_j},$$

where $m, \nu_1, \ldots, \nu_m$ are positive integers and $c_1, \ldots, c_m$ are distinct elements of $\mathbb{F}$. Then there exist $S, N \in \mathrm{End}_{\mathbb{F}}(V)$ satisfying:

1. *S is semisimple and N is nilpotent.*

2. *$SN = NS$.*

3. *$T = S + N$.*

4. *S and N are polynomials in T.*

5. *If V is finite-dimensional, then there is a basis of V relative to which the matrix of S is diagonal and the matrix of N is strictly upper-triangular.*

If each $\nu_j = 1$, that is, the roots of $p(X)$ are all distinct, then there is a basis of V relative to which the matrix of T is diagonal; if, moreover, $p(X)$ is a polynomial of minimum positive degree that vanishes on T, then the set of diagonal entries is exactly $\{c_1, \ldots, c_m\}$.

We will prove in Proposition 12.6 that the decomposition of T as $S + N$ here is unique. The last statement in Theorem 12.9 was used in the proof of Proposition 1.5; however, you can check this special case more simply without having to establish the decomposition theorem in full.

Proof. Apply Proposition 12.2 with A_j being the ideal in $A = \mathbb{F}[X]$ generated by $(X - c_j)^{\nu_j}$. Then, viewing V as an A-module by $a(X)v = a(T)v$ for all $a(X) \in A$, we see that V is the direct sum of the subspaces $V_j = \ker(T - c_j)^{\nu_j}$, and, moreover, there is a polynomial $s(X) \in A$ such that $S = s(T)$ agrees with $c_j I$ on V_j for each $j \in [m]$. Then S is semisimple. Taking $N = T - S$, we have N^{ν_j} equal to 0 on V_j for all $j \in [m]$, and so N is nilpotent. Since both S and N are polynomials in T, they commute with each other (which is clear anyway on each V_j separately).

Choose, by Proposition 12.4 applied to just the one nilpotent $N|V_j$, an ordered basis in each V_j with respect to which the matrix for $N|V_j$ is strictly upper-triangular. Stringing together all theses bases, suitably ordered, produces a basis for V relative to which S is diagonal and N is strictly upper-triangular.

If each $\nu_j = 1$, then the construction of S shows that $T = S$ on each V_j and hence on all of V. If p is a polynomial of minimum positive degree for which $p(T)$ is 0, then each $V_j \neq \{0\}$ (for otherwise $T - c_j$ is injective and hence has a left inverse, which implies that $p(X)/(X - c_j)$ vanishes on T) and so every c_j appears among the diagonal matrix entries of S. $\boxed{\text{QED}}$

The definition of a semisimple element S is awkward in that it relies on a basis for the vector space. One simple consequence, easily seen by writing everything in terms of a basis of eigenvectors, is that $\ker(S-c)^{\nu} = \ker(S-c)$ for any $c \in \mathbb{F}$ and positive integer ν. If $p(S) = 0$ for some positive-degree polynomial $p(X) \in \mathbb{F}[X]$, then every eigenvalue of S is a zero of $p(X)$ and so there are only finitely many distinct eigenvalues of S. If W is a subspace of V that is mapped into itself by S, then $p(S|W) = p(S)|W = 0$. Suppose $p(X) = \prod_{j=1}^{n}(X - c_j)^{\nu_j}$, with $c_1, \ldots, c_m$ are distinct elements of $\mathbb{F}$ and ν_j are positive integers. Then W is the direct sum of the subspaces $\ker(S-c_j)^{\nu_j}|W = V_j \cap W$, where $V_j = \ker(S - c_j)^{\nu_j} = \ker(S - c_j)$. This means that W is the direct sum of the subspaces $W_j = \ker(S - c_j)|W$. Thus, $S|W$ is semisimple: if $S \in \mathrm{End}_{\mathbb{F}}(V)$ maps a subspace W into itself, then the restriction of S to W is also semisimple.

Proposition 12.5 *Let V be a vector space over a field $\mathbb{F}$ and C be a finite subset of $\mathrm{End}_{\mathbb{F}}(V)$ consisting of semisimple elements that commute with each other. Then there is a basis of V with respect to which the matrix of every $T \in C$ is diagonal. There exists a semisimple $S \in \mathrm{End}_{\mathbb{F}}(V)$ such that every element of C is a polynomial in S. In particular, the sum of finitely many commuting semisimple elements is semisimple.*

For another, more abstract, take on this result, see Exercises 5.12, 5.13, and 5.14.

Proof. We prove this by induction on $|C|$, the case where this is 1 being clearly valid. Let $n = |C| > 1$ and assume that the result is valid for lower values of $|C|$. Pick a nonzero $S_1 \in C$; V is the direct sum of the subspaces $V_c = \ker(S_1 - cI)$ with c running over $\mathbb{F}$. Let $S_2, \ldots, S_n$ be the other elements of C. Since each S_j commutes with S_1, it maps each V_c into itself and its restriction to V_c is, as observed before, also semisimple. But then by the induction hypothesis each nonzero V_c has a basis of simultaneous eigenvectors of $S_2, \ldots, S_n$. Putting these bases together yields a basis of V that consists of simultaneous eigenvectors of $S_1, \ldots, S_n$. Thus, $V = W_1 \oplus \cdots \oplus W_m$, where each S_i is constant on each W_j, say, $S_i|W_j = c_{ij}I_{W_j}$. Now choose, for each $i \in [n]$, a polynomial $p_i(X) \in \mathbb{F}[X]$ such that $p_i(j) = c_{ij}$ for $j \in [m]$. Then $p_i(J) = S_i$, where J is the linear map equal to the constant j on W_j. $\boxed{\text{QED}}$

Now we can prove the uniqueness of the Chevalley–Jordan decomposition:

Proposition 12.6 *Let V be a vector space over a field $\mathbb{F}$. If $T \in \mathrm{End}_{\mathbb{F}}(V)$ satisfies $p(T) = 0$ for a polynomial $p(X) \in \mathbb{F}[X]$ that splits as a product of linear terms $X - \alpha$, then in a decomposition of T as $S + N$, with S semisimple and N nilpotent, and $SN = NS$, the elements S and N are uniquely determined by T.*

Proof. Remarkably, this uniqueness follows from the existence of the decomposition constructed in Theorem 12.9. If $T = S_1 + N_1$ with S_1 semisimple, N_1 nilpotent, and $S_1N_1 = N_1S_1$, then S_1 and N_1 commute with T and hence with S and N because these are polynomials in T. Then $S - S_1 = N_1 - N$ with the left side semisimple and the right side nilpotent, and hence both are 0. Hence $S = S_1$ and $T = T_1$. $\boxed{\text{QED}}$

This leads to the following sharper form of Proposition 12.5:

Proposition 12.7 *Let $V \neq 0$ be a finite-dimensional vector space over a field $\mathbb{F}$ and C be a finite subset of $\mathrm{End}_{\mathbb{F}}(V)$ consisting of elements that commute with each other. Assume also that every $T \in C$ satisfies $p(T) = 0$ for some positive-degree polynomial $p(X) \in \mathbb{F}[X]$ that is a product of linear factors of the form $X - \alpha$ with α drawn from $\mathbb{F}$. Then there is an ordered basis $b_1, \ldots, b_n$ of V such that every $T \in C$ has an upper-triangular matrix.*

Proof. We prove this by induction on $|C|$, the case where this is 1 following from Theorem 12.9. Let $n = |C| > 1$ and assume that the result is valid for lower values of $|C|$. Then V is the direct sum of the subspaces $V_j = \ker(T_1 - c_jI)^{\nu_j}$, where $p_1(T_1) = 0$ for a polynomial $p_1(X) = \prod_{j=1}^{m}(X - c_j)^{\nu_j}$, with $c_j \in \mathbb{F}$ distinct and $\nu_j \in \{1, 2, \ldots\}$. Let $T_2, \ldots, T_n$ be the other elements of C.

Since each T_j commutes with T_1, all elements of C map each V_j into itself. But then by the induction hypothesis each nonzero V_j has an ordered basis relative to which the matrices of $T_2, \ldots, T_n$ are upper-triangular. Stringing these bases together (ordered, say, with basis elements of V_i appearing before the basis elements of V_j when $i < j$) yields an ordered basis of V relative to which all the matrices of C are upper-triangular. $\boxed{\text{QED}}$

12.9 Tensor Products

In this section R is a commutative ring with multiplicative identity element 1_R. We will also use, later in the section, a possibly noncommutative ring D.

Consider R-modules $M_1, \ldots, M_n$. If N is also an R-module, a map

$$f : M_1 \times \cdots \times M_n \to N : (v_1, \ldots, v_n) \mapsto f(v_1, \ldots, v_n)$$

is called *multilinear* if it is linear in each v_j, with the other v_i held fixed:

$$f(v_1, \ldots, av_k + bv_k', \ldots, v_n) = af(v_1, \ldots, v_n) + bf(v_1, \ldots, v_k', \ldots, v_n)$$

for all $k \in \{1, \ldots, n\}$, $v_1 \in M_1, \ldots, v_k, v_k' \in M_k, \ldots, v_n \in M_n$, and $a, b \in R$.

Consider the set $S = M_1 \times \cdots \times M_n$ and the free R-module $R[S]$, with the canonical injection $j : S \to R[S]$. Inside $R[S]$ consider the submodule J spanned by all elements of the form

$$j(v_1, \ldots, av_k + bv_k', \ldots, v_n) - aj(v_1, \ldots, v_n) - bj(v_1, \ldots, v_k', \ldots, v_n)$$

with $k \in \{1, \ldots, n\}$, $v_1 \in M_1, \ldots, v_k, v_k' \in M_k, \ldots, v_n \in M_n$, and $a, b \in R$. The quotient R-module

$$M_1 \otimes \cdots \otimes M_n = R[S]/J \tag{12.52}$$

is called the *tensor product* of the modules $M_1, \ldots, M_n$. Let τ be the composite map

$$M_1 \times \cdots \times M_n \to M_1 \otimes \ldots \otimes M_n,$$

obtained by composing j with the quotient map $R[S] \to R[S]/J$. The image of $(v_1, \ldots, v_n) \in M_1 \times \cdots \times M_n$ under τ is denoted $v_1 \otimes \cdots \otimes v_n$:

$$v_1 \otimes \cdots \otimes v_n = \tau(v_1, \ldots, v_n). \tag{12.53}$$

The tensor product construction has the following "universal property": if $f : M_1 \times \cdots \times M_n \to N$ is a multilinear map, then there is a unique linear map $f_1 : M_1 \otimes \cdots \otimes M_n \to N$ such that $f = f_1 \circ \tau$, specified simply by requiring that

$$f(v_1, \ldots, v_n) = f_1(v_1 \otimes \cdots \otimes v_n)$$

for all $v_1, \ldots, v_n \in M$. Occasionally, the ring R needs to be stressed, and we then write the tensor product as

$$M_1 \otimes_R \cdots \otimes_R M_n.$$

If all the modules M_i are the same module M, then the n-fold tensor product is denoted $M^{\otimes n}$:

$$M^{\otimes n} = \underbrace{M \otimes \cdots \otimes M}_{n \text{ times}}.$$

A note of caution: tensor products can be treacherous. An infamous simple example is the tensor product of the $\mathbb{Z}$-modules $\mathbb{Q}$ and $\mathbb{Z}_2 = \mathbb{Z}/2\mathbb{Z}$:

$$\mathbb{Q} \otimes \mathbb{Z}_2 = \{0\},$$

because $1 \otimes 1 = 1/2 \otimes 2 = 0$, but $\mathbb{Z} \otimes \mathbb{Z}_2 \simeq \mathbb{Z}_2$ (induced by $\mathbb{Z} \times \mathbb{Z}_2 \to \mathbb{Z} : (m, n) \mapsto mn$) even though $\mathbb{Z}$ is a submodule of $\mathbb{Q}$.

There is a tensor product construction for two modules over a possibly noncommutative ring. We use this in two cases: (1) tensor products over division rings that arise in commutant duality; and (2) the induced representation. Let D be a ring (not necessarily commutative) with multiplicative identity element 1_D, and suppose M is a right D-module and N is a left D-module. Let J be the submodule of the $\mathbb{Z}$-module $M \otimes_{\mathbb{Z}} N$ spanned by all elements of the form $(md) \otimes n - m \otimes (dn)$, with $m \in M$, $n \in N$, and $d \in D$. The quotient is the $\mathbb{Z}$-module

$$M \otimes_D N = \mathbb{Z}[M \times N]/J. \tag{12.54}$$

This is sometimes called the *balanced* tensor product. Denote the image of $(m, n) \in M \times N$ in $M \otimes_D N$ by $m \otimes n$. The key feature now is that

$$(md) \otimes n = m \otimes (dn) \tag{12.55}$$

for all $(m, n) \in M \times N$ and $d \in D$. The universal property for the balanced tensor product

$$t : M \times N \to M \otimes_D N : (m, n) \mapsto m \otimes n \tag{12.56}$$

is that if $f : M \times N \to L$ is a $\mathbb{Z}$-bilinear map to a $\mathbb{Z}$-module L that is *balanced*, in the sense that $f(md, n) = f(m, dn)$ for all $m \in M, d \in D, n \in N$, then there is a unique $\mathbb{Z}$-linear map $f_1 : M \otimes_D N \to L$ such that $f = f_1 \circ l$.

Now suppose M is also a left R-module, for some commutative ring R with 1, such that $(am)d = a(md)$ for all $(a, m, d) \in R \times M \times D$. Then, for any $a \in R$,

$$M \times N \to M \otimes_D N : (m, n) \mapsto (am) \otimes n \qquad (12.57)$$

is $\mathbb{Z}$-bilinear and *balanced*, and so induces a unique $\mathbb{Z}$-linear map specified by

$$l_a : M \otimes_D N \to M \otimes_D N : m \otimes n \mapsto a(m \otimes n) \overset{\text{def}}{=} (am) \otimes n. \qquad (12.58)$$

The uniqueness implies that $l_{a+b} = l_a + l_b$, $l_{ab} = l_a \circ l_b$, and, of course, l_1 is the identity map. Thus, $M \otimes_D N$ is a left R-module with multiplication given by $a(m \otimes v) = (am) \otimes v$ for all $a \in R$, $m \in M$, and $m \in N$.

Despite the cautionary note and "infamous example" described earlier, there is the following comforting and useful result:

Theorem 12.10 *Let D be a ring, $\{M_i\}_{i \in I}$ be a family of right D-modules with direct sum denoted M, and $\{N_j : j \in J\}$ be a family of left D-modules with direct sum denoted N. Then the tensor product maps $t_{ij} : M_i \times N_j \to M_i \otimes N_j : (m, n) \mapsto m \otimes n$ induce an isomorphism*

$$\Theta : \bigoplus_{(i,j) \in I \times J} M_i \otimes_D N_j \to M \otimes_D N : \oplus_{i,j} t_{ij}(m_i, n_j) \mapsto \sum_{i,j} \iota_i(m_i) \otimes \iota_j(n_j), \qquad (12.59)$$

where ι_k denotes the canonical injection of the kth component module in a direct sum.

If each M_i is also a left R-module, where R is a commutative ring, satisfying

$$(am)d = a(md) \qquad (12.60)$$

for all $a \in R, m \in M_i, d \in D$, and all the balanced tensor products are given the left R-module structures, then Θ is an isomorphism of left R-modules.

If the right D-module M is free with basis $\{v_i\}_{i \in I}$ and the left D-module N is free with basis $\{w_j\}_{j \in J}$, then $M \otimes N$ is a free $\mathbb{Z}$-module with basis $\{v_i \otimes w_j\}_{(i,j) \in I \times J}$.

Note that the statement about bases applies to the D-modules M and N, not to the R-module structures.

Proof. By universality, the bilinear balanced map $M_i \times N_j \to M \otimes_D N :$ $(m, n) \mapsto \iota_i(m) \otimes \iota_j(n)$ factors through a unique $\mathbb{Z}$-linear map

$$\iota_{ij} : M_i \otimes_D N_j \to M \otimes_D N : t_{ij}(m, n) \mapsto \iota_i(m) \otimes \iota_j(n). \qquad (12.61)$$

These maps then combine to induce the $\mathbb{Z}$-linear mapping Θ on the direct sum of the $M_i \otimes_D N_j$. Since every element of M is a sum of finitely many $\iota_i(m_i)$'s, and every element of N is a sum of finitely many $\iota_j(n_j)$'s , it follows that Θ is surjective. Let π_i denote the canonical projection on the i component in a direct sum. The map

$$M \times N \to M_i \otimes_D N_j : (m, n) \mapsto \pi_i(m) \otimes \pi_j(n)$$

is $\mathbb{Z}$-bilinear and balanced and induces a $\mathbb{Z}$-linear map $\pi_{ij} : M \otimes_D N \to M_i \otimes_D N_j$. There is also the $\mathbb{Z}$-linear map ι_{ij} in (12.61). Now the composite $\pi_k \circ \iota_l$ is 0 if $k \neq l$ and is the identity map if $k = l$. Hence,

$$\pi_{ij} \circ \iota_{i'j'} = \begin{cases} \mathrm{id}_{M_i \otimes_D N_j} & \text{if } (i,j) = (i',j'), \\ 0 & \text{if } (i,j) \neq (i',j'). \end{cases} \qquad (12.62)$$

If $x \in \bigoplus_{(i,j) \in I \times J} M_i \otimes_D N_j$, then, with x_{ij} being the $M_i \otimes_D N_j$ component of x, the relations (12.62) imply $x_{ij} = \pi_{ij}(\Theta(x))$. Hence, if $\Theta(x) = 0$, then $x = 0$.

If all the modules involved are left R-modules satisfying (12.60), then Θ is R-linear as well. $\boxed{\text{QED}}$

For more on balanced tensor products, see Chevalley [12].

12.10   Extension of Base Ring

Let R be a subring of a commutative ring R_1, with the multiplicative identity 1 of R_1 lying in R. Then R_1 is an R-module in the natural way. If M is an R-module, then we have the tensor product

$$R_1 \otimes_R M,$$

which is an R-module to start with. But then it also becomes an R_1-module by means of the multiplication-by-scalar map

$$R_1 \times (R_1 \otimes M) \to R_1 \otimes M : (a, b \otimes m) \mapsto (ab) \otimes m$$

that is induced, for each fixed $a \in R_1$, from the R-bilinear map $f_a : R_1 \times M \to R_1 \otimes M : (b,m) \mapsto (ab) \otimes m$. With this R_1-module structure, we denote $R_1 \otimes_R M$ by $R_1 M$. Dispensing with $\otimes$, the typical element of $R_1 M$ looks like

$$a_1 m_1 + \cdots + a_k m_k,$$

where $a_1, \ldots, a_k \in R_1$ and $m_1, \ldots, m_k \in M$. Pleasantly confirming intuition, we find $R_1 M$ is free with finite basis if M is free with finite basis:

Theorem 12.11 *Suppose R is a subring of a commutative ring R_1 whose multiplicative identity 1 lies in R. If M is a free R-module with basis $b_1, \ldots, b_n$, then $R_1 \otimes_R M$ is a free R_1-module with basis $1 \otimes b_1, \ldots, 1 \otimes b_n$.*

Proof. View R_1^n first as an R-module. The mapping

$$R_1 \times M \to R_1^n : (a, c_1 b_1 + \cdots + c_n b_n) \mapsto (ac_1, \ldots, ac_n),$$

with $c_1, \ldots, c_n \in R$, is R-bilinear, and hence induces an R-linear mapping

$$L : R_1 \otimes_R M \to R_1^n : a \otimes (c_1 b_1 + \cdots + c_n b_n) \mapsto (ac_1, \ldots, ac_n).$$

Viewing now both $R_1 \otimes_R M$ and R_1^n as R_1-modules, w see L is clearly R_1-linear. Next we observe that the map L is invertible, with inverse given by

$$R_1^n \to R_1 \otimes_R M : (x_1, \ldots, x_n) \mapsto x_1 \otimes b_1 + \cdots + x_n \otimes b_n.$$

Thus, L is an isomorphism of $R_1 \otimes_R M$ with the free R_1-module R_1^n. The elements $(1, 0, \ldots, 0), \ldots, (0, \ldots, 1)$, forming a basis of R_1^n, are carried by L^{-1} to $1 \otimes b_1, \ldots, 1 \otimes b_n$ in $R_1 M$. This proves that $R_1 M$ is a free R_1-module and $1 \otimes b_1, \ldots, 1 \otimes b_n$ form a basis of $R_1 M$. $\boxed{\text{QED}}$

12.11   Determinants and Traces of Matrices

The *determinant* of a matrix $M = [M_{ij}]_{i,j \in [n]}$, with entries M_{ij} in a commutative ring R, is defined to be

$$\det M = \sum_{\sigma \in S_n} \text{sgn}(\sigma) M_{1\sigma(1)} \ldots M_{n\sigma(n)}. \tag{12.63}$$

As a special case, the determinant of the identity matrix I is 1. Replacing σ by σ^{-1} in (12.63) shows that the determinant of M remains unchanged if rows and columns are interchanged:

$$\det M = \det M^{\mathrm{t}}. \tag{12.64}$$

If the jth row and kth rows of M are identical, for some distinct $j, k \in [n]$, then in the sum (12.63) the term for $\sigma \in S_n$ cancels the one for $\sigma \circ (j\,k)$, and so $\det M$ is 0 in this case. Thus, a matrix with two rows or two columns has determinant 0.

Continuing with (12.63), for any $r \in [n]$, we have

$$\det M = \sum_{j=1}^{n} M_{rj} \tilde{M}_{jr}, \tag{12.65}$$

where $\tilde{M}_{jr}$ is a polynomial in the entries M_{kl}, with $k \in [n] - \{r\}$ and $l \in [n] - \{j\}$ with coefficients being ± 1; more precisely, $\tilde{M}_{rj}$ is $(-1)^{r+j}$ times the determinant of a matrix constructed by removing the rth row and the jth column from M. In fact, a little checking shows that

$$\sum_{j=1}^{n} M_{rj} \tilde{M}_{js} = (\det M)\delta_{rs} \tag{12.66}$$

for all $r, s \in [n]$. Thus, *if $\det M$ is invertible in R, then the matrix M is invertible*, with the inverse being the matrix whose (r, s) entry is $(\det M)^{-1}\tilde{M}_{rs}$.

The *trace* of a matrix $M = [M_{ij}]_{i,j\in[n]}$, with entries M_{ij} in a commutative ring R, is the sum of the diagonal entries:

$$\mathrm{Tr}\,(M) = \sum_{j=1}^{m} M_{jj}. \tag{12.67}$$

It is clear that the map Tr from the ring of $n \times n$ matrices to R is R-linear.

In the next section we will explore additional perspectives and properties of the determinant and trace.

12.12   Exterior Powers

Let E be an R-module, where R is a commutative ring. For any positive integer m and any R-module L, a map $f : E^m \to L$ is said to be *alternating* if it is multilinear and $f(v_1, \ldots, v_m)$ is 0 whenever $(v_1, \ldots, v_m) \in E^m$ has

$v_i = v_j$ for some distinct $i, j \in [m]$. We will construct an R-module $\Lambda^m E$ and an alternating map

$$w : E^m \to \Lambda^m E : (v_1, \ldots, v_m) \mapsto v_1 \wedge \ldots \wedge v_m,$$

which is *universal*, in the sense that if L is any R-module and $f : E^m \to L$ is alternating, then there is a unique R-linear map $f_* : \Lambda^m E \to L$ satisfying $f_* \circ w = f$. The construction is very similar to the construction of $E^{\otimes m}$ in Sect. 12.9. Let E_m be the free R-module on the set E^m and A_m be the subspace spanned by elements of the following forms:

$$(v_1, \ldots, v_j + v_j', \ldots, v_m) - (v_1, \ldots, v_j, \ldots, v_m) - (v_1, \ldots, v_j', \ldots, v_m),$$
$$(v_1, \ldots, av_j, \ldots, v_m) - a(v_1, \ldots, v_j, \ldots, v_m),$$
$$(v_1, \ldots, v_m), \quad \text{with } v_i = v_k \text{ for some distinct } i, k \in [m], \tag{12.68}$$

where the elements $v_1, \ldots, v_m, v_j'$ run over E and a runs over R. We define the *exterior power* $\Lambda^m E$ to be the quotient R-module

$$\Lambda^m E = E_m / A_m, \tag{12.69}$$

taken together with the map

$$w : E^m \to \Lambda^m E : (v_1, \ldots, v_m) \mapsto qj(v_1, \ldots, v_m), \tag{12.70}$$

where $j : E^m \to E_m$ is the canonical injection of E^m into the free module E_m, and $q : E_m \to E_m / A_m$ is the quotient map. The definition of A_m is designed to ensure that w is indeed an alternating map and satisfies the universal property mentioned above. If $E^{\wedge m}$ is an R-module and $w_* : E^m \to E^{\wedge m}$ is alternating and also satisfies the universal property mentioned above, then there are unique R-linear maps $i : \Lambda^m E \to E^{\wedge m}$ and $i_0 : E^{\wedge m} \to \Lambda^m E$ such that $w_* = i \circ w$ and $w = i_0 \circ w_*$, and then, examining the composites $i \circ i_0$ and $i_0 \circ i$ in light of, again, the universal property, we see that both of these are the identities on their respective domains. Thus, the universal property pins down the exterior power uniquely in this sense.

Now assume that E is a free R-module and suppose it has a finite basis consisting of distinct elements $y_1, \ldots, y_n$. Fix $m \in [n]$. Let $E^{\wedge m}$ be the free R-module spanned by the $\binom{n}{m}$ indeterminates y_I, one for each m-element subset $I \subset [n]$. Define an alternating map $w_* : E^m \to E^{\wedge m}$ by requiring that

$$w_*(y_{i_1}, \ldots, y_{i_m}) = y_I,$$

if $i_1 < \cdots < i_m$ are the elements of I in increasing order. If L is an R-module and $f : E^m \to L$ is alternating, then f is completely specified by its values on $(y_{i_1}, \ldots, y_{i_m})$ for all $i_1 < \cdots < i_m$ in $[n]$, and so $f = f_* \circ w_*$, where f_* is the linear map $E^{\wedge m} \to L$ specified by requiring that $f_*(y_I) = f(y_{i_1}, \ldots, y_{i_m})$ for all $I = \{i_1 < \cdots < i_m\} \subset [n]$. Thus, f_* is *uniquely* specified by requiring that $f = f_* \circ w_*$. Thus, $E^{\wedge m}$ is naturally isomorphic to $\Lambda^m E$, as noted in the preceding paragraph. Thus, $\Lambda^m E$ is free with a basis consisting of $\binom{n}{m}$ elements. In particular, $\Lambda^n E$ is free with a basis containing just one element.

For any endomorphism $A \in \text{End}_R(E)$, the map

$$E^m \to \Lambda^m E : (v_1, \ldots, v_m) \mapsto Av_1 \wedge \ldots \wedge Av_m$$

is alternating, and, consequently, induces a unique endomorphism $\Lambda^m A \in \text{End}_R(\Lambda^m E)$. If E is free with a basis containing n elements, so that $\Lambda^n E$ is free with a basis consisting of one element, then $\Lambda^n A$ is multiplication by a unique element $\Delta(A)$ of R:

$$(\Lambda^n A)(u) = \Delta(A)u \quad \text{for all } u \in \Lambda^n E. \tag{12.71}$$

To determine the multiplier $\Delta(A)$ we can work out the effect of $\Lambda^n A$ on $y_1 \wedge \ldots \wedge y_n$, where $y_1, \ldots, y_n$ is a basis of E,

$$Ay_1 \wedge \ldots \wedge Ay_n = \det[A_{ij}]\, y_1 \wedge \ldots \wedge y_n, \tag{12.72}$$

by a simple calculation. Hence, $\Delta(A)$ is called the determinant of the endomorphism A, and is equal to the determinant of the matrix of A with respect to any basis of E. It is denoted $\det A$:

$$\det A = \Delta(A) = \det[A_{ij}].$$

In particular, the determinant is independent of the choice of basis. Moreover, it is readily seen from (12.71) that

$$\det(AB) = \det(A)\det(B) \tag{12.73}$$

for all $A, B \in \text{End}_R(E)$, where, let us recall, E is a free R-module with finite basis. From (12.73) it follows on taking $B = A^{-1}$ that *if A is invertible, then its determinant is not* 0.

If M and N are $n \times n$ matrices with entries in a commutative ring R, then M and N naturally specify endomorphisms, also denoted by M and

N, of $E = R_1^n$, where $R_1 = R[M_{ij}, N_{kl}]_{i,j,k,l \in [n]}$, and so (12.73) implies the corresponding identity for determinants of matrices:

$$\det(MN) = \det(M)\det(N).$$

From this we see that the determinant of an invertible matrix is nonzero; earlier in the context of (12.66) we saw the converse. Thus ,*a matrix with entries in any commutative ring is invertible if and only if its determinant is invertible.*

If $A \in \text{End}_R(E)$, where E is a free R-module, where R is a commutative ring, having a basis with n elements, and t is an indeterminate, we have, for any $v_1, \ldots, v_n \in E$,

$$[\Lambda^m(tI + A)](v_1 \wedge \ldots \wedge v_n) = \sum_{k=0}^{n} t^k c_{n-k}(A)(v_1 \wedge \ldots \wedge v_n), \qquad (12.74)$$

for certain endomorphisms $c_0(A), \ldots, c_n(A) \in \text{End}_R(\Lambda^n E)$; we spare ourselves the notational change/precision needed in making (12.74) meaningful for an indeterminate t rather than for t in R. For example, $c_n(A) = \Lambda^n A$, and

$$c_1(A)(v_1 \wedge \ldots \wedge v_n) =$$
$$Av_1 \wedge v_2 \wedge \ldots \wedge v_n + v_1 \wedge Av_2 \wedge v_3 \wedge \ldots \wedge v_n + \cdots + v_1 \wedge v_2 \wedge \ldots \wedge Av_n.$$
$$(12.75)$$

Each $c_j(A)$ is multiplication by a scalar, which we also denoted by $c_j(A)$. If we take the v_i in (12.75) to form a basis of E, it follows readily from (12.75) that $c_1(A)$ is the trace of the matrix $[A_{ij}]$ of A with respect to the basis $\{v_i\}$

$$c_1(A) = \text{Tr}\,[A_{ij}]_{i,j \in [n]} = \sum_{i=1}^{n} A_{ii},$$

and so this may be called the *trace of the endomorphism* A, and is denoted $\text{Tr}\,A$:

$$\text{Tr}\,(A) \overset{\text{def}}{=} \text{Tr}\,[A_{ij}]_{i,j \in [n]} = \sum_{i=1}^{n} A_{ii}. \qquad (12.76)$$

Being equal to the multiplier, $c_1(A)$ is actually independent of the choice of basis of E. More generally, $c_j(A)$, for $j \in [n]$, is equal to $c_j([A_{rs}])$, where c_j

is defined for an $n \times n$ matrix $[M_{rs}]_{r,s \in [n]}$ with abstract indeterminate entries M_{rs} by means of the identity

$$\det(tI + M) = \sum_{k=0}^{n} t^k c_{n-k}(M),$$ (12.77)

with t being, again, an indeterminate. Note that $c_j(M)$ is a polynomial in the entries M_{rs} with integer coefficients; indeed, looking at (12.74) makes it easier to see that $c_j(M)$ is the sum of determinants of all the $j \times j$ *principal minors* (square matrices formed by removing $n - j$ columns and the corresponding rows from M).

Note that

$$c_0(M) = 1, \quad c_n(M) = \det M$$

for any $n \times n$ matrix M.

Now consider $n \times n$ matrices $[A_{ij}]$ and $[B_{ij}]$, whose entries are abstract symbols (indeterminates). Let t be another indeterminate. Then, working over the field $\mathbb{F}$ of fractions of the ring $\mathbb{Z}[A_{ij}, B_{kl}]_{i,j,k,l \in [n]}$, we have

$$\begin{aligned} \det(tI + AB) &= \det B^{-1} B(tI + AB) \\ &= \det B(tI + AB)B^{-1} \quad \text{(by (12.73))} \\ &= \det(tI + BA). \end{aligned}$$ (12.78)

This shows that

$$c_j(AB) = c_j(BA) \quad \text{for all } j \in \{0, 1, \dots, n\}.$$ (12.79)

Since this holds for matrices with entries that are indeterminates, it also holds for matrices with entries in any commutative ring (by realizing the indeterminates in this ring). Going further, since c_j of an endomorphism is equal to c_j of the matrix of the endomorphism relative to any basis, (12.79) also holds when A and B are endomorphisms of an R-module E that has a basis consisting of n elements, where n is any positive integer. Taking $j = 1$ produces the following fundamental property of the trace:

$$\mathrm{Tr}\,(AB) = \mathrm{Tr}\,(BA),$$ (12.80)

which can also be verified directly from the definition of the trace of a matrix.

If T is an upper- (or lower-) triangular matrix, then $\det A$ is the product of its diagonal entries. More generally,

$$c_j(T) = s_j(T_{11}, \ldots, T_{nn}) = \sum_{B \in \mathcal{P}_j} \prod_{i \in B} T_{ii}, \tag{12.81}$$

where $\mathcal{P}_j$ is the set of all j-element subsets of $[n]$. The polynomials s_j are called the *elementary symmetric polynomials*. They appear traditionally in studying roots of equations via the identity

$$\prod_{i=1}^{n}(X - \alpha_i) = \sum_{j=0}^{n}(-1)^{n-j}s_{n-j}(\alpha_1, \ldots, \alpha_n)X^j. \tag{12.82}$$

(Comparing with (12.77) shows the relationship with c_j for an upper-/lower-triangular matrix.)

12.13 Eigenvalues and Eigenvectors

In this section V is a finite-dimensional vector space over a field $\mathbb{F}$, with

$$n = \dim_{\mathbb{F}} V \geq 1.$$

Recall that an *eigenvalue* of an endomorphism $T \in \mathrm{End}_{\mathbb{F}}(V)$ is an element $\lambda \in \mathbb{F}$ for which there exists a nonzero $v \in V$ satisfying

$$Tv = \lambda v. \tag{12.83}$$

Thus, an element $\lambda \in \mathbb{F}$ is an eigenvalue of T if and only if $\ker(T - \lambda I) \neq 0$, which is equivalent to $T - \lambda I$ not being invertible. Hence, λ *is an eigenvalue of* T *if and only if* $\det(T - \lambda I) = 0$. Using (12.77), we obtain

$$\sum_{j=0}^{n}(-1)^j c_{n-j}(T)\lambda^j = 0. \tag{12.84}$$

If the field $\mathbb{F}$ is algebraically closed, this equation has n roots (possibly not all distinct), and so in this case every endomorphism of V has an eigenvalue. Looking at Theorem 12.9 shows that when $\mathbb{F}$ is algebraically closed there is a basis of V relative to which the matrix of T is upper-triangular.

12.14   Topology, Integration, and Hilbert Spaces

In this section we venture briefly and in a very condensed way beyond algebra, including some notions and results involving topology.

A *topology* τ on a set X is a set of subsets of X such that $\emptyset \in \tau$ and τ is closed under unions and finite intersections. The sets of τ are called the *open* sets of the topology, and their complements are called *closed* sets. The pair (X, τ), or more simply X, with τ understood, is called a *topological space*. A *neighborhood* of $p \in X$ is any open set containing p. The topology τ, or the space X, is a *Hausdorff* space if for any distinct $p, q \in X$ have disjoint neighborhoods.

In a topological space X, the intersection of all closed sets containing a given set $A \subset X$ is a closed set called the *closure* of A and denoted $\overline{A}$. A subset Y of X is *dense* if $\overline{Y} = X$.

An *open cover* of a topological space X is a set $\mathcal{U}$ of open subsets of X whose union is all of X; thus, every point of X lies in some set in $\mathcal{U}$. A topological space is *compact* if every open cover contains a finite subset that is also an open cover.

A sequence $(x_n)_{n \geq 1}$ of points in a topological space X *converges* to a *limit* $p \in X$ if for any open set U containing p there is a positive integer N such that $x_n \in U$ for all $n > N$; in this case if X is Hausdorff, then p is unique and is denoted $\lim_{n \to \infty} x_n$.

Suppose X and Y are topological spaces equipped with norms. A mapping $f : X \to Y$ is said to be *continuous* if $f^{-1}(O)$ is an open subset of X for every open set $O \subset Y$.

A *metric* d on a set X is a mapping

$$d : X \times X \to \mathbb{R} \tag{12.85}$$

such that d is nonnegative, symmetric,

$$d(x, y) = d(y, x) \quad \text{for all } x, y \in X,$$

satisfies the *triangle inequality*,

$$d(x, z) \leq d(x, y) + d(y, z) \quad \text{for all } x, y, z \in X,$$

and separates points is the sense that

$$d(x, y) = 0 \text{ if and only if } x = y.$$

The pair (X, d), or simply X, with d understood from the context, is a *metric space*. The *open ball* $B(a; r)$ in X, with *center* $a \in X$ and *radius* $r > 0$, is the set of all $x \in X$ at distance less than r from a:

$$B(a; r) = \{x \in X : \|x - a\| < r\}.$$

Let τ_d be the set of all sets that are unions of open balls; then τ_d is a topology on X. Every metric space is thus a Hausdorff topological space.

For a metric space (X, d), a sequence $(x_n)_{n \geq 1}$ of points in X is a *Cauchy sequence* if $d(x_n, x_m) \to 0$ as $n, m \to \infty$. It is easily checked that a sequence that converges is a Cauchy sequence. If, conversely, every Cauchy sequence converges to a limit, then d, or the metric space (X, d), is said to be *complete*.

A *norm* $\| \cdot \|$ on a real or complex vector space V is a function

$$V \to \mathbb{R} : v \mapsto \|v\|$$

that satisfies

$$\begin{aligned}
\|v + w\| &\leq \|v\| + \|w\|, \\
\|av\| &= |a| \, \|v\|, \\
\|v\| &\geq 0, \\
\|v\| &= 0 \quad \text{if and only if } v = 0
\end{aligned} \tag{12.86}$$

for all $v, w \in V$ and all scalars a. A vector space V equipped with a norm is called a *normed linear space*. The norm induces a metric given by

$$d(v, w) = \|v - w\|, \tag{12.87}$$

and hence a Hausorff topology, relative to which the maps

$$V \times V \to V : (v, w) \mapsto v + w$$

and

$$\mathbb{C} \times V \to V : (\alpha, v) \mapsto \alpha v$$

are continuous (with $\mathbb{C}$ replaced by $\mathbb{R}$ if V is a real vector space).

A *Hermitian inner product* on a complex vector space V is a mapping $V \times V \to \mathbb{C} : (x, y) \mapsto \langle x, y \rangle$ for which

$$\begin{aligned}
\langle av_1 + v_2, w \rangle &- \overline{a}\langle v_1, w \rangle + \langle v_2, w \rangle, \\
\langle v, aw_1 + w_2 \rangle &= a\langle v, w_1 \rangle + \langle v, w_2 \rangle, \\
\langle v, v \rangle &\geq 0 \quad \text{(in particular, } \langle v, v \rangle \text{ is real)}, \\
\langle v, v \rangle &= 0 \quad \text{if and only if } v = 0
\end{aligned} \tag{12.88}$$

for all $v, w, v_1, v_2, w_1, w_2 \in V$ and $a \in \mathbb{C}$. (Warning: a more common mathematics custom is to require that $\langle v, w \rangle$ be conjugate-linear in w.) Examining $\langle v + w, v + w \rangle$ then shows that

$$\langle v, w \rangle = \overline{\langle w, v \rangle} \quad \text{for all } v, w \in V. \tag{12.89}$$

For any positive integer n, the vector space $\mathbb{C}^n$ has the standard Hermitian inner product

$$\langle z, w \rangle = \sum_{j=1}^{n} \overline{z}_j w_j, \tag{12.90}$$

for all $z = (z_1, \ldots, z_n) \in \mathbb{C}^n$ and $w = (w_1, \ldots, w_n) \in \mathbb{C}^n$.

Vectors $v, w \in V$ are *orthogonal* if $\langle v, w \rangle = 0$.

Given a Hermitian inner product on V, let

$$\|v\| = \sqrt{\langle v, v \rangle} \quad \text{for all } v \in V. \tag{12.91}$$

Then $\| \cdot \|$ is a norm on V.

With d being as in (12.87) for a norm $\| \cdot \|$, we say that V, with the norm $\| \cdot \|$, is a *Banach space* if the metric d is complete.

A complex *Hilbert space* is a complex vector space equipped with a Hermitian inner product $\langle \cdot, \cdot \rangle$ such that the corresponding distance function d, given in (12.87) with $\| \cdot \|$ being the norm induced from $\langle \cdot, \cdot \rangle$, is complete.

An indexed family of elements $(x_\alpha)_{\alpha \in I}$ in a Hilbert space $\mathbb{H}$ is said to form an *orthonormal basis* of $\mathbb{H}$ if each x_α has norm 1 and $\langle x_\alpha, x_\beta \rangle = 0$ for all $\alpha \neq \beta$, and the closure of the linear span of $\{x_\alpha : \alpha \in I\}$ is $\mathbb{H}$. A Hilbert space $\mathbb{H}$ is *separable*, in the sense that it has a countable dense subset if and only if it has a countable orthonormal basis. If $(e_n)_{n \geq 1}$ is a countable orthonormal basis of $\mathbb{H}$, then

$$x = \sum_{n=1}^{\infty} \langle x, e_n \rangle e_n \overset{\text{def}}{=} \lim_{N \to \infty} \sum_{n=1}^{N} \langle x, e_n \rangle e_n$$

for all $x \in \mathbb{H}$.

Subspaces of a Hilbert space which are also closed sets are of great use. If L is a closed subspace of a Hilbert space $\mathbb{H}$ and x is any point in $\mathbb{H}$, then completeness implies that there is a unique point in L that is closest in distance to x; denote this point by $P_L(x)$. Then $P_L : \mathbb{H} \to \mathbb{H}$ can be shown to be a linear operator (linear mappings between Hilbert spaces are generally called linear operators). For this and all other results that we quote below, see Rudin [67] for proofs and further context.

The *norm* of a linear operator $A : \mathbb{H} \to \mathbb{H}$, where $\mathbb{H}$ is a Hilbert space, is

$$\|A\| \overset{\text{def}}{=} \sup_{v \in B(0;1)} \|Av\|. \tag{12.92}$$

A linear operator is *bounded* if its norm is finite. In this case

$$\|Av\| \leq \|A\| \|v\| \quad \text{for all } v \in \mathbb{H}.$$

From this, and with some additional work, it can be shown that a linear operator A is continuous if and only if it is bounded. The set $\mathcal{B}_{\mathbb{H}}$ of all bounded linear mappings $\mathbb{H} \to \mathbb{H}$ is an algebra, under natural addition and multiplication (composition), and is also a Banach space with respect to the operator norm $\| \cdot \|$ defined in (12.92). Moreover,

$$\|AB\| \leq \|A\| \|B\| \tag{12.93}$$

for all $A, B \in \mathcal{B}_{\mathbb{H}}$, and $\|I\| = 1$.

The *Riesz representation* theorem says that for any complex Hilbert space $\mathbb{H}$ a linear mapping $L : \mathbb{H} \to \mathbb{C}$ is continuous if and only if there is an element $z \in \mathbb{H}$ such that

$$Lv = \langle z, v \rangle \quad \text{for all } v \in \mathbb{H}. \tag{12.94}$$

A complex Banach space B that is also an algebra with multiplicative identity I, for which (12.93) holds and $\|I\| = 1$ is called a complex *Banach algebra*.

Let $C(\Delta)$ be the set of all complex-valued continuous functions on a compact Hausdorff space Δ. Then $C(\Delta)$ is clearly a complex algebra under pointwise addition and multiplication. Moreover,

$$\|f\|_{\text{sup}} = \sup_{x \in \delta} |f(x)| \tag{12.95}$$

for all $f \in C(\Delta)$ specifies a norm, called the *sup-norm*, on $C(\Delta)$, and makes $C(\Delta)$ a complex Banach algebra. Observe also that

$$\|\overline{f}\| = \|f\|$$

for all $f \in C(\Delta)$, where $\overline{f}$ is the complex conjugate of f.

The *adjoint* A^* of a bounded linear operator on a Hilbert space $\mathbb{H}$ is the bounded linear operator A^* on $\mathbb{H}$ uniquely specified by the requirement that

$$\langle A^*v, w \rangle = \langle v, Aw \rangle \quad \text{for all } v, w \in \mathbb{H}. \tag{12.96}$$

Then

$$\|A^*\| = \|A\| \tag{12.97}$$

for any bounded linear operator A.

The bounded linear operator A is said to be *self-adjoint* if $A^* = A$. A *unitary* operator on $\mathbb{H}$ is a bounded linear operator $U : \mathbb{H} \to \mathbb{H}$ for which $U^*U = UU^* = I$, the identity operator on $\mathbb{H}$.

When $\mathbb{H}$ is finite-dimensional, every linear operator on $\mathbb{H}$ is automatically continuous, and a self-adjoint operator is also called *Hermitian*. In the finite-dimensional case, the spectral theorem says that if A is a self-adjoint operator on a finite-dimensional Hilbert space $\mathbb{H}$, then there is an orthonormal basis of $\mathbb{H}$ consisting of eigenvectors of A. Equivalently, if $\lambda_1, \ldots, \lambda_m$ are the distinct eigenvalues of A, then

$$A = \sum_{j=1}^{m} \lambda_j P_j, \tag{12.98}$$

where P_j is the orthogonal projection onto the eigensubspace

$$\{v \in \mathbb{H} : Av = \lambda_j v\}.$$

Each projection P_j is a self-adjoint idempotent, satisfying

$$P_j^* = P_j \quad \text{and} \quad P_j^2 = P_j,$$

and satisfying the orthogonality condition

$$P_j P_k = 0 \quad \text{if } j \neq k,$$

and

$$\sum_{j=1}^{m} P_j = I,$$

the identity operator on $\mathbb{H}$. (Compare with the notion of a semisimple element as defined in Exercise 5.13.) The operators $P_1, \ldots, P_m$ are said to form a *resolution of the identity*, and the *spectral measure* P_A of A is given by

$$P_A(S) = \sum_{\lambda_j \in S} P_j \tag{12.99}$$

for all (Borel) subsets S of $\mathbb{R}$. Thus, the sum (12.98) can be displayed as an integral

$$A = \int_{\mathbb{R}} \lambda \, dP_A(\lambda). \tag{12.100}$$

A far-reaching generalization (see Rudin [67] for a full statement and proof) is the spectral theorem for unbounded normal operators on infinite-dimensional Hilbert spaces due to von Neumann.

Bibliography

[1] Alperin, J. L., and Bell, Rowen, B., *Groups and Representations*. Springer (1995).

[2] Artin, Emil, *Geometric Algebra*. Interscience Publishers (1957).

[3] Berkovic, Ya. G., and Zhmud', E. M., *Characters of Finite Groups*. Part I. Translated by P. Shumyatsky and V. Zobina. American Mathematical Society (1997).

[4] Birkhoff, Garrett, and Neumann, John von, *The Logic of Quantum Mechanics*, Annals of Math. 2nd Series Vol 37 Number 4, pp. 823-843 (1936).

[5] Bohm, David, *Wholeness and the Implicate Order*, Routledge (2002). (Title phrase borrowed on page 1.)

[6] Blokker, Esko, and Flodmark, Stig, *The Arbitrary Finite Group and Its Irreducible Representations*, Int. J. Quantum Chem. **4**, 463-472 (1971).

[7] Brauer, Richard, *On the Representation of a Group of Order g in the Field of g-th Roots of Unity*, Amer. J. Math. **67** (4), 461-471 (1945).

[8] Bröcker, Theodor, and Dieck, Tammo tom, *Representations of Compact Lie Groups*, Springer (1985).

[9] Burnside, William, *The theory of groups of finite order*. 2nd Edition, Cambridge University Press (1911).

[10] Ceccherini-Silberstein, Tullio, Scarabotti, Fabio, and Tolli, Filippo, *Representation Theory of the Symmetric Groups: The Okounkov-Vershik Approach, Character Formulas, and Partition Algebras*, Cambridge University Press (2010).

[11] Chalabi, Ahmed, *Modules over group algebras and their application in the study of semi-simplicity*, Math. Ann. **201** (1973), 57-63.

A.N. Sengupta, *Representing Finite Groups: A Semisimple Introduction*,
DOI 10.1007/978-1-4614-1231-1, © Springer Science+Business Media, LLC 2012

[12] Chevalley, Claude, *Fundamental Concepts of Algebra.* Academic Press, New York (1956).

[13] Cohn, Donald L., *Measure Theory*, Birkhäuser (1994).

[14] Curtis, Charles W., *Pioneers of Representation Theory: Frobenius, Burnside, Schur, and Brauer.* American Mathematical Society, London Mathematical Society (1999).

[15] Curtis, Charles W., and Reiner, Irving, *Representation theory of finite groups and associative algebras.* New York, Interscience Publishers (1962).

[16] Dante, Alighieri, *The Inferno*, Transl. Robert Pinsky, Farrar, Straus and Giroux; Bilingual edition (1997). (Alluded to on page 157.)

[17] Dedekind, Richard, *Über Zerlegungen von Zahlen durch ihre größten gemeinsamen Teiler*, in [20] (pages 103-147).

[18] Dedekind, Richard, *Über die von drei Moduln erzeugte Dualgruppe*, Mathematische Annalen **53** (1900), 371 - 403. Reprinted in [20] (pages 236-271)

[19] Dedekind, Richard, Letters to Frobenius (25th March 1896 and 6th April, 1896), in [20] (pages 420-424).

[20] Dedekind, Richard, *Gesammelte mathematische Werke*,Vol II, editors: Robert Fricke, Emmy Noether, Öystein Ore, Friedr. Vieweg & Sohn Akt.-Ges. (1931).

[21] Diaconis, Persi, *Group representations in probability and statistics*, Lecture Notes–Monograph Series, Volume 11 Hayward, CA: Institute of Mathematical Statistics (1988). Available online.

[22] Dieudonné, Jean Alexandre, and Carrell, James B.: *Invariant Theory Old and New.* Academic Press (1971).

[23] Dixon, John D., *High Speed Computation of Group Characters*, Num Math. 10 (1967), 446-450.

[24] Dixon, John D., *Computing Irreducible Representations of Groups*, Mathematics of Computation **24** (111), 707-712, July 1970.

[25] Duke, William, and Hopkins, Kimberly, *Quadratic reciprocity in a finite group*, American Mathematical Monthly 112 (2005),, no. 3, 251256.

[26] Farb, Benson, and Dennis, R. Keith, *Noncommutative Algebra*, Springer-Verlag (1993).

[27] Feit, W., *The Representation Theory of Finite Groups*, North-Holland (1982).

[28] Fontane, Theodor, *Effi Briest*, Insel-Verlag (1980); first published 1894-1895. (Alluded to on page 301.)

[29] Frobenius, Ferdinand Georg, *Über Gruppencharaktere*. Sitzungsberichte der Königlich Preußischen Akademie der Wissenschaften zu Berlin, 985-1021 (1896). In the Collected Works: Gesammelte Abhandlungen. Vol III (pages 1-37) Hrsg. von J.-P. Serre. Springer Verlag (1968).

[30] Frobenius, Ferdinand Georg, *Über die Primfactoren der Gruppendeterminante*. Sitzungsberichte der Königlich Preußischen Akademie der Wissenschaften zu Berlin, 1343-1382 (1896). In the Collected Works: Gesammelte Abhandlungen. Vol III (pages 38-77) Hrsg. von J.-P. Serre. Springer Verlag (1968).

[31] Frobenius, Ferdinand Georg, *Über Beziehungen zwischen den Primidealen eines algebraischen Körpers und den Substitutionen seiner Gruppe*, Sitzungsberichte der Königlich Preußischen Akademie der Wissenschaften zu Berlin, 689703 (1896). In the Collected Works: Gesammelte Abhandlungen. Vol II, (pages 719-733) Hrsg. von J.-P. Serre. Springer Verlag (1968).

[32] Frobenius, Ferdinand Georg, *Über die Darstellungen der endlichen Gruppen durch linearen Substitutionen*. Sitzungsberichte der Königlich Preußischen Akademie der Wissenschaften zu Berlin, 944-1015 (1897). In the Collected Works: Gesammelte Abhandlungen, Vol III (pages 82-103) Hrsg. von J.-P. Serre. Springer Verlag (1968).

[33] Frobenius, Ferdinand Georg, *Über Relationen zwischen den Charakteren einer Gruppe und denen ihrer Untergruppen*. Sitzungsberichte der Königlich Preußischen Akademie der Wissenschaften zu Berlin, 501-515 (1898). In the Collected Works: Gesammelte Abhandlungen. Vol III (pages 104-118) Hrsg. von J.-P. Serre. Springer Verlag (1968).

[34] Frobenius, Ferdinand Georg, *Über die Charaktere der symmetrischen Gruppe*. Sitzungsberichte der Königlich Preußischen Akademie der Wissenschaften zu Berlin, 516-534 (1900). In the Collected Works: Gesammelte Abhandlungen. Vol III (pages 148-166). Hrsg. von J.-P. Serre. Springer Verlag (1968).

[35] Frobenius, Ferdinand Georg, *Über die charakterischen Einheiten der symmetrischen Gruppe*. Sitzungsberichte der Königlich Preußischen Akademie der Wissenschaften zu Berlin, 328-358 (1903). In the Collected Works: Gesammelte Abhandlungen, Vol III (pages 244-274) Hrsg. von J.-P. Serre. Springer Verlag (1968).

[36] Frobenius, Ferdinand Georg, and Schur, Issai: *Über die reellen Darstellungen der endlichen Gruppen*. Sitzungsberichte der Königlich Preußischen Akademie der Wissenschaften zu Berlin, 186-208 (1906). In the Collected Works: Gesammelte Abhandlungen, Vol III (pages 355-377) Hrsg. von J.-P. Serre. Springer Verlag (1968).

[37] Fulton, William, *Young Tableaux*, Cambridge University Press (1997).

[38] Fulton, William, and Harris, Joe, *Representation Theory, A First Course*, Springer-Verlag (1991).

[39] Glass, Kenneth, and Ng, Chi-Keung, *A Simple Proof of the Hook Length Formula*. The American Mathematical Monthly, Vol. 111, No. 8 (Oct., 2004), pp. 700-704.

[40] Goodman, Roe, and Wallach, Nolan, R., *Representations and Invariants of the Classical Groups*, Cambridge University Press (1998).

[41] Hall, Brian C., *Lie Groups, Lie Algebras, and Representations An Elementary Introduction*. Springer-Verlag (2003).

[42] Hawkins, Thomas, *Emergence of the Theory of Lie Groups: An Essay in the History of Mathematics 1869-1926*. Springer-Verlag (2000).

[43] Hawkins, Thomas, *New light on Frobenius' creation of the theory of group characters*, Arch. History Exact Sci. Vol 12(1974), 217-243.

[44] Hill, Victor, E., *Groups and Characters*, Chaplan & Hall/CRC (2000).

[45] Hora, Akihito, and Obata, Nobuaki, *Quantum Probability and Spectral Analysis of Graphs*, Springer-Verlag (2007).

[46] Humphreys, James E., *Reflection Groups and Coxeter Groups*, Cambridge University Press (1990).

[47] Hungerford, Thomas, W., *Algebra*, Springer-Verlag (1974).

[48] Isaacs, J. Martin, *Character Theory of Finite Groups*, Academic Press (1976).

[49] James, Gordon D., and Liebeck, Martin, *Representations and Characters of Groups*. Cambridge University Press (2001).

[50] James, G. D., *The Representation Theory of the Symmetric Groups*, Springer-Verlag, Lecture Notes in Mathematics 682 (1978)

[51] Kock, Joachim, *Frobenius algebras and 2D topological quantum field theories*, Cambridge University Press (2004).

[52] Lam, T. Y., *A First Course on Noncommutative Rings*, Graduate Texts in Math., Vol. 131, Springer-Verlag, 1991.

[53] Lam, T. Y., *Representations of Finite Groups: A Hundred Years, Part I.* Notices of the American Mathematical Society. March 1998 Volume 45 Issue 3 (361-372).

[54] Lando, Sergei, and Zvonkin, Alexander. *Graphs on Surfaces and their Applications.* Springer Verlag (2004).

[55] Lang, Serge. *Algebra.* Springer 2nd Edition (2002).

[56] Lévy, Thierry, *Schur-Weyl duality and the heat kernel measure on the unitary group.* Adv. Math. 218 (2008), no. 2, 537–575.

[57] Littlewood, Dudley E., *The Theory of Group Characters and Matrix Representations of Groups.* Oxford at the Clarendon Press (1950).

[58] Maschke, Heinrich, *Über den arithmetischen Charakter der Coefficienten der Substitutionen endlicher linearer Substitutionensgruppen*, Math. Ann. **50**(1898), 492-498.

[59] Maschke, Heinrich, *Beweis des Satzes, daß die jenigen endlichen linearen Substitutionensgruppen, in welchen einige durchgehends verschwindende Coefficienten auftreten, intransitiv sind*, Math. Ann. **52**(1899), 363-368.

[60] Molien, Theodor, *Über Systeme höherer complexer Zahlen*, Dissertation (1891), University of Tartu. http://dspace.utlib.ee/dspace/handle/10062/110

[61] Mulase, Motohico, and Penkava, Michael, *Volume of Representation Varieties* (2002). Available online.

[62] Neusel, Mara D., *Invariant Theory*. American Mathematical Society (2006).

[63] Orlik, Peter, and Terao, Hiroaki, *Arrangement of Hyperplanes*, Springer-Verlag (1992).

[64] Okounkov, Andrei, and Vershik, Anatoly. *A New Approach to Representation Theory of Symmetric Groups.* Erwin Schrödinger International Institute for Mathematical Physics preprint ESI 333 (1996).

[65] Passman, Donald S., *The algebraic structure of group rings*. John Wiley & Sons, New York, London, Sydney, Toronto (1977).

[66] Puttaswamiah, B. M., and Dixon, John, *Modular Representations of Finite Groups*, Academic Press (1977).

[67] Rudin, Walter, *Functional Analysis*, Second Edition. McGraw-Hill (1991).

[68] Rota, Gian-Carlo, *The Many Lives of Lattice Theory*. Notices of the AMS, Volume 44, Number 11, December 1997, pp. 1440-1445.

[69] Schur, Issai, *Neue Begründing der Theorie der Gruppencharaktere*, Sitzungsberichte der Königlich Preußischen Akademie der Wissenschaften zu Berlin, 406-432 (1905). In the Collected Works: *Gesammelte Abhandlungen* Vol I (pages 143-169) Hrsg. von Alfred Brauer u. Hans Rohrbach. Springer (1973).

[70] Schur, Issai, *Über die rationalen Darstellungen der allgemeinen linearen Gruppe*, J. Reine Angew. Math. **132** (1907) 85-137; in *Gesammelte Abhandlungen* Vol I (pages 198-250) Hrsg. von Alfred Brauer u. Hans Rohrbach. Springer (1973).

[71] Schur, Issai, *Über die Darstellung der symmetrischen und der alternierenden Gruppe durch gebrochene lineare Substitutionen*, in *Gesammelte Abhandlungen* Vol III (pages 68-85) Hrsg. von Alfred Brauer u. Hans Rohrbach. Springer (1973).

[72] Sengupta, Ambar N., *The volume measure of flat connections as limit of the Yang-Mills measure*, J. Geom. Phys. Vol 47 398-426 (2003).

[73] Serre, Jean-Pierre, *Linear Representations of Finite Groups*. Translated by Leonard L. Scott. (4th Edition) Springer-Verlag (1993).

[74] Simon, Barry, *Representation Theory of Finite and Compact Groups*. American Mathematical Society (1995).

[75] Thomas, Charles B., *Representations of Finite and Lie Groups*. Imperial College Press (2004).

[76] Varadarajan, Veeravalli S., *The Geometry of Quantum Theory*. Springer; 2nd edition (December 27, 2006).

[77] Wedderburn, J. H. M., *On hypercomplex numbers*, Proc. London Math. Soc. (Ser 2) **6** (1908), 77-118.

[78] Weintraub, Steven H., *Representation Theory of Finite Groups: Algebra and Arithmetic.* American Mathematical Society (2003).

[79] Weyl, Hermann, *Group Theory and Quantum Mechanics.* Dover (1956)

[80] Witten, Edward, *On Quantum Gauge Theories in Two Dimensions,* Commun. Math. Phys. **141** (1991), 153-209.

[81] Young, Alfred. *Quantitative Substitutional Analysis I.* Proc. Lond. Math. Soc. (1) 33 (1901). Available in the Collected Works published by University of Toronto Press, c1977.

Index

Abelian group, 305
Action, group, 25, 304
Adjoint, 7, 350
Adkins, William, VIII
Albeverio, Sergio, VIII
Algebraic independence, 79
Algebraic integers, 330
Algebras
    definition, 319
    morphisms of, 319
Alperin, J.L., 25, 353
Alternating group, 303
Annihilator, 8, 11
Artin, E., 149, 353
Artin–Wedderburn structure theorem,
    144
Artinian algebras, 155
Ascending chain condition, 155
Associative
    algebras, 54, 319
    bilinear form, 156
    law, 316
    operation, 301
Atom, 155

Börchers, V., 162
Balanced map, 337
Balanced tensor product, 254
Banach algebra, 349
Banach space, 348

Basis
    cardinality, 320
    definition, 319
    existence, 320
Bell, R.B., 25, 353
Berkovic, Ya.G., 228, 353
Bilinear, 14, 32, 33, 61, 147, 204, 225,
    233, 319, 337, 339
Birkhoff, G., 25, 26, 353
Blocks, 157
Blokker, E., 353
    computing representations, 77
Bohm, D., 1, 353
Boolean algebra, 27
    and classical physics, 27
Bounded operator, 349
Bröcker, T., 282, 353
Bra-ket formalism, 10, 194, 195, 217
Braid groups, 55
Brauer, R., 6, 247, 353
Burnside, W., 24, 75, 211, 214, 353
    pq theorem, 228
    theorem on endomorphisms,
        265

Carrell, J.B., 79, 261, 354
Cauchy sequence, 347
Ceccherini-Silberstein, T., 184, 353
Center
    of $\mathbb{F}[G]$, 64, 111
    of a group, 304

A.N. Sengupta, *Representing Finite Groups: A Semisimple Introduction*,
DOI 10.1007/978-1-4614-1231-1, © Springer Science+Business Media, LLC 2012